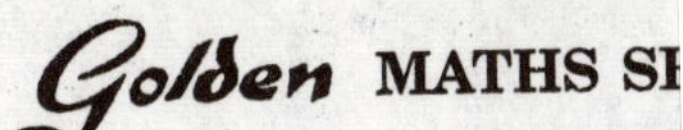

STATISTICS

For

B.A./B.Sc. STUDENTS

By

N.P. Bali

Fully Revised Edition

LAXMI PUBLICATIONS (P) LTD

BANGALORE • CHENNAI • COCHIN • GUWAHATI • HYDERABAD
JALANDHAR • KOLKATA • LUCKNOW • MUMBAI • RANCHI
NEW DELHI • BOSTON, USA

Published by :
LAXMI PUBLICATIONS (P) LTD
113, Golden House, Daryaganj,
New Delhi-110002

Phone : 011-43 53 25 00
Fax : 011-43 53 25 28

www.laxmipublications.com
info@laxmipublications.com

Price : **Rs. 105.00** *Only.* *New Edition*

OFFICES

India

✆ **Bangalore**	080-26 61 15 61
✆ **Chennai**	044-24 34 47 26
✆ **Cochin**	0484-239 70 04
✆ **Guwahati**	0361-251 38 81
✆ **Hyderabad**	040-24 65 23 33
✆ **Jalandhar**	0181-222 12 72
✆ **Kolkata**	033-22 27 37 73, 22 27 52 47
✆ **Lucknow**	0522-220 95 78
✆ **Mumbai**	022-24 91 54 15, 24 92 78 69
✆ **Ranchi**	0651-221 47 64

USA

Boston
11, Leavitt Street, Hingham,
MA 02043, USA

MGA-6718-105-G. STATISTICS (C) **C—15490/08/03**

Typeset at : Goswami Printers, Delhi. *Printed at* : Mehra Offset Press, New Delhi.

CONTENTS

GET 100 OUT OF 100 MARKS

The book has been written with a view to guide the students in their Examination.
IT ENSURES 100% MARKS.

□ □ □

The students is advised to take this book only as a guide. They should solve themselves as many questions as possible.

☞SPECIAL FEATURES

1. Solutions to all types of important questions that may be asked in the examination.
2. Questions from up-to-date papers of various Universities are inserted at proper places.
3. The book is useful to bright as well as weak students alike.
4. Approach to the subject-matter is simple and systematic.
5. The book serves triple purpose.

Text Book + Help Book + Solved University Question Papers.

IMPORTANT RESULTS AT A GLANCE

1. **Arithmetic Mean**

(*i*) For a discrete series

$$\overline{x} = \frac{1}{N} \Sigma f_i x_i$$

$$= a + \frac{1}{N} \Sigma f_i d_i, \quad \text{where} \quad d_i = x_i - a$$

(*ii*) For a continuous series having equal class intervals of width h

$$\overline{x} = a + \frac{h}{N} \Sigma f_i u_i$$

where $u_i = \dfrac{x_i - a}{h}$.

2. The algebraic sum of the deviations of all the variates from their arithmetic mean is zero.

3. Combined mean $\overline{x} = \dfrac{n_1\overline{x}_1 + n_2\overline{x}_2 + n_3\overline{x}_3 + ...}{n_1 + n_2 + n_3 + ...}$

4. **Median.** For the computation of median (or any partition value), it is necessary that the observations be arranged in ascending (or descending) order.

5. For a series of individual observations :

Median is the size of $\frac{1}{2}$ $(n + 1)$th observation.

Q_1 is the size of $\frac{1}{4}$ $(n + 1)$th observation.

Q_3 is the size of $\frac{3}{4}$ $(n + 1)$th observation.

6. For a discrete frequency distribution :

Median is the size of observation corresponding to *c.f.* just $\geq \dfrac{N + 1}{2}$,

where $N = \Sigma f_i$.

Q_1 is the size of observation corresponding to *c.f.* just

$$\geq \frac{N + 1}{4}.$$

Q_3 is the size of observation corresponding to *c.f.* just

$$\geq \frac{3(N + 1)}{4}.$$

7. For a grouped frequency distribution (the class intervals may or may not be equal).

$$\text{Median} = l + \frac{h}{f}\left(\frac{N}{2} - C\right)$$

where l = lower limit of median class

h = width of median class

f = frequency of median class, $\quad N = \Sigma f_i$

C = *c.f.* of class preceding the median class

Median class is the class corresponding to *c.f.* just $\geq \frac{N}{2}$.

$$\text{Similarly } Q_1 = l + \frac{h}{f}\left(\frac{N}{4} - C\right)$$

$$Q_3 = l + \frac{h}{f}\left(\frac{3N}{4} - C\right)$$

$$O_5 = l + \frac{h}{f}\left(\frac{5N}{8} - C\right)$$

$$D_7 = l + \frac{h}{f}\left(\frac{7N}{10} - C\right)$$

$$P_{57} = l + \frac{h}{f}\left(\frac{57N}{100} - C\right).$$

8. **Mode.** In the case of discrete frequency distribution, mode is the value of x corresponding to maximum frequency.

But in any one (or more) of the following cases :

(*i*) if the maximum frequency is repeated.

(*ii*) if the maximum frequency occurs in the very beginning or at the end of distribution.

(*iii*) if there are irregularities in the distribution, the value of mode is determined by the *method of grouping.*

9. In the case of continuous frequency distribution,

$$\text{Mode} = l + \frac{f_m - f_1}{2f_m - f_1 - f_2} \times h$$

where l = lower limit of modal class

h = width of modal class

f_m = frequency of modal class

f_1 = frequency of class preceding the modal class

f_2 = frequency of class succeeding the modal class.

For applying the above formula, the class-intervals *must* be of the same size. If they are unequal, they should first be made equal on the assumption that the frequencies are uniformly distributed throughout the class.

In case $f_m - f_1 < 0$ or $f_m - f_2 < 0$ or $2f_m - f_1 - f_2 = 0$,
use the formula

$$\text{Mode} = l + \frac{\Delta_1}{\Delta_1 + \Delta_2} \times h$$

where $\Delta_1 = |f_m - f_1|$ and $\Delta_2 = |f_m - f_2|$.

10. Empirical mode is given by
Mode = 3 Median – 2 Mean.

11. **Geometric Mean**

$$\log G = \frac{1}{N} \Sigma f_i \log x_i.$$

12. **Harmonic Mean**

$$H = \frac{N}{\sum \frac{f_i}{x_i}}$$

13. Range = L – S
where L = Largest and S = Smallest

$$\text{Co-efficient of range} = \frac{L - S}{L + S}.$$

14. Quartile Deviation = $\frac{1}{2}(Q_3 - Q_1)$

$$\text{Co-efficient of Q.D.} = \frac{Q_3 - Q_1}{Q_3 + Q_1}.$$

15. Mean Deviation $= \frac{1}{N} \Sigma f_i |x_i - A|$,

where A is the average (mean or median)

$$\text{Co-eff. of M.D.} = \frac{\text{M.D.}}{\text{Average from which it is calculated}}.$$

16. Root Mean Square Deviation

$$s = \sqrt{\frac{1}{N} \Sigma f_i (x_i - A)^2}.$$

17. Standard Deviation (S.D.)

$$\sigma = \sqrt{\frac{1}{N} \Sigma f_i (x_i - \overline{x})^2}$$

$$\text{Variance} = \sigma^2 = \frac{1}{N} \Sigma f_i x_i^2 - (\overline{x})^2$$

$$\text{S.D.} = \sqrt{\frac{1}{N} \Sigma f_i x_i^2 - \left(\frac{1}{N} \Sigma f_i x_i\right)^2}$$

If $u = \dfrac{x - a}{h}$,

then $$\text{S.D.} = h\sqrt{\frac{1}{N}\Sigma fu^2 - \left(\frac{1}{N}\Sigma fu\right)^2}$$

18. $s^2 = \sigma^2 + d^2$, where $d = \bar{x} - a$.

19. M.D. $= \frac{4}{5}$ (S.D.) ; Q.D. $= \frac{2}{3}$ (S.D.).

20. Co-efficient of dispersion based on S.D. $= \dfrac{\text{S.D.}}{\text{Mean}} = \dfrac{\sigma}{\bar{x}}$.

21. Co-efficient of variation $= \dfrac{\sigma}{\bar{x}} \times 100$.

22. Consistency ⇔ Lesser Variability.

23. **Skewness**

Bowley's co-efficient of skewness lies between – 1 and + 1

$$S_k = \frac{Q_3 + Q_1 - 2M_d}{Q_3 - Q_1}.$$

Karl Pearson's co-efficient of skewness lies between – 3 and + 3

$$S_k = \frac{\text{Mean} - \text{Mode}}{\text{S.D.}} = \frac{M - M_0}{\sigma}$$

If the mode is ill-defined, then using

$$M_0 = 3M_d - 2M,$$

$$S_k = \frac{3(M - M_d)}{\sigma}.$$

24. **Moments.** The *r*th moment of a variable *x* about any point A is denoted by μ_r' and is defined as

$$\mu_r' = \frac{1}{N}\Sigma f(x - A)^r$$

The *r*th moment of a variable *x* about the mean M is denoted by μ_r and is defined as

$$\mu_r = \frac{1}{N}\Sigma f(x - M)^r$$

In particular, $\mu_0' = \mu_0 = 1$; $\mu_1' = M - A$,

$$\mu_1 = 0,\ \mu_2 = \sigma^2$$

$$\mu_2 = \mu_2' - \mu_1'^{\,2}$$

$$\mu_3 = \mu_3' - 3\mu_2'\mu_1' + 2\mu_1'^{\,3}$$

$$\mu_4 = \mu_4' - 4\mu_3'\mu_1' + 6\mu_2'\mu_1'^{\,2} - 3\mu_1'^{\,4}$$

$$\mu_2' = \mu_2 + \mu_1'^2$$
$$\mu_3' = \mu_3 + 3\mu_2\mu_1' + \mu_1'^3$$
$$\mu_4' = \mu_4 + 4\mu_3\mu_1' + 6\mu_2\mu_1'^2 + \mu_1'^4.$$

25. β and γ co-efficients

$$\beta_1 = \frac{\mu_3^2}{\mu_2^3}, \quad \gamma_1 = \sqrt{\beta_1}$$

$$\beta_2 = \frac{\mu_4}{\mu_2^2}, \quad \gamma_2 = \beta_2 - 3.$$

26. Kurtosis

For leptokurtic curves, $\beta_2 > 3$, $\gamma_2 > 0$.

For mesokurtic curves, $\beta_2 = 3$, $\gamma_2 = 0$.

For platykurtic curves, $\beta_2 < 3$, $\gamma_2 < 0$.

27. Curve Fitting

(*i*) If $y = a + bx$ is the line of best fit, then the normal equations are

$$\Sigma y = na + b\Sigma x$$
$$\Sigma xy = a\Sigma x + b\Sigma x^2.$$

(*ii*) If $y = a + bx + cx^2$ is a second degree parabola of best fit, then the normal equations are

$$\Sigma y = na + b\Sigma x + c\Sigma x^2$$
$$\Sigma xy = a\Sigma x + b\Sigma x^2 + c\Sigma x^3$$
$$\Sigma x^2 y = a\,\Sigma x^2 + b\Sigma x^3 + c\Sigma x^4.$$

28. Most plausible solution of a system of independent equations

The normal equation for any variable is obtained by multiplying L.H.S. of each equation (R.H.S. being zero) by the co-efficient of that variable in that equation and then adding up all the resulting equations.

29. Correlation

(*i*) Karl Pearson's co-efficient of correlation between two variables x and y is defined as

$$r_{xy} = \frac{\Sigma(x_i - \bar{x})(y_i - \bar{y})}{\sqrt{\Sigma(x_i - \bar{x})^2 \,.\, \Sigma(y_i - \bar{y})^2}}$$

$$= \frac{\frac{1}{n}\Sigma(x_i - \bar{x})(y_i - \bar{y})}{\sigma_x \sigma_y}$$

$$= \frac{\text{cov}\,(x, y)}{\sigma_x \sigma_y}; \qquad | r_{xy} | \le 1.$$

(*ii*) Correlation co-efficient is independent of change of origin and scale.

If $u = \dfrac{x-a}{h}$, $v = \dfrac{y-b}{k}$, then

$$r_{xy} = r_{uv} = \frac{\frac{1}{n}\Sigma uv - \bar{u}\,\bar{v}}{\sqrt{\left(\frac{1}{n}\Sigma u^2 - \bar{u}^2\right)\left(\frac{1}{n}\Sigma v^2 - \bar{v}^2\right)}}.$$

30. **Rank Correlation.** Spearman's Rank Correlation Co-efficient is given by

$$r = 1 - \frac{6\Sigma d_i^{\,2}}{n(n^2-1)}$$

where $d_i = x_i - y_i$ = difference in ranks of ith individual

$\Sigma d_i = 0$ (always)

The above formula is used when no two individuals have same rank. For repeated ranks, use

$$r = 1 - \frac{6\left\{\Sigma d^2 + \frac{1}{12}m(m^2-1) + \frac{1}{12}m(m^2-1) + ...\right\}}{n(n^2-1)}$$

where m = the number of times a rank is repeated and every time a rank is repeated, we add the factor $\frac{1}{12}m(m^2-1)$.

31. **Regression**

Line of regression of y on x is

$$y - \bar{y} = \frac{r\sigma_y}{\sigma_x}(x - \bar{x}).$$

Line of regression of x on y is

$$x - \bar{x} = \frac{r\sigma_x}{\sigma_y}(y - \bar{y}).$$

Both the lines of regression pass through the point $(\bar{x}, \bar{y})$.

Regression co-efficient of y on x

$$= b_{yx} = \frac{r\sigma_y}{\sigma_x}.$$

Regression co-efficient of x on y

$$= b_{xy} = \frac{r\sigma_x}{\sigma_y}.$$

r, b_{yx} and b_{xy} have same sign

$$r = \sqrt{b_{yx} \times b_{xy}}.$$

32. Probability

(*i*) $$P(E) = \frac{\text{Favourable number of cases}}{\text{Exhaustive number of cases}}$$

$$P(E) + P(\bar{E}) = 1.$$

(*ii*) Addition Theorem of Probability (or Theorem of Total Probability)

$$P(A + B) = P(A) + P(B) - P(AB)$$

If A and B are mutually exclusive events, then

$$P(A + B) = P(A) + P(B)$$

(*iii*) Theorem of Compound Probability

$$P(AB) = P(A) \, . \, P(B/A)$$

If A and B are independent events, then

$$P(AB) = P(A) \, . \, P(B).$$

33. Binomial Distribution

For N sets of n trials, the frequencies of 0, 1, 2,....., n successes are given by the successive terms in the expression

$$N[q^n + {}^nC_1 q^{n-1} p + {}^nC_2 q^{n-2} p^2 + \ldots\ldots + {}^nC_r q^{n-r} p^r + \ldots\ldots + p^n]$$

which is the Bionomial expansion of $N(q + p)^n$

Mean = np, $\sigma = \sqrt{npq}$.

34. Poisson Distribution

$$P(X = r) = \frac{m^r e^{-m}}{r\,!}$$

Mean = m, $\sigma^2 = m$.

35. Normal Distribution

Standard Form of the normal curve is

$$y = \frac{1}{\sigma\sqrt{2\pi}} e^{-\frac{x^2}{2\sigma^2}}.$$

General Form of the normal curve is

$$y = \frac{N}{\sigma\sqrt{2\pi}} e^{-\frac{(x-m)^2}{2\sigma^2}}$$

where $(x - m)$ is the excess of the mean over the value chosen as origin.

Total area under the normal probability curve is unity

Mean = Mode = Median.

36. Mean and S.D. in Simple Sampling of Attributes

Expected value of (*i*) mean is np

(*ii*) S.D. is $\sqrt{npq}$

Mean of proportion of success = p

S.E. of proportion of success = $\sqrt{\frac{pq}{n}}$.

37. Test of Significance for Large Samples

If the number of successes in a large sample of size n differs from the expected value np by more than $3\sqrt{npq}$, we call the difference highly significant and the truth of the hypothesis is very improbable.

38. Chi-Square (χ^2) Test

If O_i (i = 1, 2,...., n) is a set of observed frequencies and E_i (i = 1, 2,....., n) is the corresponding set of expected frequencies, then χ^2 is defined as

$$\chi^2 = \sum_{i=1}^{n} \frac{(O_i - E_i)^2}{E_i} \quad \text{with } d.f. = n - 1.$$

39. 't' Test

The quantity t is defined as

$$t = \frac{\overline{x} - \mu}{S} . \sqrt{n}$$

where n = the number of observations in the sample

$\overline{x} = \frac{1}{n} \sum_{i=1}^{n} x_i$ is the mean of the sample

μ = the mean of the parent population from which the sample has been drawn

$S = \sqrt{\frac{1}{n-1} \sum_{i=1}^{n} (x_i - \overline{x})^2}$ is the S.D. of sample

$d.f. = n - 1$.

1

Frequency Distributions

1.1. INTRODUCTION TO STATISTICS

Statistics is as old as the human society itself. The word 'Statistics' seems to have been derived from the Latin word 'Status' or the Italian word 'Statista' or the German word 'Statistik' each of which means a 'political state'. In ancient times, the kings used to collect information about the population and wealth of the country to assess the man power of the country and to impose new taxes. With the passage of time, however, its scope began to include collection of numerical data pertaining to almost any endeavour and the presentation of data, interpretation and drawing of inferences from the data.

It is difficult to imagine any face of our life untouched by numerical data. The modern society is essentially data-oriented. It is, therefore, essential to know how to extract useful information from such data. This is the primary objective of statistics. Statistics concerns itself with the collection, presentation and drawing of inferences from numerical data which vary.

In singular sense, statistics is used to describe the principles and methods which are employed in collection, presentation, analysis and interpretation of data. These devices help to simplify the complex data and make it possible for a common man to understand it without much difficulty. Human mind is unable to assimilate complicated data at a stretch. Statistical methods make these data easy to grasp. A man is bound to be lost in figures because figures are boring. Statistical methods make these figures intelligible and readily understandable.

In plural sense, statistics is considered as a numerical description of quantitative aspect of things.

Every data is not statistics. It must fulfil certain essential characteristics to be called 'statistics'. Numerical data that are subjected to a multiplicity of causes and therefore vary from person to person, place to place, time to time etc., are statistics.

> **Def.** *Statistics is the science which deals with methods of collecting, classifying, presenting, comparing and interpreting numerical data collected to throw light on any sphere of enquiry.*

1.2. MATHEMATICAL STATISTICS

Statistics, as it is to-day, is very closely related and very much indebted to Mathematics. Wide applications of advanced mathematics have resulted into recent advancements in statistical analysis. Thus, statistics may be regarded as that branch of mathematics which systematically analyses large number of **related numerical facts.** It is that branch of Mathematics which specialises in data. Thereby arises the need for a new branch of Mathematics called Mathematical Statistics.

1.3. VARIABLE (OR VARIATE)

A quantity which can vary from one individual to another is called a **variable** or **variate**, *e.g.*, heights, weights, ages, wages of persons, rainfall records of cities etc.

Quantities which can take any numerical value within a certain range are called **continuous variables** *e.g.*, as the child grows, his/her height takes all possible values from 50 cm to 100 cm.

Quantities which are incapable of taking all possible values are called **discrete** or **discontinuous variable** *e.g.*, the number of children a man can have are +ve integers 1, 2, 3 etc. (no value between any two consecutive integers.)

1.4. FREQUENCY DISTRIBUTIONS

Consider the marks obtained by 60 students of B.A/B. Sc. Part III class in a college test in mathematics according to their roll numbers :

38, 11, 40, 0, 26, 15, 5, 45, 7, 32, 2, 18, 42, 8, 31, 27, 4, 12, 35, 15, 0, 7, 28, 46, 9, 16, 29, 34, 10, 7, 5, 1, 17, 22, 35, 8, 36, 47, 11, 30, 19, 0, 16, 14, 16, 18, 41, 38, 2, 17, 42, 45, 48, 28, 7, 21, 8, 28, 5, 20.

The data does not give any useful information. It is rather confusing to mind. These are called **raw data** or **ungrouped data.**

We would like to bring out certain salient features of this data. If we express the data in ascending or descending order of magnitude, this does not reduce the bulk of the data. We condense the data into **classes** or **groups** as below :

(*i*) Determine the **range** of the data *i.e.*, difference between largest and smallest numbers occuring in the data.
Here range = 48 – 0 = 48.

(*ii*) Decide upon the number of classes or groups into which raw data is to be grouped. There are no hard and fast rules for this. The insight of the experimentor determines this number. However, the number of classes should not, be less than 5 and not more than 30. With less number of classes accuracy is lost and with more number of classes the computations become tedious.

Let us take the number of classes = 7 here.

(*iii*) Divide the range by the desired number of classes to determine the approximate *width* or *size of class interval.* If the quotient is a fraction, take the next integer. In the above example, size of class interval is $\frac{48}{7}$ or 7.

As far as possible, classes should be of the same size.

(*iv*) Using the size of the interval, set up the class limits, making sure that the minimum and the maximum numbers occuring in the data are included in some class. As far as possible, open end classes ($a < x < b$) should be avoided since they create difficulty in analysis and interpretation. Boundaries of each class are selected in such way that there is no ambiguity as to which class a particular item of the data belongs.

(*v*) The observations corresponding to the common point of two classes should always be included in the higher class, *e.g.,* if 20 is an element of the data and 10—20 and 20—30 are two classes, then 20 is to be put in the class 20—30 and not 10—20. That is to say every class should be regarded as open to the right.

(*vi*) Take each item from the data, one at a time, and place a *tally mark* (/) opposite the class to which it belongs. Tally marks are recorded in bunches of five. Having occured four times, the fifth occurrence is represented by putting a cross-tally (\ or /) on the first four tallies (卌 or 卌). This technique facilitates the counting of the tally marks at the end.

(*vii*) *The count of tally marks in a particular class provides us with the frequency in that class.* The word 'frequency' is derived from 'how frequently' a variable occurs.

(*viii*) Marks are called the variable (x) and the number of students in a class is known as the frequency (f) or class frequency of the variable.

(*ix*) The total of all frequencies must equal the number of observations in the raw data.

(*x*) The table displaying the manner in which frequencies are distributed over various classes is called *frequency table.*

(*xi*) We are often interested in knowing, at a glance, the number of observations less than a particular value. This is done by finding cumulative frequency. The *cumulative frequency* corresponding to a class is the sum of frequencies of that class and of all classes prior to that class.

(*xii*) The table displaying the manner in which cumulative frequencies are distributed is called *cumulative frequency table.*

Using the above steps, we have the following cumulative frequency table for the example under consideration.

Class interval (*marks x*)	*Tally marks* (*number of students*)	*Frequency* (*f*)	*Cumulative Frequency*
0— 7	卌 卌	10	10
7—14	卌 卌 ||	12	22
14—21	卌 卌 ||	12	34
21—28	||||	4	38
28—35	卌 |||	8	46
35—42	卌 ||	7	53
42—49	卌 ||	7	60
Total		**60**	

Example 1. *Make a frequency table having grades of wages with class-interval of Rs. 2 each from the following data of daily wages received by 30 workers in a certain factory. Daily wages is rupees are* :

14, 16, 16, 14, 22, 13, 15, 24, 12, 23, 14, 20, 17, 21, 22, 18, 18, 19, 20, 17, 16, 15, 11, 12, 21, 20, 17, 18, 19, 23.

Sol. Minimum entry of raw data = 11

Width of a class interval = 2 (*given*)

Wages in Rs.	*No. of workers*	*Frequency*
11—13	|||	3
13—15	||||	4
15—17	卌	5
17—19	卌 |	6
19—21	卌	5
21—23	||||	4
23—25	|||	3
Total		30

Example 2. *The minimum temperature (in °C) for Delhi for the month of July, 1993 as reported by the Meterological Department is given below. Construct a frequency distribution table for it.*

30.3, 30.0, 25.8, 26.5, 24.2, 25.2, 28.0, 28.0, 29.5, 27.8, 30.0, 31.1, 27.2, 25.9, 27.6, 24.5, 24.4, 27.0, 28.1, 26.0, 25.4, 28.0, 26.9, 25.7, 27.2, 25.5, 26.6, 28.5, 28.0, 27.7, 24.0.

Sol. Range of raw data = max (31.1) – min. (24.0) = 7.1.

Let us choose the number of classes = 8

Width of class interval is a convenient number around $\frac{7.1}{8} = .88$

Take it as 1.

Min. temp. in °C	*Days of July, 1993*	*Frequency*
24.0—25.0	||||	4
25.0—26.0	~~||||~~ |	6
26.0—27.0	||||	4
27.0—28.0	~~||||~~ |	6
28.0—29.0	~~||||~~ |	6
29.0—30.0	|	1
30.0—31.0	|||	3
31.0—32.0	|	1
Total		31 (= No. of days in July)

Example 3. *The weights in grams of 50 apples picked at random from a consignment are as follows :*

106, 107, 76, 82, 109, 107, 115, 93, 187, 195, 123, 125, 111, 99, 86, 70, 126, 68, 130, 129, 139, 119, 115, 128, 100, 186, 84, 99, 113, 204, 111, 141, 136, 123, 90, 115, 98, 110, 78, 90, 107, 81, 131, 75, 84, 104, 110, 80, 118, 82.

Form the grouped frequency table by dividing the variate range into intervals of equal width, each corresponding to 20 gms. in such a way that the mid-value of the first class corresponds to 70 gms.

Sol. Mid-value of first class = 70 } *(given)*
Width of each class = 20 }

∴ The first class interval is (70 – 10) – (70 + 10) *i.e.*, 60 – 80.

Weight in grams	*No. of apples*	*Frequency*
60—80	~~||||~~	5
80—100	~~||||~~ ~~||||~~ |||	13
100—120	~~||||~~ ~~||||~~ ~~||||~~ ||	17
120—140	~~||||~~ ~~||||~~	10
140—160	|	1
160—180		0
180—200	|||	3
200—220	|	1
Total		50

Example 4. *The following are the monthly rents (in rupees) of 40 shops. Tabulate the data by grouping them in intervals of Rs. 8.*

38, 42, 49, 37, 82, 37, 75, 62, 54, 79, 84, 75, 63, 44, 74, 44, 36, 69, 54, 48, 74, 47, 52, 57, 62, 67, 72, 77, 82, 51, 31, 38, 43, 75, 67, 77, 47, 64, 84, 81.

Sol. Please try yourself.

Example 5. *Form an ordinary frequency table from the following table :*

Marks	*No. of Students*	*Marks*	*No. of Students*
Above 0	*40*	*Above 30*	*18*
" 10	*30*	*" 40*	*12*
" 20	*25*	*" 50*	*0*

Sol.

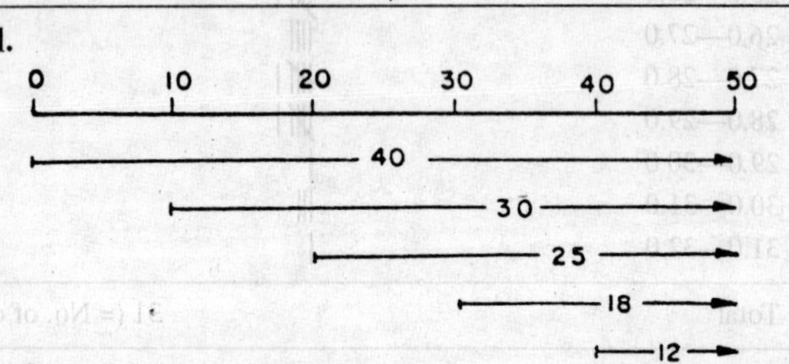

Marks	*No. of Students (f)*
0—10	40 – 30 = 10
10—20	30 – 25 = 5
20—30	25 – 18 = 7
30—40	18 – 12 = 6
40—50	12 – 0 = 12

Example 6. *Form an ordinary frequency table from the following :*

Marks	*No. of Students*	*Marks*	*No. of Students*
Below 10	*5*	*Below 40*	*22*
" 20	*7*	*" 50*	*30*
" 30	*13*	*" 60*	*38*

Sol.

Marks	*No. of Students (f)*
0—10	5
10—20	7 – 5 = 2
20—30	13 – 7 = 6
30—40	22 – 13 = 9
40—50	30 – 22 = 8
50—60	38 – 30 = 8

1.5. 'EXCLUSIVE' AND INCLUSIVE' CLASS-INTERVALS

Class-intervals of the type $\{x : a \le x < b\} = [a, b)$ are called 'exclusive' since they exclude the upper limit of the class. The following data are classified on this basis.

Income (Rs.)	*No. of persons*
50—100	88
100—150	70
150—200	52
200—250	30
250—300	23

In this method, the upper limit of one class is the lower limit of the next class. In this example, there are 88 persons whose income is from Rs. 50 to Rs. 99.99. A person whose income is Rs. 100 is included in the class Rs. 100 – Rs. 150.

Class-intervals of the type $\{x : a \le x \le b\} = [a, b]$ are called 'inclusive' since they include the upper limit of the class. The following data are classified on this basis.

Income (Rs.)	*No. of persons*
50— 99	60
100—149	38
150—199	22
200—249	16
250—299	7

However, to ensure continuity and to get correct class-limits, exclusive method of classification should be adopted. To convert inclusive class-intervals into exclusive, we have to make an adjustment.

Adjustment. Find the difference between the lower-limit of the second class and the upper limit of the first class. Divide it by 2. Subtract the value so obtained from all the lower limits and add the value to all upper limits.

In the above example, the adjustment factor is $\frac{100-99}{2} = .5$. The adjusted classes would then be as follows :

Income (Rs.)	*No. of persons*
49.5— 99.5	60
99.5—149.5	38
149.5—199.5	22
199.5—249.5	16
249.5—299.5	7

The size of the class interval is 50.

1.6. THREE TYPES OF SERIES

In this book, we will come across the following three types of series.

(*a*) **Individual Observations** (*i.e., where frequencies are not given*).

Form $x : \quad x_1, \quad x_2, \quad x_3, \ldots, x_n.$

(*b*) **Discrete Series.** It is a series of observations of the form

x : x_1, x_2, x_3,......, x_n.

f : f_1, f_2, f_3,......, f_n.

(*c*) **Continuous Series.** It is a series of observations of the form

Class Interval : $a_1 - a_2$ $a_2 - a_3$...... $a_n - a_{n+1}$

f : f_1 f_2 ...... f_n

For the purpose of further calculations in statistical work, the mid-point of each class is taken to represent the class.

Thus, if m_i is the mid-point of the ith class, then $m_i = \dfrac{a_i + a_{i+1}}{2}$ and the above series takes the form

Mid-value m : m_1, m_2, m_3,...... m_n

f : f_1, f_2, f_3, ... f_n.

The mid-value of the ith class may also be denoted by x_i. Thus, a continuous series is reduced to the form of a discrete series.

1.7. GRAPHICAL REPRESENTATION

A frequency distribution when represented by means of a graph makes the unwieldy data intelligible. A better perspective can be had by representing the frequency distribution graphically since graphs, if drawn attractively, are eye-catching and leave a more lasting impression on the mind of the observer. Graphs are a good visual aid. But graphs do not give accurate measurements of the variable as are given by the tables. Another disadvantage is that by taking different scales, the facts may be misrepresented.

Some important types of graphs are given below :

(A) **Histogram.**

In drawing the histogram of a given grouped frequency distribution :

(*a*) Mark off along the x-axis all the class intervals on a suitable scale. (If class-intervals are equal, then each = 1 cm. is quite suitable).

(*b*) Mark frequencies along y-axis on a suitable scale.

(*c*) It must not be assumed that scale for both the axes will be same. We can have different scales for the two axes. The determination of scale depends upon our convenience and the type and nature of data. The scale or scales should be so chosen as to fit the size of graph-paper and to hold all the figures of the data.

(*d*) Construct rectangles with the class-intervals as bases and heights proportional (if the class intervals are equal) to frequencies.

A diagram with all these rectangles is called a **histogram.**

Example 1. *The weights (in grams) of 40 oranges picked at random from a basket are as follows :*

45, 55, 30, 110, 75, 100, 40, 60, 65, 40, 100, 75, 70, 60, 70, 95, 85, 80, 35, 45, 40, 50, 60, 65, 55, 45, 30, 90, 85, 75, 85, 75, 70, 110, 100, 80, 70, 55, 30, 70.

Represent the data by means of a histogram.

Sol. Range = max. (110) – min. (30) = 80

Let number of class intervals = 7

Width of class interval = $\left(\frac{80}{7} \text{ or } 12.\right)$

Wts. of oranges (in gms.)	*No. of oranges*	*Frequency*
30— 42	卌 ǁ	7
42— 54	ǁǁ	4
54— 66	卌 ǀǀǀ	8
66— 78	卌 ǁǁ	9
78— 90	卌	5
90—102	卌	5
102—114	ǁ	2
Total		40

The histogram of the above frequency distribution is given below :

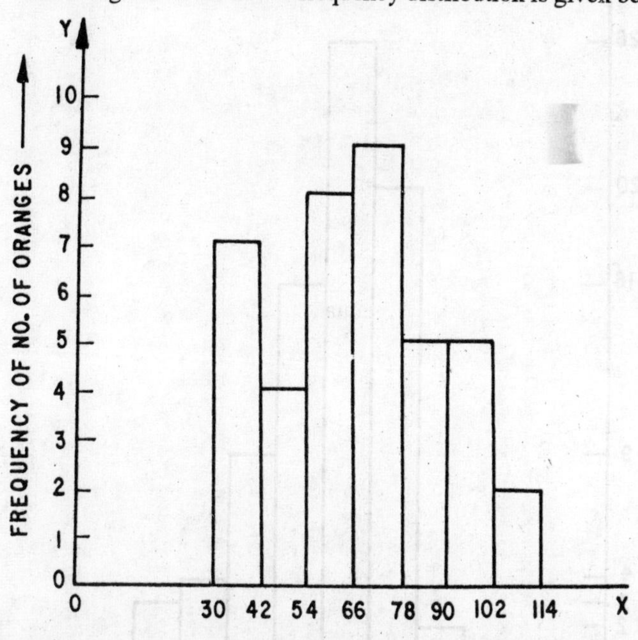

Example 2. *Draw a histogram for the following table :*

0—10	*2*	*30—40*	*8*
10—20	*4*	*40—50*	*5*
20—30	*7*	*50—60*	*3*

Sol. Please try yourself.

Example 3. *Draw a histogram representing the following frequency distribution :*

Wages (monthly) Rs.	*Frequency*	*Wages (monthly) Rs.*	*Frequency*
15	*2*	*35*	*9*
20	*20*	*40*	*4*
25	*26*	*45*	*3*
30	*16*		

Sol. Here frequency is not given in terms of class intervals. Since the difference between any two consecutive wages is 5, to have equal class intervals, the wages are the mid-points of class intervals 12.5—17.5, 17.5—22.5 etc.

The histogram is drawn below.

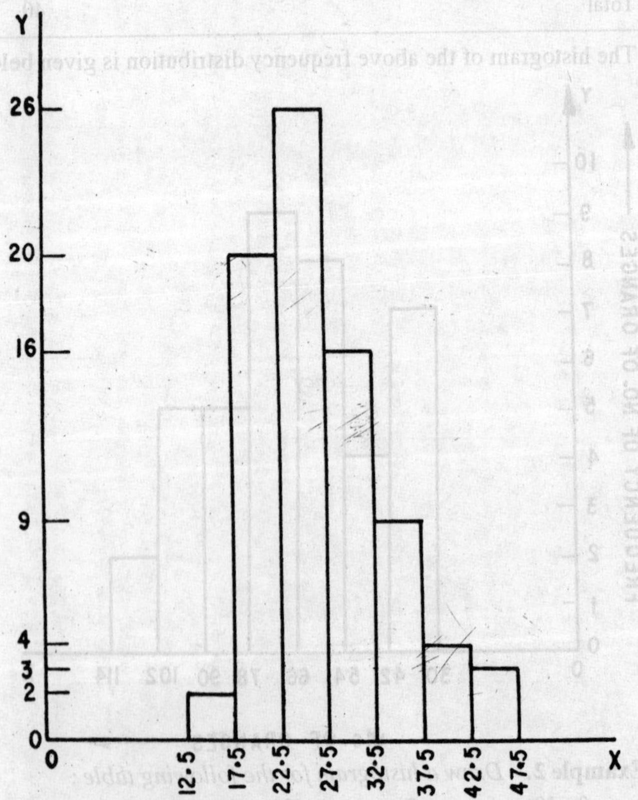

Example 4. *Draw a histogram for the following distribution :*

Degree of cloudiness (*x*)	*Frequency* (*f*)
10	*580*
9	*150*
8	*196*
7	*75*
6	*55*

Sol. Please try yourself.

(**Hint.** Take x as the central value, then class intervals are 5.5—6.5, 6.5—7.5, 7.5—8.5 etc.)

Example 5. *Represent the following data by means of a histogram* :

Weekly Wages (in Rs.)	*No. of workers*	*Weekly Wages (in Rs.)*	*No. of workers*
10—15	*7*	*30—40*	*12*
15—20	*19*	*40—60*	*12*
20—25	*27*	*60—80*	*8*
25—30	*15*		

Sol. Here the class-intervals are unequal, so we adjust the frequencies as follows :

The least size of any class-interval is 5, (first four class intervals). The size of the class-interval 30—40 is 10 = 2 × 5, we divide its frequency by 2. The sizes of the next two class-intervals are 20 = 4 × 5 each, we divide their frequencies by 4. Thus we have the following table :

Weekly Wages (in Rs.)	*Frequency*	*Adjusted Frequency*
10—15	7	7
15—20	19	19
20—25	27	27
25—30	15	15
30—40	12	6
40—60	12	3
60—80	8	2

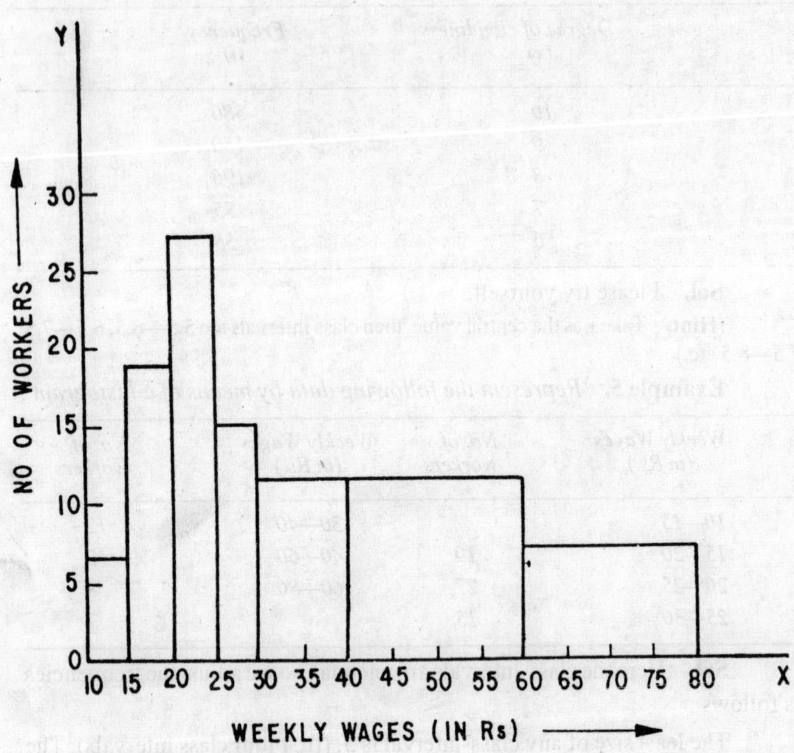

Example 6. *Draw a histogram for the following data on the size of families :*

No. of children :	*0*	*1*	*2*	*3*	*4*	*5*	*6*
No. of families :	*171*	*82*	*50*	*25*	*13*	*7*	*2*

Sol. We first arrange this data in the form of a frequency distribution with *exclusive class-intervals.*

No. of children	*No. of families*
0—1	171
1—2	82
2—3	50
3—4	25
4—5	13
5—6	7
6—7	2

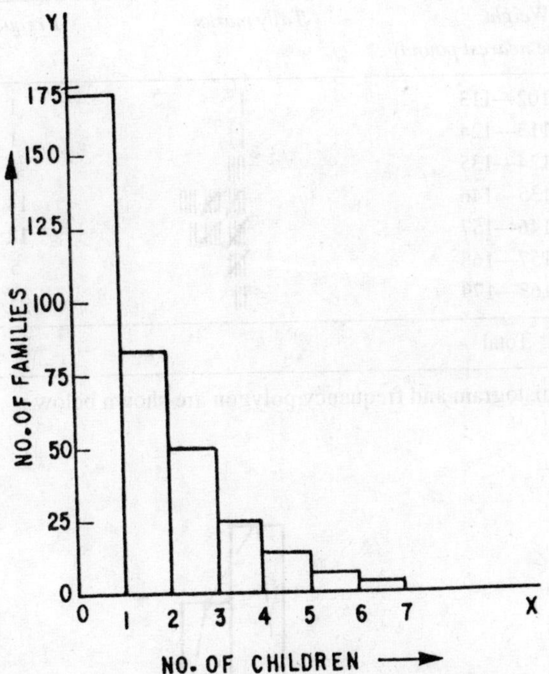

(B) **Frequency Polygon.** For a grouped frequency distribution, with equal class-intervals, a frequency polygon is obtained by joining the middle points of the upper sides (tops) of the adjacent rectangles of the histogram by means of straight lines. To complete the polygon, the mid-points at each end are joined to the immediately lower and higher mid-points at zero frequency *i.e.*, on the x-axis.

Example 7. *The following table gives the weights (to the nearest pound) of 40 male students at a university. Construct a frequency distribution with 7 classes and draw the histogram and frequency polygon.*

138, 164, 150, 132, 144, 125, 149, 157, 146, 158, 140, 147, 136, 148, 152, 144, 168, 126, 138, 176, 163, 119, 154, 165, 146, 173, 142, 147, 135, 153, 140, 135, 102, 145, 135, 142, 150, 156, 145, 128.

Sol. Range of raw data = max. (176) – min (102) = 74

Number of classes = 7

$$\therefore \quad \text{Width of class interval} = \left(\frac{74}{7} \text{ or }\right) 11.$$

Weight (*to the nearest pound*)	*Tally marks*	*Frequency*
102—113	|	1
113—124	|	1
124—135	||||	4
135—146	~~||||~~ ~~||||~~ ||||	14
146—157	~~||||~~ ~~||||~~ ||	12
157—168	~~||||~~	5
168—179	|||	3
Total		40

The histogram and frequency polygon are shown below :

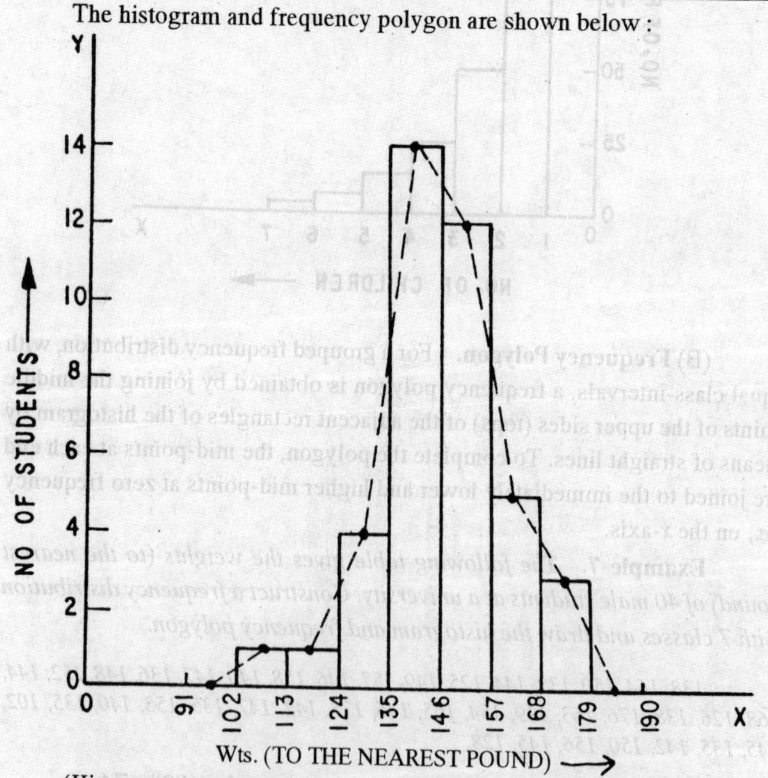

(*Histogram* :—rectangles, *Frequency polygon* :—shown dotted).

Example 8. *Draw a histogram and frequency polygon for the following data* :

0—10	*2*	*30—40*	*8*
10—20	*4*	*40—50*	*4*
20—30	*6*		

Sol. Please try yourself.

Example 9. *Represent the following distribution by a (i) histogram and (ii) frequency polygon.*

Scores	*Frequency*	*Scores*	*Frequency*
90—99	*2*	*50—59*	*14*
80—89	*12*	*40—49*	*3*
70—79	*22*	*30—39*	*1*
60—69	*20*	*20—29*	*1*

Sol. Here the grouped frequency distribution is not continuous. We first convert it into a continuous distribution as follows :

Adjustment factor $= \frac{30 - 29}{2} = .5$. Subtract it from each lower limit and add to each upper limit so as to have exclusive class intervals. Thus

Scores	*Frequency*	*Scores*	*Frequency*
19.5—29.5	1	59.5—69.5	20
29.5—39.5	1	69.5—79.5	22
39.5—49.5	3	79.5—89.5	12
49.5—59.5	14	89.5—99.5	2

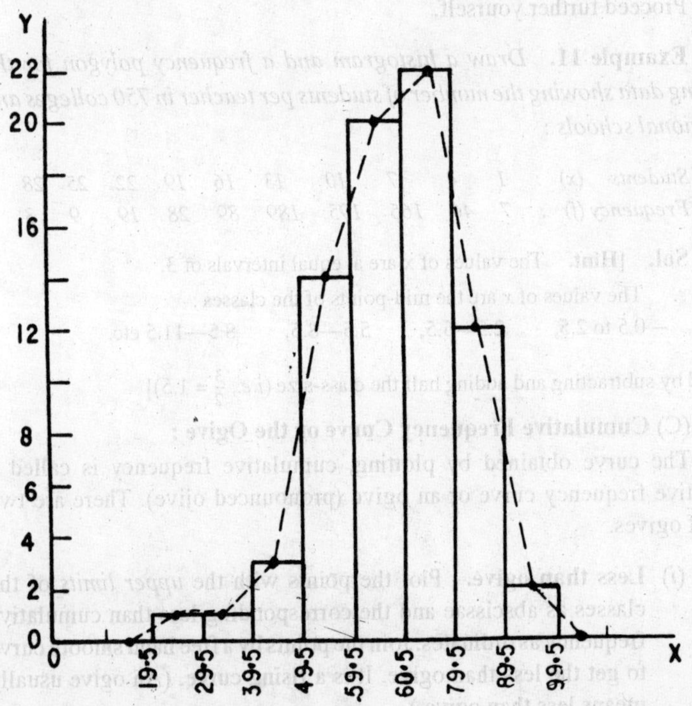

(Histogram :— rectangles, Frequency polygon : shown dotted)

Example 10. *You are given the following frequency distribution of monthly expenditure on food incurred by a sample of 100 families in a town :*

Expenditure	*Frequency*	*Expenditure*	*Frequency*
0—50	*7*	*250—400*	*27*
50—150	*24*	*400—600*	*8*
150—200	*30*	*600—800*	*4*

Draw the frequency polygon and the histogram for this distribution.

Sol. Since the class-intervals are unequal, the frequencies must be adjusted.

Expenditure	*Frequency*		*Adjusted Frequency*
0— 50	7		7
50—150	24	(÷ 2)	12
150—250	30	(÷ 2)	15
250—400	27	(÷ 3)	9
400—600	8	(÷ 4)	2
600—800	4	(÷ 4)	1

Proceed further yourself.

Example 11. *Draw a histogram and a frequency polygon for the following data showing the number of students per teacher in 750 colleges and professional schools :*

Students (x) :	*1*	*4*	*7*	*10*	*13*	*16*	*19*	*22*	*25*	*28*
Frequency (f) :	*7*	*46*	*165*	*195*	*189*	*89*	*28*	*19*	*9*	*3*

Sol. **[Hint.** The values of x are at equal intervals of 3.

∴ The values of x are the mid-points of the classes :

—0.5 to 2.5, 2.5—5.5, 5.5—8.5, 8.5—11.5 etc.

obtained by subtracting and adding half the class-size (*i.e.*, $\frac{3}{2} = 1.5$)]

(C) Cumulative Frequency Curve or the Ogive :

The curve obtained by plotting, cumulative frequency is called a cumulative frequency curve or an ogive (pronounced ojive). There are two types of ogives.

(*i*) **Less than ogive.** Plot the points with the *upper limits* of the classes as abscissae and the corresponding less than cumulative frequency as ordinates. Join the points by a free hand smooth curve to get the less than ogive. It is a rising curve. (An ogive usually means less than ogive.)

(*ii*) **More than ogive.** Plot the points with the *lower limits* of the classes as abscissae and the corresponding more than cumulative frequency as ordinates. Join the points by a free-hand smooth curve to get the more than ogive. It is a falling curve.

Consider the following frequency distribution :

Marks	*No. of students*	*Marks*	*No. of students*
10—20	4	40—50	20
20—30	6	50—60	18
30—40	10	60—70	2

Let us convert it first into a 'less than c.f.' distribution and then into a 'more than c.f.' distribution.

Marks less than	*No. of students*	*Marks more than*	*No. of students*
20	4	10	60
30	(+ 6 =) 10	20	(– 4 =) 56
40	(+ 10 =) 20	30	(– 6 =) 50
50	(+ 20 =) 40	40	(– 10 =) 40
60	(+ 18 =) 58	50	(– 20 =) 20
70	(+ 2 =) 60	60	(– 18 =) 2
		70	(– 2 =) 0

Example 12. *Draw a cumulative frequency curve for the following data :*

Marks	*No. of students*	*Marks*	*No. of students*
0— 4	*4*	*12—16*	*8*
4— 8	*6*	*16—20.*	*4*
8—12	*10*		

Sol. First we form the less than frequency distribution table.

Marks	*Frequency (no. of students)*	*Less than c.f.*
0— 4	4	4
4— 8	6	10
8—12	10	20
12—16	8	28
16—20	4	32
Total	32	

Upper limits of the classes are 4, 8, 12, 16, 20. These are the abscissae of the points to be plotted. The ordinates are 4, 10, 20, 28, 32.

∴ We plot the points (4, 4), (8, 10), (12, 20), (16, 28) and (20, 32). Joining them by free hand drawing, we get the less than cumulative frequency curve or the less than ogive.

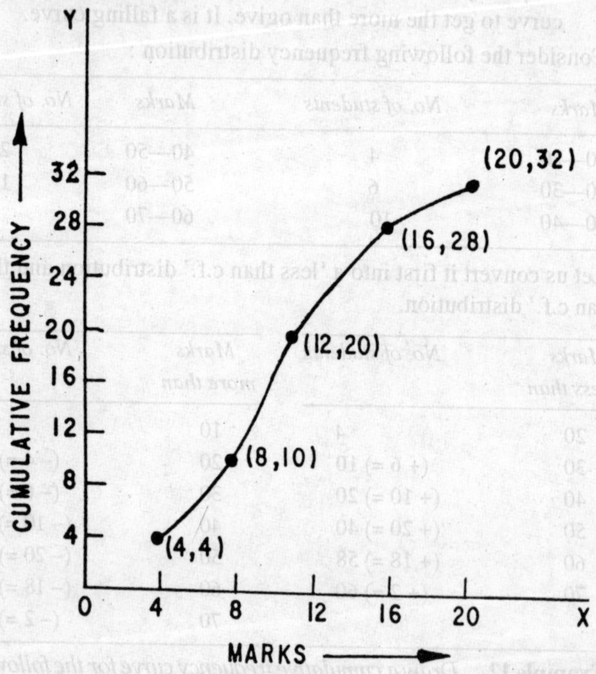

Example 13. *Represent the following distribution by an ogive.*

Marks	*No. of students*	*Marks*	*No. of students*
0—10	5	*50—60*	*4*
10—20	*13*	*60—70*	*1*
20—30	*12*	*70—80*	*3*
30—40	*11*	*80—90*	*1*
40—50	*8*	*90—100*	*2*

Sol. Please try yourself.

Example 14. *Draw the two ogives for the following distribution showing the number of marks of 59 students :*

Marks	*No. of students*	*Marks*	*No. of students*
0—10	*4*	*40—50*	*12*
10—20	*8*	*50—60*	*6*
20—30	*11*	*60—70*	*3*
30—40	*15*		

Sol.

Marks	*No. of students*	*Less than c.f.*	*More than c.f.*
0—10	4	4	59
10—20	8	12	55
20—30	11	28	47
30—40	15	38	36
40—50	12	50	21
50—60	6	56	9
60—70	3	59	3

Plotting the points (10, 4), (20, 12), (30, 23), (40, 38), (50, 50), (60, 56), (70, 59) and joining them by free-hand, the smooth rising curve so obtained is less than ogive.

Plotting the points (0, 59), (10, 55), (20, 47), (30, 36), (40, 21), (50, 9), (60, 3) and joining them by free-hand, the smooth falling curve so obtained is the more than ogive.

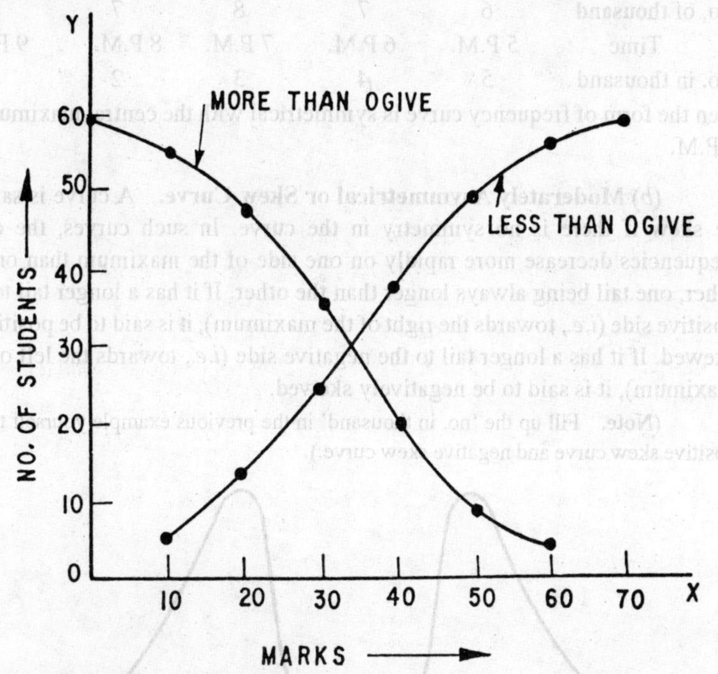

(D) **Forms of Frequency Curves.** The following are some important types of frequency curves which we generally get in the graphical representation of frequency distributions.

(*a*) **Symmetrical Curve or bell-shaped curve.** If a curve can be folded along a vertical line so that the two halves of it coincide, it is called a symmetrical curve. In such curves, the class frequencies decrease symmetrically on either side of a central maximum *i.e.*, the observations equidistant from the central maximum have the same frequency. It has a single smooth hump in the middle and tails off gradually at either end, *e.g.*, if the number of persons travelling by Haryana Roadways buses at different times in a day is given below :

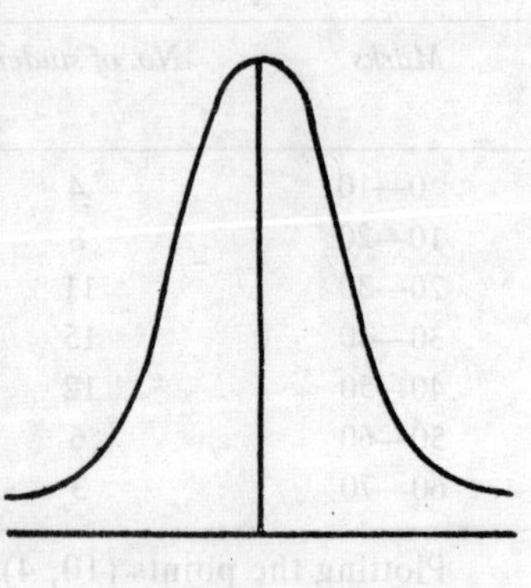

Time	7 A.M.	8 A.M.	9 A.M.	10 A.M.	11 A.M.
No. in thousand	1	2	3	4	5
Time	12 Noon	1 P.M.	2 P.M.	3 P.M.	4 P.M.
No. of thousand	6	7	8	7	6
Time	5 P.M.	6 P.M.	7 P.M.	8 P.M.	9 P.M.
No. in thousand	5	4	3	2	1

then the form of frequency curve is symmetrical with the central maximum at 2 P.M.

(*b*) **Moderately Asymmetrical or Skew Curve.** A curve is said to be skew if there is no symmetry in the curve. In such curves, the class frequencies decrease more rapidly on one side of the maximum than on the other, one tail being always longer than the other. If it has a longer tail to the positive side (*i.e.,* towards the right of the maximum), it is said to be positively skewed. If it has a longer tail to the negative side (*i.e.,* towards the left of the maximum), it is said to be negatively skewed.

(**Note.** Fill up the 'no. in thousand' in the previous example yourself to get positive skew curve and negative skew curve.)

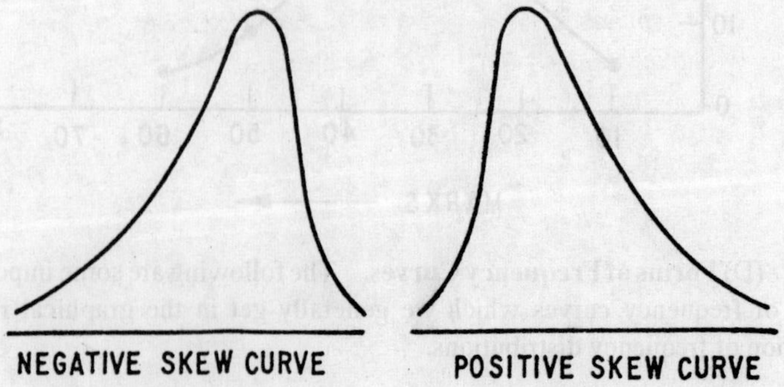

(*c*) **Extremely Asymmetrical or J-shaped curve.** When the class frequencies run upto a maximum at one end of the range, they form a J-shaped curve.

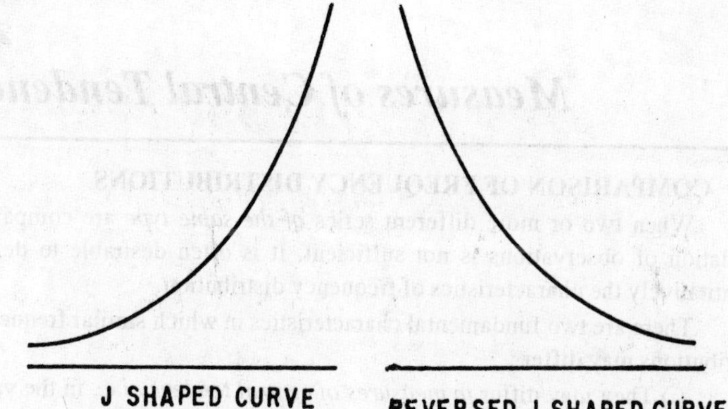

(**Note.** The form of the curve in previous example is J-shaped between 7 A.M. and 2 P.M. while it is reversed J-shaped between 2 P.M. and 9 P.M.).

(*d*) **U-shaped curve.** In this type of curve, the maximum frequency is at the ends and a minimum towards the centre.

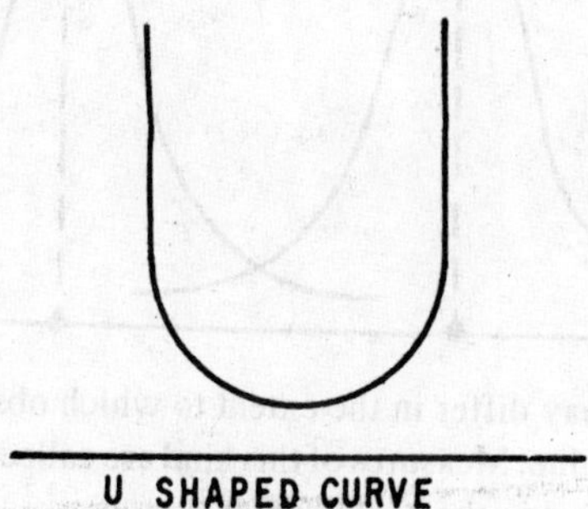

Example. *Give one illustration each of the type of data for which you would expect the frequency curve to be*

(*i*) *fairly symmetrical* (*ii*) *positively skewed*

(*iii*) *negatively skewed* (*iv*) *J-shaped.*

Sol. Please try yourself.

2

Measures of Central Tendency

2.1. COMPARISON OF FREQUENCY DISTRIBUTIONS

When two or more different series *of the same type* are compared, tabulation of observations is not sufficient. It is often desirable to define quantitatively the characteristics of frequency distribution.

There are two fundamental characteristics in which similar frequency distributions may differ :

(*i*) They may differ *in measures of central tendency i.e.,* in the value of the variate x round which they centre. The figure shows two frequency distributions which differ in central value. Measures of this kind are generally known as *Averages.*

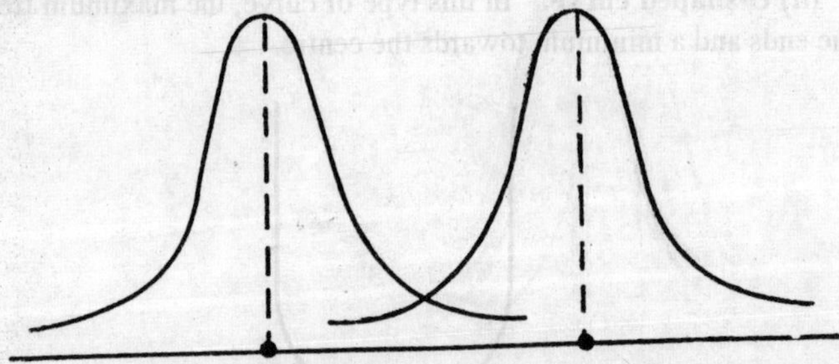

(*ii*) They may differ in the extent to which observations are scattered about the central value. Measures of this kind are called measures of dispersion and will be discussed in the next chapter.

2.2. MEASURES OF CENTRAL TENDENCY (OR AVERAGES)

Tabulation and graphical representation have considerably advanced our knowledge of the real significance of statistical data, but they have by no means completed it. Tabulation arranges facts in a logical order and helps their understanding and comparison. But often, the groups tabulated are still too large for their characteristics to be readily grasped. What is desired is a numerical expression which summarises the characteristic of the group. Measures of central tendency (also popularly called Averages) serve this purpose.

A figure which is used to represent a whole series should neither have the lowest value nor the highest in the series, but a value somewhere between these two limits, possibly in the centre, where most of the items of the series cluster. Such figures are called *Measures of Central Tendency* (or Averages).

According to Professor Bowley, "Averages are Statistical Constants which enable us to comprehend in a single effort the significance of the whole."

There are five types of averages in common use :

1. Arithmetic Average or Mean
2. Median
3. Mode
4. Geometric Mean
5. Harmonic Mean

We shall take them one by one.

1. ARITHEMTIC MEAN

In the case of Individual Observations (*i.e.*, where frequency is not given).

(*i*) **Direct Method.** If $x : x_1, x_2, \ldots\ldots, x_n$ then A.M. $\bar{x}$ is given by

$$\bar{x} = \frac{x_1 + x_2 + \ldots\ldots + x_n}{n} = \frac{1}{n} \Sigma x.$$

(*ii*) **Short cut Method** (Shift of origin). Shifting the origin to an *arbitrary point a*, the formula

$$\bar{x} = \frac{1}{n} \Sigma x \qquad \text{becomes} \qquad \bar{x} - a = \frac{1}{n} \Sigma(x - a)$$

or
$$\bar{x} = a + \frac{1}{n} \Sigma d_x \qquad \text{where} \qquad d_x = x - a$$

Thus
$$\bar{\mathbf{x}} = \mathbf{a} + \frac{\mathbf{1}}{\mathbf{n}} \Sigma \mathbf{d_x}$$

where a = an arbitrary number, called *Assumed Mean*

$\Sigma d_x = \Sigma(x - a) = (x_1 - a) + (x_2 - a) + \ldots\ldots + (x_n - a)$

= sum of the deviations of the variate x from a

n = number of observations.

In the case of Discrete Series.

(*i*) **Direct Method.** If the frequency distribution is

$x : \; x_1, \; x_2, \ldots\ldots, x_n$

$f : \; f_1, \; f_2, \ldots\ldots, f_n$, then

$$\bar{x} = \frac{f_1 x_1 + f_2 x_2 + \ldots\ldots + f_n x_n}{f_1 + f_2 + \ldots\ldots + f_n} = \frac{\Sigma fx}{\text{N}}$$

where $\text{N} = f_1 + f_2 + \ldots\ldots + f_n = \Sigma f.$

(*ii*) **Short cut Method.** (Shift of origin). Shifting the origin to an arbitrary point 'a', the formula

$$\bar{x} = \frac{1}{\text{N}} \Sigma fx \qquad \text{becomes} \qquad \bar{x} - a = \frac{1}{\text{N}} \Sigma f(x - a)$$

or $$\bar{x} = a + \frac{1}{N}\Sigma fd_x \quad \text{where} \quad d_x = x - a$$

Thus $$\bar{x} = a + \frac{1}{N}\Sigma \mathbf{fd_x}$$

where a = any assumed number

$$\Sigma fd_x = \Sigma f(x - a)$$
$$= f_1(x_1 - a) + f_2(x_2 - a) + \ldots\ldots + f_x(x_n - a)$$

= sum of the products of f and the deviation of the corresponding variate x from a

$$N = f_1 + f_2 + \ldots\ldots + f_n = \Sigma f.$$

Note. If the frequencies are given in terms of class intervals, the mid-values of class intervals are considered as x and then the above formulae are applied.

(*iii*) **In the case of Continuous Series having equal class intervals,** say of width h, we use a different formula (Shift of origin and change of scale ; Step Deviation Method).

Let $$u = \frac{x - a}{h} \quad \text{then} \quad x = a + hu$$

$$\therefore \quad \Sigma fx = \Sigma f(a + hu) = a\,\Sigma f + h\Sigma fu$$

Dividing both sides by $N = \Sigma f$, we get

$$\frac{\Sigma fx}{N} = a + \frac{h\Sigma fu}{N}$$

or $$\bar{x} = \mathbf{a} + \mathbf{h}\frac{\Sigma \mathbf{fu}}{N} \quad \text{where} \quad u = \frac{x - a}{h}.$$

Weighted Arithmetic Mean. If the variate-values are not of equal importance, we may attach to them 'weights' $w_1, w_2, \ldots\ldots, w_n$ as measures of their importance.

The weighted mean $\bar{x}_w$ is defined as

$$\bar{x}_w = \frac{w_1x_1 + w_2x_2 + \ldots\ldots + w_nx_n}{w_1 + w_2 + \ldots\ldots + w_n} = \frac{\Sigma wx}{\Sigma w}$$

(*i.e.*, write w for f).

Example 1. *Marks obtained by 9 students in statistics are given below :*

52, 75, 40, 70, 43, 40, 65, 35, 48.

Calculate their arithmetic mean.

Sol. Direct Method

$$\bar{x} = \frac{\Sigma x}{n} = \frac{52 + 75 + 40 + 70 + 43 + 40 + 65 + 35 + 48}{9}$$

$$= \frac{468}{9} = 52 \text{ marks.}$$

Short-cut Method

$$\bar{x} = a + \frac{\Sigma d_x}{n}.$$

Let assumed mean $a = 50$ (some no. round the middle)

Variate (x)	*Deviation from assumed mean 50* $d_x = x - a = x - 50$
52	(52 – 50 =) 2
75	(75 – 50 =) 25
40	(40 – 50 =) – 10
70	(70 – 50 =) 20
43	(43 – 50 =) – 7
40	(40 – 50 =) – 10
65	(65 – 50 =) 15
35	(35 – 50 =) –15
48	(48 – 50 =) – 2
	$\Sigma d_x = 62 - 44 = 18$

$$\therefore \quad \bar{x} = a + \frac{\Sigma d_x}{n} = 50 + \frac{18}{9} = 50 + 2 = 52 \text{ marks.}$$

(**Note.** The answer must be the same by both the methods.)

Example 2. *The following table gives the marks obtained by the students in an examination. Calculate the arithmetic mean by direct and short-cut methods* :

65, 85, 99, 100, 120, 122, 125, 132, 133, 140, 140, 146.

Sol. Please try yourself. [**Ans.** 117.25 marks]

Example 3. *Calculate the arithmetic mean of the distribution* :

Variate x :	5	10	15	20	25	30	35	40	45	50
Frequency f :	20	43	75	67	72	45	39	9	8	6

Sol. Direct Method

Variate (x)	*Frequency* (f)	*Product* fx
5	20	100
10	43	430
15	75	1125
20	67	1340
25	72	1800
30	45	1350

35	39	1365
40	9	360
45	8	360
50	6	300
	N = Σf = 384	Σfx = 8530

$$\therefore \quad \bar{x} = \frac{\Sigma fx}{N} = \frac{8530}{384} = 22.2.$$

Short-cut Method

Let the assumed mean = 30.

Variate (x)	*Frequency* (f)	*Dev. from ass. mean* $d_x = x - 30$	fd_x
5	20	− 25	− 500
10	43	− 20	− 860
15	75	− 15	− 1125
20	67	− 10	− 670
25	72	− 5	− 360
30	45	0	0
35	39	5	195
40	9	10	90
45	8	15	120
50	6	20	120
	N = Σf = 384		Σfd_x = − 3515 + 525 = − 2990

$$\bar{x} = a + \frac{\Sigma fd_x}{N} = 30 - \frac{2990}{384} = 30 - 7.8 = 22.2.$$

Example 4. *The following two rows give the variable and the corresponding frequency. Find the arithmetic mean by direct and short-cut methods.*

Variable x :	*1*	*2*	*3*	*4*	*5*
Frequency f :	*3*	*5*	*9*	*6*	*2*

Sol. Please try yourself. **[Ans.** 2.96]

Example 5. *Compute the mean of the following by both direct and short-cut methods :*

Height in cms :	*219*	*216*	*213*	*210*	*207*	*204*	*201*	*198*	*195*
No. of persons :	*2*	*4*	*6*	*10*	*11*	*7*	*5*	*4*	*1*

Sol. Please try yourself. **[Ans.** 207.54 cm]

Example 6. *Find the arithmetic mean for the following data :*

Marks	*No. of students*
0—10	*5*
10—20	*10*
20—30	*40*
30—40	*20*
40—50	*25*

Sol. First Method

$$\overline{x} = \frac{\Sigma fx}{N}$$

Marks	*Mid-values* (*x*)	*No. of students* (*f*)	*fx*
0—10	5	5	25
10—20	15	10	150
20—30	25	40	1000
30—40	35	20	700
40—50	45	25	1125
		$N = \Sigma f = 100$	$\Sigma fx = 3000$

$$\overline{x} = \frac{\Sigma fx}{N} = \frac{3000}{100} = 30 \text{ marks.}$$

Second Method

$$\overline{x} = a + \frac{\Sigma fd_x}{N}$$

Let assumed mean = 25.

Mid. values of class intervals (*x*)	*No. of students* (*f*)	*Dev. from ass. mean* $d_x = x - a$	fd_x
5	5	– 20	– 100
15	10	– 10	– 100
25	40	0	0
35	20	10	200
45	25	20	500
	$N = \Sigma f = 100$		$\Sigma fd_x = 500$

$$\overline{x} = a + \frac{\Sigma fd_x}{N} = 25 + \frac{500}{100}$$

$$= 25 + 5 = 30 \text{ marks.}$$

Third Method

$$\bar{x} = a + h\frac{\Sigma fu}{N} \qquad \text{where} \qquad u = \frac{x-a}{h}.$$

Let assumed mean $a = 25$

Here h = Width of equal class intervals = 10.

Mid-values of class intervals (x)	*No. of students (f)*	$u = \frac{x-a}{h}$	*fu*
5	5	– 2	– 10
15	10	– 1	– 10
25	40	0	0
35	20	1	20
45	25	2	50
	$N = \Sigma f = 100$		$\Sigma fu = 50$

$$\therefore \quad \bar{x} = a + h\frac{\Sigma hu}{N} = 25 + 10 \times \frac{50}{100} = 25 + 5 = 30 \text{ marks.}$$

Example 7. *Calculate the arithmetic mean of the marks from the following table :*

Marks :	*0—10*	*10—20*	*20—30*	*30—40*	*40—50*	*50—60*
No. of students :	*12*	*18*	*27*	*20*	*17*	*6*

Sol. Please try yourself. **[Ans.** 28 marks]

Example 8. *Make a frequency table having grades of wages with class-intervals of Rs. 2 each from the following data of daily wages received by 30 workers in a certain factory and then compute the average daily wage paid to workers :*

Daily wages in Rs. :

14, 16, 16, 14, 22, 13, 15, 24, 12, 23, 14, 20, 17, 21, 18, 18, 19, 20, 17, 16, 15, 11, 12, 21, 20, 17, 18, 19, 22, 23.

Sol. Tabular layout of data for computing average daily wages :

Wages in Rs.	*Mid-values (x)*		*Frequency f*	*fx*
11—13	12	|||	3	36
13—15	14	||||	4	56
15—17	16	~~||||~~	5	80
17—19	18	~~||||~~ |	6	108
19—21	20	~~||||~~	5	100
21—23	22	||||	4	88
23—25	24	|||	3	72
			$N = \Sigma f = 30$	$\Sigma fx = 540$

Average daily wage $\bar{x} = \frac{\Sigma fx}{N} = \frac{540}{30}$ = Rs. 18.

Example 9. *Given the following frequency distribution, calculate the arithmetic mean :*

Monthly wages	*No. of Workers*
12.5—17.5	2
17.5—22.5	22
22.5—27.5	19
27.5—32.5	14
32.5—37.5	3
37.5—42.5	4
42.5—47.5	6
47.5—52.5	1
52.5—57.5	1

Sol. Please try yourself. **[Ans.** Rs. 27.85]

Example 10. *The number of asthma sufferers whose first attacks come at various ages is given in the following table. Calculate the mean age at the first attack by any method :*

Age at first attack	*Number of cases*
0— 5	298
5—10	113
10—15	64
15—20	61
20—25	70
25—30	81
30—35	77
35—40	64
40—45	53
45—50	40
50—55	35
55—60	24
60—65	20

Sol. Please try yourself. **[Ans.** 20.8 years]

Example 11. *Following is the frequency distribution of yield of cane in tons per acre. Calculate the mean.*

Class-intervals	*Frequency*
35—40	7
40—45	8
45—50	12
50—55	26
55—60	32
60—65	42
65—70	42
70—75	15
75—80	17
80—85	9

Sol. Please try yourself. **[Ans.** 61.84 tons]

Example 12. *Find the average marks of students from the following table :*

Marks	*No. of students*	*Marks*	*No. of students*
Above 0	*80*	*Above 60*	*28*
Above 10	*77*	*Above 70*	*16*
Above 20	*72*	*Above 80*	*10*
Above 30	*65*	*Above 90*	*8*
Above 40	*55*	*Above 100*	*0*
Above 50	*43*		

Sol. The frequency table can be written as :

Marks	*Mid-values* (*x*)	*f*	$x - a$ (*a* = 55)	$u = \frac{x-a}{h}$ (*h* = 10)	*fu*
0—10	5	3	− 50	− 5	− 15
10—20	15	5	− 40	− 4	− 20
20—30	25	7	− 30	− 3	− 21
30—40	35	10	− 20	− 2	− 20
40—50	45	12	− 10	− 1	− 12
50—60	55	15	0	0	0
60—70	65	12	10	1	12
70—80	75	6	20	2	12
80—90	85	2	30	3	6
90—100	95	8	40	4	32
		N = Σ*f* = 80			Σ*fu* = − 26

Mean $\bar{x} = a + h\frac{\Sigma fu}{N} = 55 + 10 \times -\frac{26}{80}$

$= 55 - 3.25 = 51.75$ marks.

Example 13. *Find the mean from the following data :*

Marks	*No. of students*	*Marks*	*No. of students*
Below 10	5	*Below 60*	60
Below 20	9	*Below 70*	70
Below 30	17	*Below 80*	78
Below 40	29	*Below 90*	83
Below 50	45	*Below 100*	85

Sol. The frequency distribution table can be written as :

Marks	*Mid-values* (*x*)	*f*	$u = \frac{x-55}{10}$	*fu*
0— 10	5	5	−5	−25
10— 20	15	4	−4	−16
20— 30	25	8	−3	−24
30— 40	35	12	−2	−24
40— 50	45	16	−1	−16
50— 60	55	15	0	0
60— 70	65	10	1	10
70— 80	75	8	2	16
80— 90	85	5	3	15
90—100	95	2	4	8
	N = Σ*f* = 85			−56

Mean $\bar{x} = a + h\frac{\Sigma fu}{N} = 55 + 10 \times \frac{-56}{85}$ [Here $a = 55, h = 10$]

$= 55 - \frac{112}{17} = 55 - 6.59 = 48.41$ marks.

Example 14. *Find the average marks of students from the following cumulative frequency table :*

Marks	*No. of students*	*Marks*	*No. of students*
Below 10	15	*Below 50*	96
Below 20	35	*Below 60*	127
Below 30	60	*Below 70*	198
Below 40	84	*Below 80*	250

Sol. Please try yourself. **[Ans.** 50.4 marks]

Example 15. *Two hundred people were interviewed by a public opinion polling agency. The frequency distribution gives the ages of people interviewed.*

Age groups (years)	*Frequency*	*Age groups (years)*	*Frequency*
80—89	*2*	*40—49*	*56*
70—79	*2*	*30—39*	*40*
60—69	*6*	*20—29*	*42*
50—59	*20*	*10—19*	*32*

Calculate the value of mean.

Sol.

Age Group (years)	*Mid-values (x)*	*f*	$u = \frac{x - 44.5}{10}$	*fu*
80—89	84.5	2	4	8
70—79	74.5	2	3	6
60—69	64.5	6	2	12
50—59	54.5	20	1	20
40—49	44.5	56	0	0
30—39	34.5	40	– 1	– 40
20—29	24.5	42	– 2	– 84
10—19	14.5	32	– 3	– 96
		N = 200		– 174

$$\text{Mean } \overline{x} = a + h\frac{\Sigma fu}{N} = 44.5 + 10 \times \frac{-174}{200} = 44.5 - 8.7 = 35.8 \text{ years.}$$

Note. Here the class-intervals are inclusive and $h = 10$ (not 9) since 80—89 includes both 80 and 89.

Second Method. Let us convert inclusive class-intervals into exclusive class intervals so as to have continuous series.

Adjustment factor $= \frac{80 - 79}{2} = 0.5$. Subtracting 0.5 from lower class limits and adding 0.5 to upper class limits, the frequency distribution takes the form.

Age Groups (years)	*Mid-values (x)*	*f*	$u = \frac{x - 44.5}{10}$	*fu*
79.5—89.5	84.5	2	4	8
69.5—79.5	74.5	2	3	6
59.5—69.5	64.5	6	2	12
49.5—59.5	54.5	20	1	20
39.5—49.5	44.5	56	0	0
29.5—39.5	34.5	40	– 1	– 40
19.5—29.5	24.5	42	– 2	– 84
9.5—19.5	14.5	32	– 3	– 96
		N = 200		– 174

Mean $\quad \bar{x} = a + h\dfrac{\Sigma fu}{N} = 44.5 + 10 \times \dfrac{-174}{200}$

$= 44.5 - 8.7 = 35.8$ years.

☞**Observation.** *When the frequency distribution is given in terms of inclusive class-intervals, there is no need to convert it into a continuous series for calculating arithmetic mean because the mid-points remain the same whether or not the adjustment is made.*

☞**Caution.** The same is not true for median and mode. In their case the adjustment is a must *i.e.,* inclusive class-intervals must be converted into exclusive class intervals so as to have a continuous series.

Example 16. *The following are the monthly salaries in rupees of 30 employees of a firm :*

91	*139*	*126*	*119*	*100*	*87*	*65*	*77*	*99*	*95*
108	*127*	*86*	*148*	*116*	*76*	*69*	*88*	*112*	*118*
89	*116*	*97*	*105*	*95*	*80*	*86*	*106*	*93*	*135*

The firm gave bonuses of 10, 15, 20, 25, 30, 35, 40, 45 and 50 to employees in the respective salary groups, exceeding 60 but not exceeding 70, exceeding 70 but not exceeding 80; and so on upto exceeding 140 but not exceeding 150. Find the average bonus paid per employee.

Sol. (*i*) **Classification of data** (in terms of inclusive class-interval).

Monthly Salaries (in Rs.)	*Tally Marks*	*No. of employees (f)*
61—70	||	2
71—80	|||	3
81—90	||||/	5
91—100	||||/ ||	7
101—110	|||	3
111—120	||||/	5
121—130	||	2
131—140	||	2
141—150	|	1

(*ii*) Calculation of average bonus paid.

Bonus (in Rs.) (x)	*No. of employees* (f)	$u = \frac{x-30}{5}$	*fu*
10	2	−4	−8
15	3	−3	−9
20	5	−2	−10
25	7	−1	−7
30	3	0	0
35	5	1	5
40	2	2	4
45	2	3	6
50	1	4	4
	N = 30		−15

$$\text{Mean } \bar{x} = a + h\frac{\Sigma fu}{N} = 30 + 5 \times \frac{-15}{30} = 30 - 2.5 = 27.5$$

∴ Average bonus paid per employee = Rs. 27.5.

Example 17. *From the following data of income distribution, calculate the arithmetic mean. It is given that (i) the total income of persons in the highest income group is Rs. 435 and (ii) none is earning less than Rs. 20.*

Income (Rs.)	*No. of persons*	*Income (Rs.)*	*No. of persons*
Below 30	*16*	*Below 70*	*87*
" 40	*36*	*" 80*	*95*
" 50	*61*	*80 and above*	*5*
" 60	*76*		

Sol.

Income (Rs.)	*Mid-values* (x)	*No. of persons* (f)	*fx*
20—30	25	16	400
30—40	35	20	700
40—50	45	25	1125
50—60	55	15	825
60—70	65	11	715
70—80	75	8	600
80 and above	—	5	435 (given)
		N = 100	4800

$\therefore$ Mean $\bar{x} = \frac{\Sigma fx}{N} = \frac{4800}{100} = 48.$

Hence average income = Rs. 48.

Example 18. *The (arithmetic) mean height of 50 students of a college is 5′ – 8″. The height of 30 of these is given in the frequency distribution below. Find the (arithmetic) mean height of the remaining 20 students.*

Height :	*5′—4″*	*5′—6″*	*5′—8″*	*5′—10″*	*6′—0″*
Frequency :	*4*	*12*	*4*	*8*	*2*

Sol. Calculation of (arithmetic) mean height of 30 students :

' ′ ' stands for feet and ' ″ ' stand for inches. (1 foot = 12 inches.)

Height (in inches) (*x*)	*Frequency* (*f*)	$u = \frac{x-68}{2}$	*fu*
64	4	– 2	– 8
66	12	– 1	– 12
68	4	0	0
70	8	1	8
72	2	2	4
	N = 30		– 8

Mean $\bar{x} = a + h\frac{\Sigma fu}{N} = 68 + 2 \times \frac{-8}{30} = 68 - \frac{8}{15} = \frac{1012}{15}$

$\Rightarrow$ Mean height of 30 students $\frac{1012}{15}$ inches

$\therefore$ Total height of 30 students = $\frac{1012}{15} \times 30 = 2024$ inches ...(*i*)

Also mean height of 50 students = 5′ – 8″ = 68 inches

$\therefore$ Total height of 50 students = 68 × 50 = 3400 inches ...(*ii*)

Subtracting (*i*) from (*ii*)

Total height of remaining 20 students = 3400 – 2024

= 1376 inches

$\therefore$ Mean height of remaining 20 students = $\frac{1376}{20}$ = 68.8″ = 5′ 8.8″

Example 19. *Find the class intervals if the arithmetic mean of the following distribution is 33 and assumed mean is 35.*

Step deviation (u) :	*– 3*	*– 2*	*– 1*	*0*	*+ 1*	*+ 2*
Frequency (f) :	*5*	*10*	*25*	*30*	*20*	*10*

(M.D.U. 1982 S)

Sol.

Step deviation (u)	*Frequency* (f)	fu
− 3	5	− 15
− 2	10	− 20
− 1	25	− 25
0	30	0
+ 1	20	20
+ 2	10	20
	N = 100	− 20

Here $\overline{x} = 33$, assumed mean 'a' = 35 (given)

$\therefore \quad \overline{x} = a + h\dfrac{\Sigma fu}{N}$

$\therefore \quad 33 = 35 + h \times \dfrac{-20}{100} \Rightarrow -2 = -\dfrac{1}{5}h \Rightarrow h = 10$

$\therefore$ Size of each class-interval = 10

Now, $u = \dfrac{x-a}{h}$, where x is the mid-value of corresponding class-interval.

$\Rightarrow \quad -3 = \dfrac{x-35}{10} \Rightarrow -30 = x - 35 \Rightarrow x = 5$

$\therefore$ 5 is the mid-value of the first class-interval.

Hence the class-intervals are

0—10, 10—20, 20—30, 30—40, 40—50, 50—60.

Example 20. *For a frequency distribution of marks in Statistics for 100 students, the arithmetic average was found to be 50. Later on it was discovered that marks '48' were misread as '84'. Find the correct mean.*

Sol. Here incorrect value of $\overline{x} = 50$, $n = 100$

Since $\overline{x} = \dfrac{\Sigma x}{n} \quad \therefore \quad \Sigma x = n\overline{x}$

Using incorrect value of $\overline{x}$

Incorrect $\Sigma x = 100 \times 50 = 5000$

(Less) incorrect marks = 84

= 4916

(Add) correct marks = 48

= 4964

$\therefore$ Correct value of $\Sigma x = 4964$

$\therefore$ Correct mean $= \dfrac{\text{correct value of } \Sigma x}{n} = \dfrac{4964}{100} = 49.64$ marks.

Example 21. *The mean of 200 items was 50. Later on it was discovered that two items were misread as 92 and 8 instead of 192 and 88. Find out the correct mean.*

Sol. Here incorrect value of $\overline{x} = 50$, $n = 200$

Since $\overline{x} = \frac{\Sigma x}{n}$ $\therefore$ $\Sigma x = n\overline{x}$

Using incorrect value of $\overline{x}$,

Incorrect $\Sigma x = 200 \times 50 = 10000$

(Less) incorrect items $= 100$ $(\because\ 92 + 8 = 100)$

$= 9900$

(Add) correct items $= 280$ $(\because\ 192 + 88 = 280)$

$= 10180$

$\therefore$ Correct value of $\Sigma x = 10180$

$\therefore$ Correct mean $= \frac{\text{correct value of } \Sigma x}{n} = \frac{10180}{200} = 50.9.$

Example 22. *The average weight of a group of 25 boys was calculated to be 78.4 lbs. It was later discovered that one value was misread as 69 lbs. instead of the correct value 96 lbs. Calculate the correct average.*

Sol. Please try yourself **[Ans.** 79.48 lbs.]

Example 23. *A candidate obtains the following percentages in an examination. English 60 ; Hindi 75 ; Mathematics 63 ; Physics 59 and Chemistry 55. Find the weighted mean if weights 2, 1, 5, 5, 3 are allotted to the subjects. Also calculate the simple average.*

Sol.

Subjects	*Marks %* x	*Weight* w	wx
English	60	2	120
Hindi	75	1	75
Mathematics	63	5	315
Physics	59	5	295
Chemistry	55	3	165
	312	16	970

Weighted mean $\overline{x}_w = \frac{\Sigma wx}{\Sigma w} = \frac{970}{16} = 60.625\%$ marks

Simple average $\overline{x} = \frac{\Sigma x}{n} = \frac{312}{5} = 62.4\%$ marks.

Example 24. *A candidate obtains the following percentage of marks in an examination. English 75 ; Statistics 60 ; Mathematics 59 ; Physics 55 ; Chemistry 63.*

Find the weighted mean if weights 2, 1, 3, 3 and 1 respectively are allotted to the subjects. Calculate also the simple mean.

Sol. Please try yourself. [**Ans.** Weighted mean = 61.5% marks
Simple mean = 62.4% marks]

Example 25. *From the following data calculate the missing frequency :*

No. of tablets :	*4—8*	*8—12*	*12—16*	*16—20*	*20—24*	*24—28*	*28—32*
No of persons cured :	*11*	*13*	*16*	*14*	*?*	*9*	*17*

No. of tablets :	*32 – 36*	*36 – 40*
No. of persons cured :	*6*	*4*

The average number of tablets to cure a person is 20.

Sol.

Class	*Mid-value* x	*Frequency* f	$u = \frac{x-22}{4}$	*fu*
4— 8	6	11	– 4	– 44
8—12	10	13	– 3	– 39
12—16	14	16	– 2	– 32
16—20	18	14	– 1	– 14
20—24	22	α	0	0
24—28	26	9	1	9
28—32	30	17	2	34
32—36	34	6	3	18
36—40	38	4	4	16
		N = 90 + α		– 52

Using $\overline{x} = a + h\dfrac{\Sigma fu}{N}$, we have

$$20 = 22 + 4 \times \frac{-52}{90+\alpha} \qquad \therefore \quad \alpha = 14.$$

Example 26. *Show that the A.M. of the first n natural numbers is* $\frac{1}{2}(n+1)$.

Sol. Let $\overline{x}$ be the A.M. of the first n natural numbers.

Then, $$\overline{x} = \frac{1+2+3+\ldots+n}{n} = \frac{\frac{n(n+1)}{2}}{n} = \tfrac{1}{2}(n+1).$$

Example 27. *Show that the weighted arithmetic mean of first n natural numbers whose weights are equal to the corresponding numbers is equal to* $\frac{1}{3}(2n+1)$.

Sol. First n natural numbers are 1, 2, 3,......, n (x)

Corresponding weights are 1, 2, 3,......, n (w)

$$\therefore \quad \text{Weighted A.M. } \bar{x}_w = \frac{\Sigma wx}{\Sigma w} = \frac{1^2 + 2^2 + 3^2 + \ldots + n^2}{1 + 2 + 3 + \ldots + n}$$

$$= \frac{\Sigma n^2}{\Sigma n} = \frac{\frac{1}{6} n(n+1)(2n+1)}{\frac{1}{2}n(n+1)} = \frac{1}{3}(2n+1).$$

Example 28. *The arithmetic mean of n numbers of a series is $\bar{x}$. The sum of the first (n – 1) numbers is k. Show that the nth number is $n\bar{x} - k$.*

Sol. Let the n numbers be $x_1, x_2, x_3, \ldots, x_n$. Then

$$\bar{x} = \frac{\Sigma x}{n} = \frac{x_1 + x_2 + \ldots + x_{n-1} + x_n}{n} = \frac{k + x_n}{n}$$

or $\quad n\bar{x} = k + x_n \quad \therefore \quad x_n = n\bar{x} - k.$

Example 20. *If*

$$\bar{x}_1 = \frac{1}{n} \sum_{1}^{n} x_i, \quad \bar{x}_2 = \frac{1}{n} \sum_{2}^{n+1} x_i \quad \text{and} \quad \bar{x}_3 = \frac{1}{n} \sum_{3}^{n+2} x_i$$

then show that

$$(a) \quad \bar{x}_2 = \bar{x}_1 + \frac{1}{n}(x_{n+1} - x_1) \text{ and } \quad (b) \quad \bar{x}_3 = \bar{x}_2 + \frac{1}{n}(x_{n+2} - x_2).$$

Sol. $\bar{x}_1 = \frac{1}{n}\sum_{1}^{n} x_i = \frac{1}{n}(x_1 + x_2 + x_3 + \ldots + x_n)$

$$\bar{x}_2 = \frac{1}{n}\sum_{2}^{n+1} x_i = \frac{1}{n}(x_2 + x_3 + x_4 + \ldots + x_n + x_{n+1})$$

$$\bar{x}_3 = \frac{1}{n}\sum_{3}^{n+2} x_i = \frac{1}{n}(x_3 + x_4 + x_5 + \ldots + x_{n+1} + x_{n+2})$$

Now $\quad \bar{x}_2 - \bar{x}_1 = \frac{1}{n}(x_{n+1} - x_1)$

$$\therefore \quad \bar{x}_2 = \bar{x}_1 + \frac{1}{n}(x_{n+1} - x_1)$$

And $\quad \bar{x}_3 - \bar{x}_2 = \frac{1}{n}(x_{n+2} - x_2)$

$$\therefore \quad \bar{x}_3 = \bar{x}_2 + \frac{1}{n}(x_{n+2} - x_2).$$

Example 30. *The frequencies of values 0, 1, 2,..., n of a variable are given by*

$$q^n, {}^nC_1\, q^{n-1}p, {}^nC_2 q^{n-2}p^2 \ldots, p^n$$

where $p + q = 1$. Show that the mean is np.

Sol. Here $x: \quad 0, \quad 1, \quad 2, \ldots\ldots, n$

$f: \quad q^n, \quad {}^nC_1 q^{n-1}p, \quad {}^nC_2 q^{n-2}p^2, \quad \ldots\ldots, p^n$

$$\therefore \quad \text{Mean } \bar{x} = \frac{\Sigma fx}{\Sigma f}$$

$$= \frac{q^n(0) + {}^nC_1 q^{n-1}p(1) + {}^nC_2 q^{n-2}p^2(2) + \ldots + p^n(n)}{q^n + {}^nC_1 q^{n-1}p + {}^nC_2 q^{n-2}p^2 + \ldots + p^n}$$

$$= \frac{{}^nC_1 q^{n-1}p + 2\,{}^nC_2 q^{n-2}p^2 + \ldots + np^n}{(q+p)^n}$$

$$= nq^{n-1}p + 2 \cdot \frac{n(n-1)}{2.1} q^{n-2}p^2 + \ldots + np^n \quad [\because p + q = 1]$$

$$= np[q^{n-1} + (n-1)q^{n-2}p + \ldots + p^{n-1}]$$

$$= np(q+p)^{n-1} = np. \quad [\because p + q = 1]$$

Example 31. *What is the A.M. of the following data ?*

Variate (x)	:	0	1	2	......	n
Frequency (f)	:	${}^nC_0$	${}^nC_1$	${}^nC_2$	......	${}^nC_n$.

Sol. $\bar{x} = \dfrac{\Sigma fx}{\Sigma f}$

Now $\Sigma f = {}^nC_0 + {}^nC_1 + {}^nC_2 + \ldots\ldots + {}^nC_n$

$$= (1+1)^n = 2^n$$

$$\Sigma fx = 0 \,.\, {}^nC_0 + 1 \,.\, {}^nC_1 + 2 \,.\, {}^nC_2 + 3 \,.\, {}^nC_3 + \ldots + n \,.\, {}^nC_n$$

$$= n + 2 \cdot \frac{n(n-1)}{2.1} + 3 \cdot \frac{n(n-1)(n-2)}{3.2.1} + \ldots + n \,.\, 1$$

$$= n\left[1 + (n-1) + \frac{(n-1)(n-2)}{2.1} + \ldots + 1\right]$$

$$= n[{}^{n-1}C_0 + {}^{n-1}C_1 + {}^{n-1}C_2 + \ldots + {}^{n-1}C_{n-1}]$$

$$= n(1+1)^{n-1} = n \,.\, 2^{n-1}$$

$$\therefore \quad \bar{x} = \frac{n \,.\, 2^{n-1}}{2^n} = \frac{n}{2}.$$

2.3. PROPERTIES OF ARITHMETIC MEAN

Property I. *The algebraic sum of the deviations of all the variates from their arithmetic mean is zero.* (M.D.U. 1981 S)

Proof. Let d_x be the deviation of the variate x from the mean $\bar{x}$, then

$$d_x = x - \bar{x}$$

$$\therefore \quad \Sigma f d_x = \Sigma f(x - \bar{x}) = \Sigma fx - \bar{x}\Sigma f$$

$$= N\bar{x} - N\bar{x} = 0 \qquad \Big| \because \bar{x} = \frac{\Sigma fx}{N}, \text{ where } N = \Sigma f.$$

Property II. *The sum of the squares of the deviations of a set of values is minimum when taken about mean.*

Proof. Let the frequency distribution be x_i/f_i, $i = 1, 2, ..., n$. Let z be the sum of the squares of the deviations of the given values from an arbitrary point a (say).

$$\Rightarrow \quad \text{Let} \qquad z = \sum_{i=1}^{n} f(x-a)^2$$

We have to show that z is minimum when $a = \bar{x}$

z will be minimum when

$$\frac{\partial z}{\partial a} = 0 \qquad \text{and} \qquad \frac{\partial^2 z}{\partial a^2} > 0$$

Now $$\frac{\partial z}{\partial a} = \sum_{i=1}^{n} 2f(x-a).(-1) = -2\sum_{i=1}^{n} f(x-a)$$

$\therefore$ $$\frac{\partial z}{\partial a} = 0 \quad \Rightarrow \quad -2\Sigma f(x-a) = 0$$

$$\Rightarrow \quad \Sigma fx - a\Sigma f = 0$$

$$\Rightarrow \quad N\bar{x} - aN = 0 \qquad \left[\because \quad \bar{x} = \frac{\Sigma fx}{N}, \Sigma f = N\right]$$

$$\Rightarrow \quad \bar{x} - a = 0 \qquad (\because \quad N = \Sigma f \neq 0)$$

$$\Rightarrow \quad a = \bar{x}$$

Also $$\frac{\partial^2 z}{\partial a^2} = -2\sum_{i=1}^{n} f(-1) = 2\Sigma f = 2N > 0$$

Hence z is minimum when $a = \bar{x}$.

Property III. (*Mean of the composite series*)

If $\bar{x}_i$ $(i = 1, 2, ..., k)$ be the arithmetic means of k distributions with respective frequencies n_i $(i = 1, 2, ..., k)$ then mean $\bar{x}$ of the whole distribution obtained by combining the k distributions is given by

$$\bar{x} = \frac{n_1\bar{x}_1 + n_2\bar{x}_2 + ... + n_k\bar{x}_k}{n_1 + n_2 + ... + n_k} = \frac{\sum_i n_i\bar{x}_i}{\sum_i n_i}$$

Proof. Let $x_{11}, x_{12}, x_{13}, ..., x_{1n_1}$ be the variables of the first distribution, $x_{21}, x_{22}, ..., x_{2n_2}$ be the variables of the second distribution, and so on. Then by def.

$$\left.\begin{aligned} \bar{x}_1 &= \frac{1}{n_1}(x_{11} + x_{12} + ... + x_{1n_1}) \\ \bar{x}_2 &= \frac{1}{n_2}(x_{21} + x_{22} + ... + x_{2n_2}) \\ &\cdots\cdots\cdots\cdots\cdots\cdots\cdots\cdots \\ \bar{x}_k &= \frac{1}{n_k}(x_{k_1} + x_{k_2} + ... + x_{kn_k}) \end{aligned}\right\} \quad (A)$$

The mean $\overline{x}$ of the whole distribution of size $(n_1 + n_2 + ... + n_k)$ is given by

$$\overline{x} = \frac{(x_{11} + x_{12} + \ldots + x_{1n_1}) + (x_{21} + x_{22} + \ldots x_{2n_2}) + \ldots + (x_{k_1} + x_{k_2} + \ldots + x_{kn_k})}{n_1 + n_2 + ... + n_k}$$

$$= \frac{n_1\overline{x}_1 + n_2\overline{x}_2 + \ldots + n_k\overline{x}_k}{n_1 + n_2 + ... + n_k} \qquad \text{[Using (A)]}$$

$$= \frac{\sum_i n_i\overline{x}_i}{\sum_i n_i}.$$

Example 1. *The mean wage of 200 workers working in a factory is Rs. 50. The mean wage of 75 workers of the first shift is Rs. 60. Find the mean wage of the rest.*

Sol. No. of workers in first shift $(n_1) = 75$

$\therefore$ No. of other workers $(n_2) = 200 - 75 = 125$

Mean wage of workers in first shift $(\overline{x}_1)$ = Rs. 60

Mean wage of all the workers $(\overline{x})$ = Rs. 50

Let the mean wage of the rest be $\overline{x}_2$

Using $\overline{x} = \dfrac{n_1\overline{x}_1 + n_2\overline{x}_2}{n_1 + n_2}$, we have

$$50 = \frac{75 \times 60 + 125\,\overline{x}_2}{200}$$

or $$10000 - 4500 = 125\,\overline{x}_2$$

$$\therefore \quad \overline{x}_2 = \frac{5500}{125} = \text{Rs. } 44.$$

Example 2. *The mean wage of 500 workers in a factory running two shifts of 360 and 140 workers respectively is Rs. 70. The mean wage of 360 workers working in the day shift is Rs. 75. Find the mean wage of 140 workers working in night shift.*

Sol. Please try yourself. **[Ans.** Rs. 57.14]

Example 3. *A distribution consists of three components with frequencies 300, 200 and 600 having their means 16, 8 and 4 respectively. Find the mean of the combined distribution.*

Sol. Here $n_1 = 300, \quad \overline{x}_1 = 16, \quad n_2 = 200, \quad \overline{x}_2 = 8,$

$n_3 = 600, \quad \overline{x}_3 = 4$

Let $\overline{x}$ be the mean of the combined distribution, then

$$\overline{x} = \frac{n_1\overline{x}_1 + n_2\overline{x}_2 + n_3\overline{x}_3}{n_1 + n_2 + n_3}$$

$$= \frac{4800 + 1600 + 2400}{300 + 200 + 600} = \frac{8800}{1100} = 8.$$

Example 4. *The mean of marks obtained in an examination by a group of 100 students was found to be 49.96. The mean of the marks obtained in the same examination by another group of 200 students was 52.32. Find the mean of the marks obtained by both the groups of students taken together.*

Sol. Please try yourself. **[Ans.** 51.53]

Example 5. *The mean marks obtained by 300 students in the subject of Statistics are 45. The mean of the top 100 of them was found to be 70 and the mean of the last 100 was known to be 20. What is the mean of the remaining 100 students ?*

Sol. Let the mean of the remaining 100 students be $\bar{x}_3$

Here $\bar{x} = 45,\quad n_1 = 100,\quad \bar{x}_1 = 70$

$n_2 = 100,\quad \bar{x}_2 = 20,\quad n_3 = 100$

Using $\bar{x} = \dfrac{n_1\bar{x}_1 + n_2\bar{x}_2 + n_3\bar{x}_3}{n_1 + n_2 + n_3}$, we have

$$45 = \frac{7000 + 2000 + 100\bar{x}_3}{300}$$

or $13500 - 9000 = 100\,\bar{x}_3$

$$\therefore \quad \bar{x}_3 = \frac{4500}{100} = 45.$$

Example 6. *The mean weight of 150 students in a certain class is 60 kg. The mean weight of boys in the class is 70 kg. and that of the girls is 55 kg. Find the number of boys and the number of girls in the class.*

Sol. Let number of boys = n_1 and number of girls = n_2

Mean weight of boys $(\bar{x}_1)$ = 70 kg.

Mean weight of girls $(\bar{x}_2)$ = 55 kg.

Mean weight of all students $(\bar{x})$ = 60 kg.

Also $n_1 + n_2 = 150$ ...(1)

Using $\bar{x} = \dfrac{n_1\bar{x}_1 + n_2\bar{x}_2}{n_1 + n_2}$, we have

$$60 = \frac{70n_1 + 55n_2}{150}$$

or $70n_1 + 55n_2 = 9000$

or $14n_1 + 11n_2 = 1800$

or $14n_1 + 11(150 - n_1) = 1100$ [Using (1)]

or $3n_1 = 1800 - 1650 = 150$

$\therefore \quad n_1 = 50, \quad n_2 = 150 - 50 = 100$

Hence No. of boys = 50, No. of girls = 100.

Example 7. *The mean annual salary paid to all employees of a company was Rs. 5000. The mean annual salaries paid to male and female* *employees were Rs. 5200 and Rs. 4200 respectively. Determine the percentage of males and females employed by the company.*

Sol. Let p_1 and p_2 represent percentage of males and females respectively.

Then $p_1 + p_2 = 100$ ...(1)

Mean annual salary of all employees $(\bar{x})$ = Rs. 5000

Mean annual salary of males $(\bar{x}_1)$ = Rs. 5200

Mean annual salary of females $(\bar{x}_2)$ = Rs. 4200

Using $\bar{x} = \dfrac{p_1\bar{x}_1 + p_2\bar{x}_2}{p_1 + p_2}$, we get

$$5000 = \frac{5200p_1 + 4200p_2}{100}$$

or $5200p_1 + 4200p_2 = 500000$

or $26p_1 + 21p_2 = 2500$

or $26p_1 + 21(100 - p_1) = 2500$ | Using (1)

or $5p_1 = 2500 - 2100 = 400$

$\therefore$ $p_1 = 80$ and $p_2 = 100 - 80 = 20$

Hence the percentage of males and females is 80 and 20 respectively.

Example 8. *In a certain examination, the average grade of all students in class A is 68.4 and that of all students in class B is 71.2. If the average of both classes combined is 70, find the ratio of the number of students in class A to the number in class B.*

Sol. Let the number of students in class A be n_1 and in class B be n_2.

Average grade of all students $(\bar{x}) = 70$

Average grade of all students in class A = $\bar{x}_1 = 68.4$

Average grade of all students in class B = $\bar{x}_2 = 71.2$

Using $\bar{x} = \dfrac{n_1\bar{x}_1 + n_2\bar{x}_2}{n_1 + n_2}$, we have

$$70 = \frac{68.4n_1 + 71.2n_2}{n_1 + n_2}$$

$\Rightarrow$ $70n_1 + 70n_2 = 68.4n_1 + 71.2n_2$

$\Rightarrow$ $1.6n_1 = 1.2n_2$

$\Rightarrow$ $\dfrac{n_1}{n_2} = \dfrac{1.2}{1.6} = \dfrac{12}{16} = \dfrac{3}{4}$

$\therefore$ Reqd. ratio = $n_1 : n_2 = 3 : 4$.

2. MEDIAN

1. Median is the central value of the variable **when the values are arranged in ascending or descending order of magnitude.** When the observations are arranged in the order of their size, the median is the value of that item which has equal number of observations on either side. Median divides the distribution into two equal parts. Median is, thus, a potential average.

☞For the computation of median, it is necessary that the items be arranged in ascending or descending order.

2. For an ungrouped frequency distribution, if the n values of the variate are arranged in ascending or descending order of magnitude.

(*a*) *When n is odd,* the middle value *i.e.,* $\frac{1}{2}(n+1)$th value gives the median.

(*b*) *When n is even,* there are two middle values $\frac{1}{2}n$th and $(\frac{1}{2}n+1)$th. The arithmetic mean of these two values gives the median.

3. For discrete frequency distribution, median is obtained by considering cumulative frequencies. Find $\frac{N+1}{2}$, where $N = \sum_i f_i$. See the cumulative frequency just greater than $\frac{N+1}{2}$. The corresponding value of x is the median.

4. For a grouped frequency distribution, the median is given by the formula,

$$\textbf{Median} = l + \frac{h}{f}\left(\frac{N}{2} - C\right)$$

where, l = lower limit of median class, where median class is the class corresponding to cumulative frequency just greater than $\frac{N}{2}$

h = the width of median class

f = the frequency of the median class.

N = Σf

C = cumulative frequency of the class **preceding** the median class.

5. Partition Values. These are the values of the variate which divide the total frequency into a number of equal parts. Median being that value of the variate which divides the total frequency into two equal parts.

(*a*) **Quartiles.** Quartiles are those values of the variate which divide the total frequency into four equal parts. When the lower half before the median is divided into two equal parts, the value of the dividing variate is called Lower Quartile and is denoted by Q_1. The value of the variate dividing the upper half into two equal parts is called the Upper Quartile and is denoted by Q_3. (Q_2 being the median). The formulae for computation are :

$$Q_1 = l + \frac{h}{f}\left(\frac{N}{4} - C\right)$$

$$Q_3 = l + \frac{h}{f}\left(\frac{3N}{4} - C\right).$$

(*b*) **Deciles.** Deciles are those values of the variate which divide the total frequency into 10 equal parts. If D_1, D_2...... denote respectively the first, second,... deciles, then

$$D_1 = l + \frac{h}{f}\left(\frac{N}{10} - C\right),$$

$$D_2 = l + \frac{h}{f}\left(\frac{2N}{10} - C\right),$$

$$D_3 = l + \frac{h}{f}\left(\frac{3N}{10} - C\right)$$

(The fifth decile D_5 is the median).

(*c*) **Percentiles.** Percentiles are those values of the variate which divide the total frequency into 100 equal parts. If P_1, P_2,...... denote respectively the first, second...... percentiles, then

$$P_1 = l + \frac{h}{f}\left(\frac{N}{100} - C\right),$$

$$P_2 = l + \frac{h}{f}\left(\frac{2N}{100} - C\right) \text{ etc.}$$

(The 50th percentile P_{50} is the median.)

In the above formulae for Quartiles, Deciles and Percentiles, the letters *l*, *i*, *f*, N, C have been used in the same sense in which they have been used in the formula for the median.

Example 1. *Obtain the median and quartiles from the following data :* *5, 7, 9, 11, 13, 15, 17.*

Sol. The data is arranged in ascending order. Number of items $n = 7$ (odd).

Median = size of middle item

$= \text{size of } \frac{n+1}{2}$ th, *i.e.*, 4th item = 11

$Q_1 = \text{size of } \frac{n+1}{4}$ th, *i.e.*, 2nd item = 7

$Q_3 = \text{size of } \frac{3(n+1)}{4}$ th, *i.e.*, 6th item = 15.

Example 2. *Find the median of the following :* *20, 18, 22, 27, 25, 12, 15.*

Sol. Arranging the data in ascending order

12, 15, 18, 20, 22, 25, 27.

Number of items $n = 7$ (odd)

$$\text{Median} = \text{size of } \frac{n+1}{2} \text{ th, } i.e., \text{ 4th item} = 20.$$

Example 3. *Fifteen students took a test. The marks obtained by 10 student, who have cleared the test, are given below :*

7, 8, 10, 9, 18, 15, 12, 11, 14, 17.

Find out the median marks.

Sol. The five students who did not clear the test get marks less than 7. Arranging the marks in ascending order

1 2 3 4 5	6	7	8	9	10	11	12	13	14	15
Less than 7	7,	8,	9,	10,	11,	12,	14,	15,	17,	18

Number of students = 15 (odd)

$$\text{Median marks} = \text{marks of } \frac{15+1}{2} \text{ th, } i.e., \text{ 8th student} = 9.$$

Example 4. *Find out the median from the following data :*

2, 30, 12, 25, 20, 8, 10, 4, 15.

Sol. Please try yourself. **[Ans.** 12]

Example 5. *According to the census of 1981, following are the population figures, in thousand, of 10 cities :*

2000, 1180, 1785, 1500, 560, 782, 1200, 385, 1123, 222.

Find the median.

Sol. Arranging the data in ascending order

222, 385, 560, 782, 1123, 1180, 1200, 1500, 1785, 2000.

Number of items = 10 (even)

$$\therefore \quad \text{Median} = \text{A.M. of size of } \frac{n}{2} \text{ th and } \left(\frac{n}{2}+1\right) \text{ th items}$$

$$= \text{A.M. of 5th and 6th items}$$

$$= \frac{1123 + 1180}{2} = 1151.5 \text{ thousands.}$$

Example 6. *Determine the median from the following data :*

30, 12, 25, 38, 2, 10, 4, 8, 15, 20.

Sol. Please try yourself. **[Ans.** 13.5]

Example 7. *Below are given the marks obtained by a batch of 20 students in a certain class in Mathematics and Physics :*

Roll Nos. :	*1*	*2*	*3*	*4*	*5*	*6*	*7*	*8*	*9*	*10*
Marks in Maths. :	*53*	*54*	*52*	*32*	*30*	*60*	*47*	*46*	*35*	*28*
Marks in Physics :	*58*	*55*	*25*	*32*	*26*	*85*	*44*	*80*	*33*	*72*

Roll. Nos.	:	*11*	*12*	*13*	*14*	*15*	*16*	*17*	*18*	*19*	*20*
Marks in Maths.	:	*25*	*42*	*33*	*48*	*72*	*51*	*45*	*33*	*65*	*29*
Marks in Physics	:	*10*	*42*	*15*	*46*	*50*	*64*	*39*	*38*	*30*	*36*

In which subject is the level of knowledge of the students higher ?

Sol. To find out the subject in which the level of knowledge of the students is higher, we find out the medians of both the series. The subject for which the median value is higher will be the subject in which the level of knowledge of the students is higher. Let us arrange the marks in ascending order of magnitude.

S. No.	*Marks in Maths.*	*Marks in Physics*	*S. No.*	*Marks in Maths.*	*Marks in Physics*
1	25	10	11	46	42
2	28	15	12	47	44
3	29	25	13	48	46
4	30	26	14	51	50
5	32	30	15	52	55
6	33	32	16	53	58
7	33	33	17	54	64
8	35	36	18	60	72
9	42	38	19	65	80
10	45	39	20	72	85

Number of items in each case

= 20 (even)

Median marks in Mathematics

$$= \text{A.M. of size of } \left(\frac{20}{2}\right) \text{th and } \left(\frac{20}{2}+1\right) \text{th items}$$

= A.M. of sizes of 10th and 11th items

$$= \frac{45+46}{2} = 45.5.$$

Median marks in Physics

= A.M. of sizes of 10th and 11th items

$$= \frac{39+42}{2} = 40.5.$$

Since the median marks in Mathematics are greater than the median marks in Physics, the level of knowledge in Mathematics is higher.

Example 8. *Below are given the marks obtained by a batch of 25 students in a certain test in Mathematics and English :*

Roll Nos.	:	*1*	*2*	*3*	*4*	*5*	*6*	*7*	*8*	*9*	*10*
Marks in Maths.	:	*29*	*65*	*33*	*45*	*51*	*72*	*48*	*33*	*42*	*25*
Marks in English	:	*36*	*30*	*38*	*39*	*64*	*50*	*46*	*15*	*42*	*10*

Roll Nos.	:	*11*	*12*	*13*	*14*	*15*	*16*	*17*	*18*	*19*	*20*
Marks in Maths.	:	*28*	*35*	*46*	*47*	*60*	*30*	*32*	*52*	*54*	*56*
Marks in English	:	*72*	*33*	*80*	*44*	*85*	*20*	*32*	*25*	*55*	*28*
Roll Nos.	:	*21*	*22*	*23*	*24*	*25*					
Marks in Maths.	:	*58*	*49*	*38*	*40*	*46*					
Marks in English	:	*53*	*35*	*40*	*62*	*58*					

In which subject is the level of knowledge of the students higher ?

Sol. Please try yourself. **[Ans.** Mathematics]

Example 9. *Obtain the median for the following frequency distribution :*

x :	*1*	*2*	*3*	*4*	*5*	*6*	*7*	*8*	*9*
f :	*8*	*10*	*11*	*16*	*20*	*25*	*15*	*9*	*6*

Sol. The cumulative frequency distribution table is given below :

x	f	*c.f.*
1	8	8
2	10	18
3	11	29
4	16	45
5	20	65
6	25	90
7	15	105
8	9	114
9	6	120

Here N = 120 $\therefore \quad \frac{N+1}{2} = 60.5$

Cumulative frequency just greater than $\frac{N+1}{2}$ is 65 and the value of x corresponding to *c.f.* 65 is 5. Hence median is 5.

Example 10. *Find the median from the following table :*

x :	*5*	*7*	*9*	*11*	*13*	*15*	*17*	*19*
f :	*1*	*2*	*7*	*9*	*11*	*8*	*5*	*4*

Sol. Please try yourself. **[Ans.** 13]

Example 11. *Find the median from the following table :*

Marks	*No. of students*	*Marks*	*No. of students*
0—10	*2*	*40—50*	*35*
10—20	*18*	*50—60*	*20*
20—30	*30*	*60—70*	*6*
30—40	*45*	*70—80*	*3*

Sol. Let us calculate cumulative frequencies.

Marks	*No. of students*	*c.f.*	*Marks*	*No. of students*	*c.f.*
0—10	2	2	40—50	35	130
10—20	18	20	50—60	20	150
20—30	30	50	60—70	6	156
30—40	45	95	70—80	3	159

Here N = 159

Median class is the class corresponding to *c.f.* just greater than

$$\frac{N}{2} = 79.5.$$

Hence 30—40 is the median class

l = lower limit of median class = 30

h = width of median class = 40 – 30 = 10

f = frequency of median class = 45

C = *c.f.* of the class preceding the median class = 50

$$\therefore \quad \text{Median} = l + \frac{h}{f}\left(\frac{N}{2} - C\right)$$

$$= 30 + \frac{10}{45}(79.5 - 50) = 30 + 6.56 = 36.56.$$

Example 12. *Find the median for the following distribution :*

Wages in Rs. :	*0—10*	*10—20*	*20—30*	*30—40*	*40—50*
No. of workers :	*22*	*38*	*46*	*35*	*20*

Sol. Please try yourself. **[Ans.** Rs. 24.46]

Example 13. *Calculate the mean and median from the following table :*

Class Intervals	*Frequency*
6.5— 7.5	*5*
7.5— 8.5	*12*
8.5— 9.5	*25*
9.5—10.5	*48*
10.5—11.5	*32*
11.5—12.5	*6*
12.5—13.5	*1*

Sol. Let assumed mean 'a' = 10

Class Intervals	*Mid-values* (x)	*Frequency* (f)	$d_x = x - 10$	fd_x	*c.f.*
6.5—7.5	7	5	− 3	− 15	5
7.5—8.5	8	12	− 2	− 24	17
8.5—9.5	9	25	− 1	− 25	42
9.5—10.5	10	48	0	0	90
10.5—11.5	11	32	1	32	122
11.5—12.5	12	6	2	12	128
12.5—13.5	13	1	3	3	129
		N = Σf = 129		− 17	

Mean $\bar{x} = a + \frac{\Sigma f d_x}{N} = 10 - \frac{17}{129} = 10 - .133 = 9.867$

$$\frac{N}{2} = \frac{129}{2} = 64.5 \quad \therefore \quad \text{Median class is } 9.5 - 10.5$$

$$l = 9.5, \quad h = 1, \quad f = 48, \quad C = 42$$

$$\therefore \quad \text{Median} = l + \frac{h}{f}\left(\frac{N}{2} - C\right) = 9.5 + \frac{1}{48}(64.5 - 42)$$

$$= 9.5 + \frac{22.5}{48} = 9.5 + .47 = 9.97.$$

Example 14. *Calculate the median from the following data :*

Mid. values	*Frequency*	*Mid. values*	*Frequency*
15	*6*	*65*	*7*
25	*8*	*75*	*9*
35	*10*	*85*	*4*
45	*2*	*95*	*6*
55	*4*		

Sol. We are given mid-values. The difference between two consecutive mid-values is 10. We reconstruct the table with upper and lower limits of various classes.

Class Intervals	*Frequency*	*Cumulative Frequency*
10—20	6	6
20—30	8	14
30—40	10	24
40—50	2	26
50—60	4	30
60— 70	7	37
70— 80	9	46
80— 90	4	50
90—100	6	56

Here $N = 56$

$\frac{N}{2} = 28$ $\therefore$ Median class is 50—60

$l = 50,\quad h = 10,\quad f = 4,\quad C = 26$

$$\therefore \text{ Median} = l + \frac{h}{f}\left(\frac{N}{2} - C\right)$$

$$= 50 + \frac{10}{4}(28 - 26) = 55.$$

Example 15. *Compute the median from the following data :*

Mid-value	*Frequency*	*Mid-value*	*Frequency*
115	*6*	*165*	*60*
125	*25*	*175*	*38*
135	*48*	*185*	*22*
145	*72*	*195*	*3*
155	*116*		

Sol. Please try yourself. **[Ans.** 153.8]

Example 16. *Find the median, lower and upper quartiles from the following table :*

Marks	*No. of students*
Below 10	15
,, 20	35
,, 30	60
,, 40	84
,, 50	94
,, 60	127
,, 70	198
,, 80	249

Sol. From the above table, we reconstruct the *c.f.* table with class intervals

Marks	*No. of students (f)*	*c.f.*
0—10	15	15
10—20	20	35
20—30	25	60
30—40	24	84
40—50	10	94
50—60	33	127
60—70	71	198
70—80	51	249

Here $\quad N = 249$

(*i*) **Calculation of Median**

$$\frac{N}{2} = 124.5 \quad \therefore \quad \text{Median class is } 50\text{—}60$$

$$l = 50, \quad h = 10, \quad f = 33, \quad C = 94$$

$$\therefore \quad \text{Median} = l + \frac{h}{f}\left(\frac{N}{2} - C\right) = 50 + \frac{10}{33}(124.5 - 94)$$

$$= 50 + \frac{305}{33} = 50 + 9.24 = 59.24 \text{ marks.}$$

(*ii*) **Calculation of lower quartile Q_1**

$$\frac{N}{4} = 62.25 \quad \therefore \quad \text{lower quartile class is } 30\text{—}40$$

$$l = 30, \quad h = 10, \quad f = 24, \quad C = 60$$

$$\therefore \quad Q_1 = l + \frac{h}{f}\left(\frac{N}{4} - C\right) = 30 + \frac{10}{24}(62.25 - 60)$$

$$= 30 + \frac{22.5}{24} = 30 + .94 = 30.94 \text{ marks}$$

(*iii*) **Calculation of upper quartile Q_3**

$$\frac{3N}{4} = \frac{747}{4} = 186.75 \quad \therefore \quad \text{upper quartile class is } 60\text{—}70$$

$$l = 60, \quad h = 10, \quad f = 71, \quad C = 127$$

$$\therefore \quad Q_3 = l + \frac{h}{f}\left(\frac{3N}{4} - C\right) = 60 + \frac{10}{71}(186.75 - 127)$$

$$= 60 + \frac{597.5}{71} = 60 + 8.41 = 68.41 \text{ marks.}$$

Example 17. *Find the median and quartiles for the following distribution :*

Variable :	*100—200*	*200—300*	*300—400*	*400—500*	*500—600*
Frequency :	*15*	*18*	*30*	*0*	*17*

Sol. Please try yourself. **[Ans.** 356.7, 255.5, 460]

Example 18. *Find the median, quartiles, 7th decile and 82nd percentile for the following distribution :*

Wages in Rs.	*No. of Workers*
0—10	*22*
10—20	*38*
20—30	*46*
30—40	*35*
40—50	*20*

Sol. The cumulative frequency table is

Class	*Frequency*	*Cumulative Frequency*
0—10	22	22
10—20	38	60
20—30	46	106
30—40	35	141
40—50	20	161

Here N = 161

(*i*) $\frac{N}{2} = 80.5 \quad \Rightarrow \quad$ Median class is 20—30

$\therefore \quad \text{Median} = l + \frac{h}{f}\left(\frac{N}{2} - C\right)$

$= 20 + \frac{10}{46}(80.5 - 60)$

$= 20 + \frac{205}{46} = 20 + 4.46 = \text{Rs. } 24.46.$

(*ii*) $\frac{N}{4} = 40.25 \quad \Rightarrow \quad$ Class of Q_1 is 10—20

$\therefore \quad Q_1 = l + \frac{h}{f}\left(\frac{N}{4} - C\right) = 10 + \frac{10}{38}(40.25 - 22)$

$= 10 + \frac{182.5}{38} = 10 + 4.8 = \text{Rs. } 14.8.$

(*iii*) $\frac{3N}{4} = 120.75 \quad \Rightarrow \quad$ Class of Q_3 is 30—40

$\therefore \quad Q_3 = l + \frac{h}{f}\left(\frac{3N}{4} - C\right) = 30 + \frac{10}{35}(120.75 - 106)$

$= 30 + \frac{147.5}{35} = 30 + 4.21 = \text{Rs. } 34.21.$

(*iv*) *To calculate* D_7

$\frac{7N}{10} = \frac{1127}{10} = 112.7 \quad \Rightarrow \quad$ Class of D_7 is 30—40

$$D_7 = l + \frac{h}{f}\left(\frac{7N}{10} - C\right) = 30 + \frac{10}{35}(112.7 - 106)$$

$$= 30 + \frac{67}{35} = 30 + 1.91 = \text{Rs. } 31.91.$$

(*v*) *To calculate* P_{82}

$\frac{82N}{100} = 132.02 \quad \Rightarrow \quad$ Class of P_{82} is 30—40

$$\therefore \qquad P_{82} = l + \frac{h}{f}\left(\frac{82N}{100} - C\right) = 30 + \frac{10}{35}(132.02 - 106)$$

$$= 30 + \frac{260.2}{35} = 30 + 7.43 = \text{Rs. } 37.43.$$

Example 19. *Calculate the quartiles, fourth decile and 60th percentile for the following distribution :*

Marks Group	*No. of students*
5—10	*5*
10—15	*6*
15—20	*15*
20—25	*10*
25—30	*5*
30—35	*4*
35—40	*2*
40—45	*2*

(G.N.D.U. 1982)

Sol. Please try yourself.

[**Ans.** $Q_1 = 15.41$, $Q_3 = 25.75$, $D_4 = 17.86$, $P_{60} = 21.7$]

Example 20. *Find the median, quartiles, 7th decile and 85th percentile from the following data :*

Monthly Rent (Rs.)	*No. of families*	*Monthly Rent (Rs.)*	*No. of families*
20— 40	*6*	*120—140*	*15*
40— 60	*9*	*140—160*	*10*
60— 80	*11*	*160—180*	*8*
80—100	*14*	*180—200*	*7*
100—120	*20*		

(K.U. 1982)

Sol. Please try yourself.

[**Ans.** (Rs.) 110, 78.18, 140, 133.33, 160]

Example 21. *From the following data, find the value of median :*

Income (Rs.)	:	*100*	*150*	*80*	*200*	*250*	*180*
No. of persons	:	*24*	*26*	*16*	*20*	*6*	*30*

Sol. Here the values of the variate (*i.e.,* income in rupees) is not arranged in ascending or descending order of magnitude. We first arrange the values in ascending order.

Income (Rs.) arranged in ascending order	No. of persons (f)	c.f.
80	16	16
100	24	40
150	26	66
180	30	96
200	20	116
250	6	122

Here N = 122, *c.f.* just greater than $\frac{N+1}{2} = 61.5$ is 66.

The value of the variate corresponding to *c.f.* 66 is 150

Hence median income = Rs. 150.

Example 22. *Calculate the median from the following data :*

Marks :	*0—10*	*10—30*	*30—60*	*60—80*	*80—90*
No. of students :	*5*	*16*	*30*	*8*	*2*

Sol. Here the class intervals are unequal. *For the computation of median, there is no need to make the class-intervals equal* as is illustrated below :

(*i*) Computation of median from the given data

Marks	*No. of students (f)*	*c.f.*
0—10	5	5
10—30	16	21
30—60	30	51
60—80	8	59
80—90	2	61

Here N = 61, *c.f.* just greater than $\frac{N}{2} = 30.5$ is 51.

∴ Median class is 30—60.

$\Rightarrow \quad l = 30, \quad f = 30, \quad h = 60 - 30 = 30, \quad C = 21$

$$\therefore \quad \text{Median} = l + \frac{h}{f}\left(\frac{N}{2} - C\right) = 30 + \frac{30}{30}(30.5 - 21)$$

$$= 30 + 9.5 = 39.5 \text{ marks.}$$

(*ii*) Computation of median after making the class intervals equal, each of size 10.

Marks		*No. of students* (*f*)	*c.f.*
0—10		5	5
10—20	(16 ÷ 2 =)	8	13
20—30		8	21
30—40	(30 ÷ 3 =)	10	31
40—50		10	41
50—60		10	51
60—70	(8 ÷ 2 =)	4	55
70—80		4	59
80—90		2	61

Here N = 61, *c.f.* just greater than $\frac{N}{2} = 30.5$ is 31.

Median class is 30—40.

$\Rightarrow \quad l = 30, \quad f = 10, \quad h = 10, \quad C = 21$

$$\therefore \quad \text{Median} = l + \frac{h}{f}\left(\frac{N}{2} - C\right) = 30 + \frac{10}{10}(30.5 - 21)$$

$$= 30 + 9.5 = 39.5 \text{ marks.}$$

Example 23. *Find the median, lower and upper quartiles, 4th decile and 60th percentile for the following distribution :*

Marks	*No. of students*	*Marks*	*No. of students*
0— 4	*10*	*18—20*	*8*
4— 8	*12*	*20—25*	*4*
8—12	*18*	*25 and above*	*6*
12—14	*7*		
14—18	*5*		

(K.U. 1981 S)

Sol. Here the class-intervals are not all equal. To find any partition value, there is no need to make them equal. (See Ex. 22)

Marks	*No. of students* (*f*)	*c.f.*
0— 4	10	10
4— 8	12	22
8—12	18	40
12—14	7	47
14—18	5	52
18—20	8	60
20—25	4	64
25 and above	6	70
	N = 70	

(*i*) **Calculation of median**

$\frac{N}{2} = 35 \quad \therefore \quad$ median class is 8 – 12

$l = 8, \quad h = 4, \quad f = 18, \quad C = 22$

$$\therefore \quad \text{Median} = l + \frac{h}{f}\left(\frac{N}{2} - C\right)$$

$$= 8 + \frac{4}{18}(35 - 22) = 10.89.$$

(*ii*) **Calculation of lower quartile Q_1**

$\frac{N}{4} = 17.5 \quad \therefore \quad$ lower quartile class is 4—8

$l = 4, \quad h = 4, \quad f = 12, \quad C = 10$

$$Q_1 = l + \frac{h}{f}\left(\frac{N}{4} - C\right) = 4 + \frac{4}{12}(17.5 - 10) = 6.5.$$

(*iii*) **Calculation of upper quartile Q_3**

$\frac{3N}{4} = 52.5 \quad \therefore \quad$ upper quartile class is 18 – 20

$l = 18, \quad h = 2, \quad f = 8, \quad C = 52.$

$$\therefore \quad Q_3 = l + \frac{h}{f}\left(\frac{3N}{4} - C\right)$$

$$= 18 + \frac{2}{8}(52.5 - 52) = 18.125.$$

(*iv*) **Calculation of 4th decile D_4**

$\frac{4N}{10} = 28 \quad \therefore \quad$ Class of D_4 is 8 – 12

$l = 8, \quad h = 4, \quad f = 18, \quad C = 22$

$$\therefore \quad D_4 = l + \frac{h}{f}\left(\frac{4N}{10} - C\right)$$

$$= 8 + \frac{4}{18}(28 - 22) = 9.33.$$

(*v*) **Calculation of 60th percentile P_{60}**

$\frac{60N}{100} = 42 \quad \therefore \quad$ Class of P_{60} is 12 – 14

$l = 12, \quad h = 2, \quad f = 7, \quad C = 40$

$$\therefore \quad P_{60} = l + \frac{h}{f}\left(\frac{60N}{100} - C\right)$$

$$= 12 + \frac{2}{7}(42 - 40) = 12.57.$$

Example 24. *Calculate the median, quartiles, 7th octile, 4th decile and 15th percentile of the following series of marks obtained by 10 candidates in an examination :*

24, 23, 28, 15, 10, 40, 42, 32, 48, 8.

Sol. ☞ Arranging the marks in ascending order, we have

S. No. :	1	2	3	4	5	6	7	8	9	10
Marks :	8	10	15	23	24	28	32	40	42	48

(*i*) Median = size of $\left(\frac{n+1}{2}\right)$th = $\left(\frac{10+1}{2}\text{th} = \right)$ 5.5th item

= size of 5th item + .5 (size of 6th item – size of 5th item)

= 24 + .5 (28 – 24) = 26 marks

[*OR* 5.5 lies mid-way between 5 and 6

∴ size of 5.5th item = $\frac{1}{2}$ (size of 5th item + size of 6th item)

i.e., mean of 5th and 6th items

= $\frac{1}{2}$ (24 + 28) = 26]

(*ii*) Q_1 = size of $\left(\frac{n+1}{4}\right)$th = $\left(\frac{11}{4}\text{th} = \right)$ 2.75th item

= size of 2nd item + .75 (size of 3rd item – size of 2nd item)

= 10 + .75 (15 – 10) = 13.75 marks

(*iii*) Q_3 = size of $\frac{3(n+1)}{4}$th = $\left(\frac{33}{4}\text{th} = \right)$ 8.25th item

= size of 8th item + .25 (size of 9th item – size of 8th item)

= 40 + .25 (42 – 40) = 40.5 marks

(*iv*) O_7 = size of $\frac{7(n+1)}{8}$th = $\left(\frac{77}{8}\text{th} = \right)$ 9.6th item

= size of 9th item + .6 (size of 10th item – size of 9th item)

= 42 + .6 (48 – 42) = 45.6 marks

(*v*) D_4 = size of $\frac{4(n+1)}{10}$th = $\left(\frac{44}{10}\text{th} = \right)$ 4.4th item

= size of 4th item + .4 (size of 5th item – size of 4th item)

= 23 + .4 (24 – 23) = 23.4 marks

(*vi*) P_{15} = size of $\frac{15(n+1)}{100}$th = $\left(\frac{165}{100}\text{th} = \right)$ 1.65th item

= size of 1st item + .65 (size of 2nd item – size of 1st item)

= 8 + .65 (10 – 8) = 9.3 marks.

Example 25. *The following data relates to the sizes of shoes sold at a store during a given week. Find the median size of the shoes. Also calculate the quartiles, 3rd quintile, 3rd decile and 85th percentile.*

Size of shoes	*Frequency*	*Size of shoes*	*Frequency*
5	2	7.5	40
5.5	8	8	25
6	20	8.5	10
6.5	30	9	3
7	70	9.5	1

Sol.

Size of shoes	*Frequency*	*c.f.*
5	2	2
5.5	8	10
6	20	30
6.5	30	60
7	70	130
7.5	40	170
8	25	195
8.5	10	205
9	3	208
9.5	1	209
		N = 209

(*i*) *c.f.* just greater than $\frac{N+1}{2} = 105$ is 130

∴ Median = size of shoe corresponding to *c.f.* 130 = 7

(*ii*) *c.f.* just greater than $\frac{N+1}{4} = 52.5$ is 60

∴ Q_1 = size of shoe corresponding to *c.f.* 60 = 6.5

(*iii*) *c.f* just greater than $\frac{3(N+1)}{4} = 157.5$ is 170

∴ Q_3 = size of shoe corresponding to *c.f.* 170 = 7.5

(*iv*) Quintiles divide the total frequency into 5 equal parts.

c.f. just greater than $\frac{3(N+1)}{5} = 126$ is 130

∴ 3rd Quintile = size of shoe corresponding to *c.f.* 130 = 7

(*v*) *c.f.* just greater than $\frac{3(N+1)}{10} = 63$ is 130

D_3 = size of shoe corresponding to *c.f.* 130 = 7

(*vi*) *c.f.* just greater than $\dfrac{85(N+1)}{100} = 178.5$ is 195

P_{85} = size of shoe corresponding to *c.f.* 195 = 8.

Example 26. *The following table gives the distribution of males in an Indian town. Find the median age.*

Age groups	*Males*	*Age groups*	*Males*
0— 9	*2756*	*50—59*	*610*
10—19	*2124*	*60—69*	*245*
20—29	*1677*	*70—79*	*67*
30—39	*1481*	*80—89*	*6*
40—49	*1021*	*90—99*	*3*

Sol. Please try yourself. **[Ans.** 20.185 years**]**

(Here N = 9990, Median class is 20—29. But the class intervals are given by inclusive method. Adjustment factor = .5. Real limits of median class are 19.5—29.5 $\therefore l = 19.5$. Also $h = 10$ and not 9).

Example 27. *Calculate the quartiles from the following data :*

Weekly wages in Dollars : *35—36—37—38—39—40—41—42*
No. of wage earners : *14 20 42 54 45 18 7*

Sol. $x \in 35$— $\Rightarrow 35 \leq x < 36$

Weekly wages in Dollars	*No. of wage earners (f)*	*c.f.*
35—36	14	14
36—37	20	34
37—38	42	76
38—39	54	130
39—40	45	175
40—41	18	193
41—42	7	200

Here N = 200

For computation of Q_1

c.f. just greater than $\dfrac{N}{4} = 50$ is 76,

$\Rightarrow$ class of Q_1 is 37—38. As usual

$$Q_1 = l + \frac{h}{f}\left(\frac{N}{4} - C\right) = 37 + \frac{1}{42}(50-34)$$

$$= 37 + \frac{8}{21} = 37.38 \text{ dollars.}$$

Similarly (please find), $Q_3 = 39.44$ dollars.

Example 28. *An incomplete frequency distribution is given as follows :*

Variable	*Frequency*	*Variable*	*Frequency*
10—20	*12*	*50—60*	*?*
20—30	*30*	*60—70*	*25*
30—40	*?*	*70—80*	*18*
40—50	*65*	*Total :*	*229*

Given that the median value is 46, determine the missing frequencies using the median formula. (D.U.1984)

Sol. Let the frequency of the class 30—40 be f_1 and that of 50—60 be f_2.

Since $\quad N = \Sigma f = 229$

$\therefore \quad 12 + 30 + f_1 + 65 + f_2 + 25 + 18 = 229$

$\Rightarrow \quad f_1 + f_2 = 79$ ...(1)

Since the median is given to be 46 $\quad\therefore\quad$ Median class is 40—50

$l = 40, \quad h = 10, \quad f = 65, \quad C = 12 + 30 + f_1 = 42 + f_1$

Using median formula,

$$\text{Median} = l + \frac{h}{f}\left(\frac{N}{2} - C\right), \text{ we have}$$

$$46 = 40 + \frac{10}{65}(114.5 - 42 - f_1)$$

or $$6 \times \frac{65}{10} = 72.5 - f_1$$

or $$f_1 = 72.5 - 39 = 33.5$$

Since frequency is never a fraction, $\quad\therefore\quad f_1 = 34$

from (1) $\quad f_2 = 79 - 34 = 45.$

3. MODE

1. Mode. Mode is the value which occurs most frequently in a set of observations and around which the other items of the set cluster densely. It is the point of maximum frequency or the point of greatest density. In other words, the mode or modal value of the distribution is that value of the variate for which frequency is maximum.

2. Calculation of Mode

(*a*) In the case of discrete frequency distribution, mode is the value of x corresponding to maximum frequency.

But in any one (or more) of the following cases :

(*i*) if the maximum frequency is repeated

(*ii*) if the maximum frequency occurs in the very beginning or at the end of the distribution (*iii*) if there are irregularities in the distribution, the value of mode is determined by the *method of grouping* (illustrated in examples below).

(*b*) In the case of continuous frequency distribution, mode is given by the formula :

$$\text{Mode} = l + \frac{f_m - f_1}{2f_m - f_1 - f_2} \times h$$

where l is the lower limit, h the width and f_m the frequency of the modal class, f_1 and f_2 are the frequencies of the classes preceding and succeeding the modal class respectively.

☞ While applying the above formula, it is necessary to see that the class-intervals are of the same size. **If they are unequal, they should first be made equal** on the assumption that the frequencies are equally distributed throughout the class.

☞ In case $f_m - f_1 < 0$ or $2f_m - f_1 - f_2 = 0$, use the formula

$$\text{Mode} = l + \frac{\Delta_1}{\Delta_1 + \Delta_2} \times h$$

where $\Delta_1 = |f_m - f_1|$ and $\Delta_2 = |f_m - f_2|$ (See Example 17)

(*c*) **For a symmetrical distribution, mean, median and mode co-incide.**

(*d*) Where mode is **ill-defined** *i.e.*, where the method of grouping also fails, its value can be ascertained by the formula

Mode = 3 Median – 2 Mean

This measure is called the **empirical mode.**

Example 1. *If seven men are receiving daily wages of Rs. 5, 6, 7, 7, 7, 8 and 10, find the modal wage.*

Sol. Since 7 occurs thrice and no other item occurs three times or more than three times, hence modal wage is Rs. 7.

Example 2. *Determine the mode from the following figures :*

25 15 23 40 27 25 23 25 20

Sol. Please try yourself. **[Ans.** 25]

Example 3. *Find the mode from the following frequency distribution :*

x :	*1*	*2*	*3*	*4*	*5*	*6*	*7*	*8*
f :	*4*	*9*	*16*	*25*	*22*	*16*	*8*	*3*

Sol. Here maximum frequency is 25 and the corresponding value of x is 4. Hence mode is 4.

Example 4. *Calculate the mode from the following frequency distribution :*

Size (x)	:	4	5	6	7	8	9	10	11	12	13
Frequency (f)	:	2	5	8	9	12	14	14	15	11	13

Sol. **Method of Grouping**

Size (x)	*Frequency* I	II	III	IV	V	VI
4	2	7		15		
5	5		13		22	
6	8	17				29
7	9		21	35		
8	12	26			**40**	
9	14		**28**			**43**
10	14	**29**		**40**		
11	**15**		26		39	
12	11	24				
13	13					

Explanation :

In column I, *original* frequencies are written.

In column II, frequencies of column I are combined *two by two.*

In column III, *leave the first frequency of column I* and combine the others *two by two.*

In column IV, frequencies of column I are combined *three by three.*

In column V, *leave the first frequency of column I* and combine the others *three by three.*

In column IV, *leave the first two frequencies in column I* and combine the others *three by three.*

In all these columns, the maximum frequency is written in bold black type.

☞(**Note.** All operations are done on column I).

Now we frame another table in which against every maximum item of columns I to VI, we write down the corresponding size or sizes. The size (x) which occurs maximum number of times is the mode.

Columns	*Size of item having max. frequency*
I	11
II	10, 11
III	9, 10
IV	10, 11, 12
V	8, 9, 10
VI	9, 10, 11

Since the item 10 occurs maximum number of times (*i.e.*, 5), hence the mode is 10.

Example 5. *Find the mode of the following frequency distribution* :

Size (x)	:	*1*	*2*	*3*	*4*	*5*	*6*	*7*	*8*	*9*	*10*	*11*	*12*
Frequency (f)	:	*3*	*8*	*15*	*23*	*35*	*40*	*32*	*28*	*20*	*45*	*14*	*6*

Sol. By Method of Grouping

Size (x)	*Frequency*					
	I	*II*	*III*	*IV*	*V*	*VI*
1	3	11		26		
2	8		23		46	
3	15	38				73
4	23		58	98		
5	35	75			107	
6	40		72			100
7	32	60		80		
8	28		48		93	
9	20	65				79
10	**45**		59	65		
11	14	20				
12	6					

Analysis

Columns	*Size of item having max. frequency*
I	10
II	5, 6
III	6, 7
IV	4, 5, 6
V	5, 6, 7
VI	6, 7, 8

Since the item 6 occurs maximum number of times (*i.e.,* 5), hence the mode is 6.

Example 6. *Compute the mode of the following frequency distribution :*

Size (x) :	*2*	*3*	*4*	*5*	*6*	*7*	*8*	*9*	*10*	*11*	*12*	*13*
f :	*3*	*8*	*10*	*12*	*16*	*14*	*10*	*8*	*17*	*5*	*4*	*1*

Sol. Please try yourself. **[Ans.** 6]

Example 7. *Find out the mode of the following series :*

Size	*f*	*Size*	*f*
5	48	13	52
6	52	14	41
7	56	15	57
8	60	16	63
9	63	17	52
10	57	18	48
11	55	19	40
12	50		

Sol. Please try yourself. **[Ans.** 9]

Example 8. *Calculate the modal wage from the following data :*

Wages (Rs.)	*No. of workers*
Below 100	*8*
100—200	*12*
200—300	*25*
300—400	*15*
400—500	*10*
Above 500	*6*

Sol. Since the highest frequency is 25, the modal class is 200—300

$\therefore$ l = lower limit of modal class = 200

f_m = frequency of modal class = 25

f_1 = frequency of class preceding the modal class = 12

f_2 = frequency of class succeeding the modal class = 15

h = width of modal class = 100

$$\therefore \quad \text{Mode} = l + \frac{f_m - f_1}{2f_m - f_1 - f_2} \times h$$

$$= 200 + \frac{25 - 12}{50 - 12 - 15} \times 100 = 200 + \frac{13}{23} \times 100$$

$$= 200 + 56.52 = \text{Rs. } 256.52.$$

Example 9. *Find the mode of the following :*

Marks :	*1—5*	*6—10*	*11—15*	*16—20*	*21—25*
No. of candidates :	*7*	*10*	*16*	*32*	*24*
Marks :	*26—30*	*31—35*	*36—40*	*41—45*	
No. of candidates :	*18*	*10*	*5*	*1*	

Sol. Here the greatest frequency 32 lies in the class 16—20. Hence modal class is 16—20. But the actual limits of this class are 15.5—20.5.

$$l = 15.5, \quad f_m = 32, \quad f_1 = 16, \quad f_2 = 24, \quad h = 5$$

$$\therefore \quad \text{Mode} = l + \frac{f_m - f_1}{2f_m - f_1 - f_2} \times h = 15.5 + \frac{32 - 16}{64 - 16 - 24} \times 5$$

$$= 15.5 + \frac{16}{24} \times 5 = 15.5 + \frac{10}{3} = 15.5 + 3.33$$

$$= 18.83 \text{ marks.}$$

Example 10. *Find the mode of the following distribution :*

Class :	*0—7*	*7—14*	*14—21*	*21—28*	*28—35*	*35—42*	*42—49*
Frequency :	*19*	*25*	*36*	*72*	*51*	*43*	*28*

Sol. Please try yourself. **[Ans.** 25.4]

Example 11. *Find the mode and median from the following table :*

Marks	*No. of students*	*Marks*	*No. of students*
0—10	*2*	*40—50*	*35*
10—20	*18*	*50—60*	*20*
20—30	*30*	*60—70*	*6*
30—40	*45*	*70—80*	*3*

Sol. Please try yourself. **[Ans.** 36, 36.6]

Example 12. *Calculate the median and mode of the following data :*

No. of days absent	*No. of students*
Less than 5	29
” ” *10*	224
” ” *15*	465
” ” *20*	582
” ” *25*	634
” ” *30*	644
” ” *35*	650
” ” *40*	653
” ” *45*	655

Sol.

No. of days absent	*No. of students* (*f*)	*c.f.*
0—5	29	29
5—10	195	224
10—15	241	465
15—20	117	582
20—25	52	634
25—30	10	644
30—35	6	650
35—40	3	653
40—45	2	655

$\frac{N}{2} = \frac{655}{2} = 327.5$ ∴ Median class is 10—15

$l = 10, \quad h = 5 \quad f = 241, \quad C = 224$

$$\therefore \quad \text{Median} = l + \frac{h}{f}\left(\frac{N}{2} - C\right)$$

$$= 10 + \frac{5}{241}(327.5 - 224) = 10 + \frac{5 \times 103.5}{240}$$

$$= 10 + \frac{517.5}{240} = 10 + 2.16 = 12.16$$

Max. frequency 241 lies in 10—15, which is, therefore, the modal class.

$l = 10, \quad f_m = 241, \quad f_1 = 195, \quad f_2 = 117, \quad h = 5$

$$\therefore \quad \text{Mode} = l + \frac{f_m - f_1}{2f_m - f_1 - f_2} \times h$$

$$= 10 + \frac{241 - 195}{482 - 195 - 117} \times 5 = 10 + \frac{46}{170} \times 5$$

$$= 10 + 1.35 = 11.35.$$

Example 13. *Find out the mean, median and mode of the following distribution :*

Marks	*Frequency*	*Marks*	*Frequency*
10—25	*6*	*55—70*	*26*
25—40	*20*	*70—85*	*3*
40—55	*44*	*85—100*	*1*

Sol. Let assumed mean 'a' = 45.5

Marks	*Mid-value* (*x*)	*Frequency* (*f*)	*c.f.*	$d_x = x - 45.5$	fd_x
10—25	17.5	6	6	− 28	− 168
25—40	32.5	20	26	− 13	− 260
40—55	47.5	44	70	2	88
55—70	62.5	26	96	17	442
70—85	77.5	3	99	32	96
85—100	92.5	1	100	47	47
		N = 100			245

$$\text{Mean } \bar{x} = a + \frac{\Sigma fd_x}{\text{N}} = 45.5 + \frac{245}{100} = 45.5 + 2.45 = 47.95 \text{ marks}$$

$$\text{N} = 100 \quad \therefore \quad \frac{\text{N}}{2} = 50 \quad \therefore \quad \text{Median class is } 40\text{—}55$$

$$l = 40, \qquad h = 15, \qquad f = 44, \qquad \text{C} = 26$$

$$\text{Median} = l + \frac{h}{f}\left(\frac{\text{N}}{2} - \text{C}\right) = 40 + \frac{15}{44}(50 - 26)$$

$$= 40 + \frac{15 \times 24}{44} = 40 + 8.18 = 48.18 \text{ marks}$$

Max. frequency 44 lies in 40—55 which is, therefore, the modal class.

$$l = 40, \qquad f_m = 44, \qquad f_1 = 20, \qquad f_2 = 26, \qquad h = 15$$

$$\therefore \quad \text{Mode} = l + \frac{f_m - f_1}{2f_m - f_1 - f_2} \times h$$

$$= 40 + \frac{44 - 20}{88 - 20 - 26} \times 15 = 48.57 \text{ marks.}$$

Example 14. *Calculate mode from the following data :*

Marks :	*0—10*	*10—20*	*20—40*	*40—50*	*50—70*
Frequency :	*5*	*15*	*40*	*32*	*28*

Sol.

☞ Here the class-intervals are unequal. We must make them equal before we start computing the mode. To make the class intervals equal, we assume that the frequencies are uniformly distributed throughout the class.

Marks	f	Marks	f
0—10	5	40—50	32
10—20	15	50—60	14
20—30	20	60—70	14
30—40	20		

Modal class is 40—50. Proceeding as usual, mode = 44 marks.

Example 15. *Calculate mode from the following :*

Marks	*No. of students*	*Marks*	*No. of students*
0— 2	*8*	*25— 30*	*45*
2— 4	*12*	*30— 40*	*60*
4—10	*20*	*40— 50*	*20*
10—15	*10*	*50— 60*	*13*
15—20	*16*	*60— 80*	*15*
20—25	*25*	*80—100*	*4*

Sol. Here the class-intervals are unequal. We must make them equal before computing mode.

Marks	*No. of students* (*f*)
0— 20	8 + 12 + 20 + 10 + 16 = 66
20— 40	25 + 45 + 60 = 130
40— 60	20 + 13 = 33
60— 80	15
80—100	4

Modal class is 20—40. Proceeding as usual, mode = 27.95 marks.

Example 16. *Calculate the value of mean, median and mode of the following distribution*

Class-interval	*f*
0— 4	*4*
4— 6	*6*
6— 8	*8*
8—12	*12*
12—18	*7*
18—20	*2*

(M.D.U. 1981)

Sol. Here, the class-intervals are unequal. For computation of mean and median, there is no need to adjust them.

Class-intervals	*Mid-values* (*x*)	*f*	*fx*	*c.f.*
0— 4	2	4	8	4
4— 6	5	6	30	10
6— 8	7	8	56	18
8—12	10	12	120	30
12—18	15	7	105	37
18—20	19	2	38	39
		N = 39	357	

Mean $\bar{x} = \frac{\Sigma fx}{N} = \frac{357}{39} = 9.154$ ...(1)

$\frac{N}{2} = 19.5 \quad \Rightarrow \quad$ median class is 8—12

As usual, median $= l + \frac{h}{f}\left(\frac{N}{2} - C\right) = 8 + \frac{4}{12}(19.5 - 18) = 8.5$ ...(2)

For finding the mode, we make the class-intervals of equal size.

Class	*f*
0— 4	4
4— 8	(6 + 8 =)14
8—12	12
12—16	5
16—20	4

[Size of class 12 –18 is 6 and frequency is 7

$\therefore$ If size of class is 4, frequency $= \frac{7}{6} \times 4 = 4.7$

i.e., 5 (next integer)]

Here modal class is 4 – 8 ; $l = 4$, $f_m = 14$, $f_1 = 4$, $f_2 = 12$, $h = 4$

$\therefore \quad \text{Mode} = l + \frac{f_m - f_1}{2f_m - f_1 - f_2} \times h = 4 + \frac{10}{12} \times 4 = 7.3.$

Example 17. *Calculate the mode of the distribution given below :*

Monthly wages in Rs.	*No. of workers*
50— 70	*4*
70— 90	*44*
90—110	*38*
110—130	*28*
130—150	*6*
150—170	*8*
170—190	*12*
190—210	*2*
210—230	*2*

Sol. The distribution is bimodal. We use the method of grouping to find the modal class.

Class	*Frequency* I	II	III	IV	V	VI
50— 70	4	48		86		
70— 90	**44**		**82**		**110**	
90—110	38	**66**				**72**
110—130	28		34	42		
130—150	6	14			26	
150—170	8		20			22
170—190	12	14		16		
190—210	2		4			
210—230	2					

Analysis

Column	*Class*
I	70—90
II	90—110, 110—130
III	70—90, 90—110
IV	50—70, 70—90, 90—110
V	70—90, 90—110, 110—130
VI	90—110, 110—130, 130—150
	1 4 5 3 1

Since the class 90—110 occurs maximum number of times (*i.e.*, 5), the modal class is 90—110.

$l = 90, \quad f_m = 38, \quad f_1 = 44, \quad f_2 = 28, \quad h = 20$

$f_m - f_1 < 0$

$\therefore \quad \Delta_1 = |f_m - f_1| = |-6| = 6$

$\Delta_2 = |f_m - f_2| = |10| = 10$

$$\text{Mode} = l + \frac{\Delta_1}{\Delta_1 + \Delta_2} \times h = 90 + \frac{6}{6+10} \times 20 = \text{Rs. } 97.50.$$

Example 18. *The marks (out of maximum 100) obtained by candidates in an examination are shown in the following frequency table. Calculate the Mode and the Arithmetic Average :*

Marks	*No. of candidates*
17.5—22.5	*2*
22.5—27.5	*8*
27.5—32.5	*33*
32.5—37.5	*80*
37.5—42.5	*170*
42.5—47.5	*243*
47.5—52.5	*213*
52.5—57.5	*145*
57.5—62.5	*67*
62.5—67.5	*35*
67.5—72.5	*4*

(K.U. 1981)

Sol. Please try yourself. **[Ans.** 46, 46.965]

Example 19. *From the following table, find the mean, median and modal ages of married women at first child births :*

Age at the birth of first child :

13 14 15 16 17 18 19 20 21 22 23 24 25

Number of married women :

7 162 343 390 256 433 161 355 65 85 49 46 40

Sol.

(Age at the birth of first child)	*(Number of married women) f*	*d = x – 19*	*fd*	*c.f.*
13	37	– 6	– 222	37
14	162	– 5	– 810	199
15	343	– 4	– 1372	542
16	390	– 3	– 1170	932
17	256	– 2	– 512	1188
18	433	– 1	– 433	1621
19	161	0	0	1782
20	355	1	355	2137
21	65	2	130	2202
22	85	3	255	2287
23	49	4	196	2336
24	46	5	230	2382
25	40	6	240	2422
	N = 2422		– 3113	

$$\text{Mean } \bar{x} = a + \frac{\Sigma fd}{N} = 19 - \frac{3113}{2422} = 17.8 \text{ years}$$

$\frac{N}{2} = \frac{2422}{2} = 1211$, *c.f.* just greater than 1211 is 1621 and the value of x corresponding to this *c.f.* is 18.

∴ Median = 18 years.

Computing the mode by the method of grouping (left for the student).

Mode = 18 years.

Example 20. *In a moderately asymmetrical distribution, the mode and mean are 32.1 and 35.4 respectively. Compute the median.*

Sol. Mode = 32.1, Mean = 35.4

Using Mode = 3 Median – 2 Mean, we have

$$32.1 = 3 \text{ Median} - 70.8 \quad \text{or} \quad 3 \text{ Median} = 102.9$$

∴ Median = 34.3.

Example 21. *An incomplete distribution of families according to their expenditure per week is given below. The median and mode for the distribution are Rs. 25 and Rs. 24 respectively. Calculate the missing frequencies*

Expenditure :	*0—10*	*10—20*	*20—30*	*30—40*	*40—50*
No. of families :	*14*	*?*	*27*	*?*	*15*

Sol. Let the missing frequencies be x and y respectively.

Expenditure	*No. of families* (*f*)	*c.f.*
0—10	14	14
10—20	x	$14 + x$
20—30	27	$41 + x$
30—40	y	$41 + x + y$
40—50	15	$56 + x + y$
	$N = 56 + x + y$	

(*i*) Median = Rs. 25 ⇒ Median class is 20—30

$l = 20, \quad h = 10, \quad f = 27, \quad C = 14 + x$

Using $\text{Median} = l + \frac{h}{f}\left(\frac{N}{2} - C\right)$, we have

$$25 = 20 + \frac{10}{27}\left(\frac{56 + x + y}{2} - 14 - x\right)$$

or $$\frac{135}{10} = \frac{28 - x + y}{2} \quad \text{or} \quad 27 = 28 - x + y$$

∴ $$y = x - 1 \qquad \text{...(1)}$$

(*ii*) Mode = Rs. 24 $\Rightarrow$ Modal class is 20—30

$l = 20, \quad h = 10, \quad f_m = 27, \quad f_1 = x, \quad f_2 = y$

Using Mode $= l + \dfrac{f_m - f_1}{2f_m - f_1 - f_2} \times h$, we have

$$24 = 20 + \frac{27 - x}{54 - x - y} \times 10$$

or $\quad 4(54 - x - y) = 10(27 - x)$

or $\quad 2[54 - x - (x - 1)] = 5(27 - x)$ | Using (1)

or $\quad 110 - 4x = 135 - 5x \quad \therefore \quad x = 25, \quad y = 24.$

Example 22. *The expenditure of 100 families is given below :*

Expenditure	:	*0—10*	*10—20*	*20—30*	*30—40*	*40—50*
No. of families	:	*14*	*?*	*27*	*?*	*15*

Mode for the distribution is 24. Calculate missing frequencies.

(D.U. 1981)

Sol. Proceeding as in example 21,

$N = 100 = 56 + x + y \quad \therefore \quad x + y = 44$

Also $\quad 4(54 - x - y) = 10(27 - x)$

or $\quad 2[54 - x - (44 - x)] = 5(27 - x)$

or $\quad 20 = 135 - 5x \quad \therefore \quad x = 23, \quad y = 21.$

Example 23. *A distribution $x_1, x_2, ..., x_r, ..., x_n$ with frequencies, $f_1, f_2, ..., f_r, ..., f_n$ is transformed into the distribution $y_1, y_2, ..., y_r, ..., y_n$ with the same corresponding frequencies by the relation $y_r = ax_r + b$, where a and b are constants. Show that the mean, median and mode of the new distribution are given in terms of the first distribution by the same transformation.*

Sol. The given distribution is

$$\left.\begin{array}{l} x : x_1 \quad , \quad x_2 \quad , \quad x_3 \,......, x_n \\ f : f_1 \quad , \quad f_2 \quad , \quad f_3 \,......, f_n \end{array}\right\} \quad ...(1)$$

The transformed distribution is

$$\left.\begin{array}{l} y : y_1 \quad , \quad y_2 \quad , \quad y_3 \,......, y_n \\ f : f_1 \quad , \quad f_2 \quad , \quad f_3 \,......, f_n \end{array}\right\} \quad ...(2)$$

where $\quad y = ax + b$

(*a*) If $\bar{x}$ is the mean of (1) and $\bar{y}$, the mean of (2), then

$$\bar{x} = \frac{1}{N} \sum fx$$

$$\bar{y} = \frac{1}{N} \sum fy = \frac{1}{N} \sum f(ax + b)$$

$$= a \cdot \frac{1}{N} \sum fx + b \cdot \frac{1}{N} \sum f$$

$$= a\bar{x} + b \qquad (\because \ \Sigma f = N)$$

☞ $y = ax + b \quad \Rightarrow \quad \bar{y} = a\bar{x} + b$

(*b*) If f_r is the *c.f.* just greater than $\frac{N}{2}$ and x_i is the value of x corresponding to it, then median of (1) is $M_x = x_i$ and median of (2) is

$$M_y = y_i = ax_i + b = aM_x + b$$

$$\therefore \quad M_y = aM_x + b$$

(*c*) If f_r is the maximum frequency in each distribution, then the modes (*i.e.*, corresponding values of variate) are x_r and y_r respectively which are connected by $y_r = ax_r + b.$

Hence the mean, median and mode of the new distribution are given in terms of the first distribution by the same transformation.

4. GEOMETRIC MEAN

Geometric Mean. (*a*) Geometric Mean (G.M.) of n individual observations $x_1, x_2, ..., x_n$ $(x_i \neq 0)$ is the nth root of their product.

Thus $$G = (x_1 x_2 ...... x_n)^{1/n}$$

Taking logarithms of both sides

$$\log G = \frac{1}{n}(\log x_1 + \log x_2 + ... + \log x_n)$$

$$= \frac{1}{n}\sum_{i=1}^{n} \log x_i$$

$$\therefore \quad G = \text{antilog}\left[\frac{1}{n}\sum_{i=1}^{n} \log x_i\right].$$

(*b*) If $x_1, x_2, ..., x_n$ occur $f_1, f_2, ..., f_n$ times respectively and $N = \sum_{1}^{n} f_i$, then G.M. is given by

$$G = (x_1^{f_1} x_2^{f_2} ..... x_n^{f_n})^{1/N}$$

Taking logarithms of both sides

$$\log G = \frac{1}{N}(f_1 \log x_1 + f_2 \log x_2 + ... + f_n \log x_n)$$

$$= \frac{1}{N}\sum_{i=1}^{n} f_i \log x_i$$

$$G = \text{antilog}\left[\frac{1}{N}\sum_{i=1}^{n} f_i \log x_i\right]$$

(*c*) In the case of continuous frequency distribution, x is taken to be the value corresponding to the mid-points of the class-intervals.

Example 1. *Find the geometric mean of the series*

1, 2, 4, 8,......, 2^n.

Sol. $x : 1, 2^1, 2^2, 2^3, \ldots\ldots, 2^n$

Number of observations $= n + 1$

$$\text{G.M.} = (1 \,.\, 2^1 \,.\, 2^2 \,.\, 2^3 \ldots\ldots 2^n)^{\frac{1}{n+1}}$$

$$= (2^{1+2+3 \ldots\ldots + n})^{\frac{1}{n+1}} = \left[2^{\frac{n(n+1)}{2}}\right]^{\frac{1}{n+1}} = 2^{n/2}.$$

Example 2. *Compute the geometric mean from the following data :*

10, 110, 120, 50, 80, 60, 52, 37.

Sol.

Size of item (*x*)	*log x*
10	1.0000
110	2.0414
120	2.0792
50	1.6990
80	1.9031
60	1.7782
52	1.7160
37	1.5682
No. of items (n) = 8	13.7851

$$\therefore \qquad \log G = \frac{1}{n} \sum \log x_i = \frac{1}{8}(13.7851) = 1.723$$

$$G = \text{antilog}\,(1.723) = 52.84.$$

Example 3. *The marks obtained by seven students are 5, 10, 15, 20, 25, 30, 35. Find the geometric mean.*

Sol. Please try yourself. **[Ans.** 16.9]

Example 4. *Calculate the geometric mean of the following :*

2574, 475, 75, 5, 0.8, 0.08, 0.005, 0.0009.

Sol. Please try yourself. **[Ans.** 1.841]

Example 5. *Compute the geometric mean of the following data :*

x :	*10*	*15*	*18*	*20*	*25*
f :	*2*	*3*	*5*	*6*	*4*

Sol.

x	*f*	*log x*	*f log x*
10	2	1.0000	2.0000
15	3	1.1761	3.5283
18	5	1.2553	6.2765
20	6	1.3010	7.8060
25	4	1.3979	5.5916
	$N = 20$		25.2024

$$\log G = \frac{1}{N} \Sigma f \log x = \frac{1}{20} \times 25.2024 = 1.2601$$

$$G = \text{antilog } 1.2601 = 18.20.$$

Example 6. *The marks obtained by 25 students in a test are given below :*

Marks	:	*11*	*12*	*13*	*14*	*15*
No. of students	:	*3*	*7*	*8*	*5*	*2*

Find their geometric mean.

Sol. Please try yourself. **[Ans.** 12.8]

Example 7. *Compute the geometric mean from the following data :*

Marks	*No. of students*
0—10	*10*
10—20	*5*
20—30	*8*
30—40	*7*
40—50	*20*

Sol.

Marks	*Mid-values* (*x*)	*No. of students* (*f*)	*log x*	*f log x*
0—10	5	10	0.6990	6.9900
10—20	15	5	1.1761	5.8805
20—30	25	8	1.3979	11.1832
30—40	35	7	1.5441	10.8087
40—50	45	20	1.6532	33.0640
		$N = 50$		67.9264

$$\log G = \frac{1}{N} \sum f \log x = \frac{67.9264}{50} = 1.3585$$

$$G = \text{antilog } 1.3585 = 22.83.$$

Example 8. *Find the geometric mean of the following distribution :*

Marks	:	*0—10*	*10—20*	*20—30*	*30—40*
No. of students	:	*5*	*8*	*3*	*4*

Sol. Please try yourself. **[Ans.** 14.58]

Example 9. *In a frequency table, the upper boundary of each class interval has a constant ratio to the lower boundary. Show that the geometric mean G may be expressed by the formula*

$$\log G = x_0 + \frac{C}{N} \sum f_i(i - 1)$$

where x_0 is the logarithm of the mid-value of the first interval and C is the logarithm of the ratio between the upper and the lower boundaries.

Sol. Let r be the constant ratio of the upper boundary of any class to the lower boundary of the same class. Let a be the mid-value of the first interval. Then the mid-values of the n intervals (say) are

$$a, ar, ar^2, \ldots\ldots, ar^{n-1}$$

$$\log G = \frac{1}{N} \sum_{i=1}^{n} f_i \log x_i \qquad \text{where } x_i \text{ is mid-value of } i\text{th class.}$$

$$= \frac{1}{N} [f_1 \log a + f_2 \log (ar) + f_3 \log (ar^2) + \ldots\ldots$$

$$\ldots + f_i \log (ar^{i-1}) + \ldots\ldots + f_n \log (ar^{n-1})]$$

$$= \frac{1}{N} [f_1 \log a + (f_2 \log a + f_2 \log r) + (f_3 \log a + 2f_2 \log r)$$

$$+ \{ f_i \log a + (i-1) f_i \log r \} + \ldots\ldots$$

$$+ \{ f_n \log a + (n-1) f_n \log r \}]$$

$$= \frac{1}{N} [(f_1 + f_2 + \ldots\ldots + f_n) \log a + \log r \, \Sigma f_i (i-1)]$$

$$= \frac{1}{N} [N x_0 + C \, \Sigma f_i (i-1)] \qquad | \because \quad \log a = x_0 ; \log r = C \text{ (given)}$$

$$= x_0 + \frac{C}{N} \Sigma f_i (i-1).$$

Example 10. *If n_1 and n_2 are the sizes, G_1 and G_2 the geometric means of two series respectively, then the geometric mean G of the combined series is given by*

$$\log G = \frac{n_1 \log G_1 + n_2 \log G_2}{n_1 + n_2}. \qquad \text{(D.U. 1989)}$$

Sol. Let x_{1i} $(i = 1, 2, \ldots\ldots, n_1)$ and x_{2j} $(j = 1, 2, \ldots\ldots, n_2)$ be n_1 and n_2 items of the two series respectively, then

$$\left.\begin{aligned} G_1 &= \left(x_{11} x_{12} \ldots\ldots x_{1n_1} \right)^{\frac{1}{n_1}} \quad \therefore \quad \log G_1 = \frac{1}{n_1} \sum_{i=1}^{n_1} \log x_{1i} \\ G_2 &= \left(x_{21} x_{22} \ldots\ldots x_{2n_2} \right)^{\frac{1}{n_2}} \quad \therefore \quad \log G_2 = \frac{1}{n_2} \sum_{j=1}^{n_2} \log x_{2j} \end{aligned}\right\} \ldots (A)$$

Geometric mean G of the combined series is given by

$$G = \left(x_{11} x_{12} \ldots\ldots x_{1n_1} \cdot x_{21} x_{22} \ldots\ldots x_{2n_2} \right)^{\frac{1}{n_1 + n_2}}$$

$$\therefore \quad \log G = \frac{1}{n_1 + n_2} \log (x_{11} x_{12} \ldots\ldots x_{1n_1} \cdot x_{21} x_{22} \ldots\ldots x_{2n_2})$$

$$= \frac{1}{n_1 + n_2} [\log (x_{11} x_{12} \ldots\ldots x_{1n_1}) + \log (x_{21} x_{22} \ldots\ldots x_{2n_2})]$$

$$= \frac{1}{n_1 + n_2} \left[\sum_{i=1}^{n_1} \log x_{1i} + \sum_{j=1}^{n_2} \log x_{2j} \right]$$

$$= \frac{n_1 \log G_1 + n_2 \log G_2}{n_1 + n_2} \qquad [using\ (A)]$$

Example 11. *Show that in finding the arithmetic mean of a set of readings on a thermometer, it does not matter whether we measure temperature in Centigrade or Fahrenheit, but that in finding the geometric mean, it does matter which scale we use.*

Sol. Let $C_1, C_2, C_3, \ldots, C_n$ be the n readings on the Centigrade thermometer. Their arithmetic mean M_C is given by

$$M_C = \frac{1}{n}(C_1 + C_2 + C_3 + \ldots + C_n)$$

If F and C be the readings in Fahrenheit and Centigrade respectively, then

$$\frac{F - 32}{180} = \frac{C}{100} \qquad \Rightarrow \qquad F = 32 + \frac{9}{5} C$$

$\therefore$ Fahrenheit equivalents of $C_1, C_2, \ldots, C_n$ are

$$F_1 = 32 + \frac{9}{5} C_1, \quad F_2 = 32 + \frac{9}{5} C_2, \ldots, F_n = 32 + \frac{9}{5} C_n$$

respectively. Their arithmetic mean M_F is given by

$$M_F = \frac{1}{n}(F_1 + F_2 + \ldots + F_n)$$

$$= \frac{1}{n}\left[32n + \frac{9}{5}(C_1 + C_2 + \ldots + C_n) \right]$$

$$= 32 + \frac{9}{5} M_C$$

which is the Fahrenheit equivalent of M_C.

Hence in finding the arithmetic mean of a set of n readings on a thermometer, it is immaterial whether we measure temperature in Centigrade or Fahrenheit.

Geometric mean G_C of n readings in Centigrade is

$$G_C = (C_1 C_2 \ldots C_n)^{\frac{1}{n}}$$

Geometric mean G_F of n readings in Fahrenheit is

$$G_F = (F_1 F_2 \ldots F_n)^{\frac{1}{n}}$$

$$= \left[\left(32 + \frac{9}{5} C_1\right)\left(32 + \frac{9}{5} C_2\right) \ldots \left(32 + \frac{9}{5} C_n\right) \right]^{\frac{1}{n}}$$

$$= \left[32\left(1+\frac{9}{160}C_1\right).32\left(1+\frac{9}{160}C_2\right)......32\left(1+\frac{9}{160}C_n\right)\right]^{\frac{1}{n}}$$

$$= \left(32^n\right)^{\frac{1}{n}} \cdot \left(1+\frac{9}{160}C_1\right)^{\frac{1}{n}} \left(1+\frac{9}{160}C_2\right)^{\frac{1}{n}} ......\left(1+\frac{9}{160}C_n\right)^{\frac{1}{n}}$$

$$= 32\left(1+\frac{9}{160n}C_1+.....\right)\left(1+\frac{9}{160n}C_2+......\right)$$

$$.....\left(1+\frac{9}{160n}C_n+...\right)$$

$$= 32\left[1+\frac{9}{160n}(C_1+C_2+......+C_n)+......\right]$$

$$= 32+\frac{9}{5n}(C_1+C_2+......+C_n)+......$$

But the Fahrenheit equivalent of G_C is

$$32+\frac{9}{5}G_C = 32+\frac{9}{5}\left(C_1 C_2 ..... C_n\right)^{\frac{1}{n}} \neq G_F.$$

Hence in finding the geometric mean of n readings on a thermometer, the scale matters.

5. HARMONIC MEAN

Harmonic Mean. Harmonic mean of a number of observations is the reciprocal of the arithmetic mean of the reciprocals of the given values. Thus, the harmonic mean H of n observations $x_1, x_2, ..., x_n$ is

$$H = \frac{1}{\frac{1}{n}\sum_{i=1}^{n}\frac{1}{x_i}} = \frac{n}{\frac{1}{x_1}+\frac{1}{x_2}+...+\frac{1}{x_n}}.$$

If $x_1, x_2, ..., x_n$ (none of them being zero) have the frequencies $f_1, f_2, ...,$ f_n respectively, then harmonic mean is given by

$$H = \frac{1}{\frac{1}{N}\sum_{i=1}^{n}\frac{f_i}{x_i}}$$

$$= \frac{N}{\frac{f_1}{x_1} + \frac{f_2}{x_2} + \ldots + \frac{f_n}{x_n}}, \quad N = \sum_{i=1}^{n} f_i$$

In the case of class-intervals, x is taken to be the mid-value of the class-interval.

Example 1. *Calculate the harmonic mean of the following :*

5, 10, 20.

Sol.

$$\text{H.M.} = \frac{3}{\frac{1}{x_1} + \frac{1}{x_2} + \frac{1}{x_3}} = \frac{3}{\frac{1}{5} + \frac{1}{10} + \frac{1}{20}}$$

$$= \frac{60}{7} = 8.57.$$

Example 2. *The marks obtained by five students are :*

10, 20, 40, 60, 120.

Find the harmonic mean.

Sol. Please try yourself. **[Ans.** 25]

Example 3. *Find out the harmonic mean of the following data :*

Marks (out of 150)	*No. of students*
10	*2*
20	*3*
40	*6*
60	*5*
120	*4*

Sol.

x	f	$\frac{1}{x}$	$\frac{f}{x}$
10	2	.100	.200
20	3	.050	.150
40	6	.025	.150
60	5	.017	.085
120	4	.008	.032
	N = 20		.617

$$\text{H.M.} = \frac{N}{\sum \frac{f}{x}} = \frac{20}{.617} = \frac{20 \times 1000}{617} = 32.4.$$

Example 4. *The marks obtained by 25 students in a test are given below :*

Marks	:	*11*	*12*	*13*	*14*	*15*
No. of students	:	*3*	*7*	*8*	*5*	*2*

Find their harmonic mean.

Sol. Please try yourself. **[Ans.** 12.7]

Example 5. *Calculate the harmonic mean of the following data :*

Marks	*No. of students*
0—10	*4*
10—20	*6*
20—30	*10*
30—40	*7*
40—50	*3*

Sol.

Marks	*Mid-value* (x)	*No. of students* (f)	$\frac{1}{x}$	$\frac{f}{x}$
0—10	5	4	.200	.800
10—20	15	6	.067	.402
20—30	25	10	.040	.400
30—40	35	7	.029	.203
40—50	45	3	.022	.066
		N = 30		1.871

$$\text{H.M.} = \frac{\text{N}}{\sum \frac{f}{x}} = \frac{30}{1.871} = 16.03.$$

Example 6. *Calculate the H.M. of the following distribution :*

Class :	*2—4*	*4—6*	*6—8*	*8—10*
f :	*20*	*40*	*30*	*10*

Sol. Please try yourself. **[Ans.** 4.985]

Example 7. *Calculate A.M., G.M. and H.M. of the following observations and show that :*

$$A.M. > G.M. > H.M.$$

$$32, 35, 36, 37, 39, 41, 43.$$

Sol. Please try yourself.

Example 8. *A variate takes values* $a, ar, ar^2, \ldots\ldots, ar^{n-1}$ *each with frequency unity. If A, G, H be respectively A.M., G.M. and H.M. show that*

$$A = \frac{a(1 - r^n)}{a(1 - r)}, \qquad G = ar^{\frac{n-1}{2}}$$

$$H = \frac{an(1 - r)r^{n-1}}{1 - r^n}.$$

Prove further that $AH = G^2$ (or A, G, H are in G.P.)

Prove also that $A > G > H$ unless $n = 1$ when all the three means coincide.

Sol. $A = \dfrac{\Sigma fx}{N}$

$$= \frac{a + ar + ar^2 + \ldots + ar^{n-1}}{n} \qquad |\because \quad N = \Sigma f = n$$

$$= \frac{a(1 - r^n)}{n(1 - r)} \qquad \text{...(I)}$$

$$G = (x_1^{f_1} . x_2^{f_2} \ldots\ldots x_n^{f_n})^{\frac{1}{N}}$$

$$= (a.ar.ar^2 \ldots\ldots ar^{n-1})^{\frac{1}{n}} \qquad |\because \quad f_i = 1, i = 1, 2, \ldots\ldots, n$$

$$= [a^n \, r^{1 + 2 \ldots\ldots + (n-1)}]^{\frac{1}{n}}$$

$$= \left[a^n . r^{\frac{n(n-1)}{2}} \right]^{\frac{1}{n}} = ar^{\frac{n-1}{2}} \qquad \text{...(II)}$$

$$H = \frac{N}{\sum \frac{f}{x}} = \frac{n}{\frac{1}{a} + \frac{1}{ar} + \frac{1}{ar^2} + \ldots + \frac{1}{ar^{n-1}}}$$

$$= \frac{n}{\frac{1}{ar^{n-1}}(r^{n-1} + r^{n-2} + \ldots + 1)} = \frac{na\, r^{n-1}}{1 + r + \ldots + r^{n-1}}$$

$$= \frac{na\, r^{n-1}}{\frac{1(1 - r^n)}{1 - r}} = \frac{an(1 - r)\, r^{n-1}}{1 - r^n} \qquad \text{...(III)}$$

$$\text{Now } AH = \frac{a(1 - r^n)}{n(1 - r)} . \frac{an(1 - r)\, r^{n-1}}{1 - r^n}$$

$$= a^2\, r^{n-1} = G^2.$$

$$\text{Since } \frac{x_1 + x_2 + \ldots + x_n}{n} \geq (x_1\, x_2 \ldots x_n)^{\frac{1}{n}},$$

the equality holding when $n = 1$ or when all x's are equal, *i.e.*, when $r = 1$ which is not possible because, then, the variate becomes a constant.

$\therefore \qquad A \geq G,$ the equality holding when $n = 1$

Also $\qquad AH = G^2$

or $$\frac{A}{G} = \frac{G}{H}$$

$\because$ $A \geq G$ *(proved)* $\therefore$ $\frac{A}{G} \geq 1$

$\Rightarrow$ $\frac{G}{H} \geq 1$ $\therefore$ $G \geq H$

Hence $A \geq G \geq H$,

the equality holding when $n = 1$.

Example 9. *If G_x be the G.M. of N x's and G_y be the G.M. of N y's ; then prove that G, the G.M. of the 2N values is given by*

$$G^2 = G_x G_y. \qquad \text{(D.U. 1984, 87)}$$

Sol. By def. $$G_x = (x_1 x_2 \ldots\ldots x_N)^{\frac{1}{N}}$$

$$G_y = (y_1 y_2 \ldots\ldots y_N)^{\frac{1}{N}}$$

$$G = (x_1 x_2 \ldots\ldots x_N \; y_1 y_2 \ldots\ldots y_N)^{\frac{1}{2N}}$$

$\therefore$ $$G^2 = (x_1 x_2 \ldots\ldots x_N \,.\, y_1 y_2 \ldots\ldots y_N)^{\frac{1}{N}}$$

$$= (x_1 x_2 \ldots\ldots x_N)^{\frac{1}{N}} \,.\, (y_1 y_2 \ldots\ldots y_N)^{\frac{1}{N}}$$

$$= G_x G_y.$$

Example 10. *A man motors from A to B. A part of the distance is up hill and he gets a mileage of only 10 miles per gallon of gasoline. On the return trip, he makes 15 miles per gallon. Find the harmonic mean of his mileage. Verify the fact that this is the proper average to use by assuming that the distance from A to B is 60 miles.*

Sol. $$\text{H.M.} = \frac{2}{\frac{1}{10} + \frac{1}{15}} = 12 \text{ miles per gallon.}$$

Distance from A to B is 60 miles.

In the onward journey, consumption $\frac{60}{10} = 6$ gallons and in the return journey consumption is $\frac{60}{15} = 4$ gallons.

Total consumption = 6 + 4 = 10 gallons.

Average consumption = $\frac{120}{10}$ = 12 gallons which is given by the H.M.

2.4. REQUISITES FOR AN IDEAL MEASURE OF CENTRAL TENDENCY

An ideal measure of central tendency should satisfy the following characteristics : (M.D.U. 1982) (*i*) It should be rigidly defined and its value should be definite.

(*ii*) It should be based upon all the observations.

(*iii*) It should be readily comprehensible and easy to calculate.

(*iv*) It should be suitable for further mathematical treatment, *i.e.*, if two or more series of observations on similar material are given, we should be able to calculate the average of the composite series obtained on combining the given series.

(*v*) It should be least affected by fluctuations of sampling *i.e.*, if different samples are taken from the same universe, the averages of these samples should not differ much from each other.

2.5. MERITS AND DEMERITS OF VARIOUS AVERAGES

(1) ARITHMETIC MEAN

Merits. (*i*) It is rigidly defined.

(*ii*) It is easy to understand and easy to calculate.

(*iii*) It is never indefinite, *i.e.*, it is determinate.

(*iv*) It is based upon all the observations.

(*v*) It is least affected by fluctuations of sampling.

(*vi*) It lends itself to algebraic treatment. The mean of the composite series can be readily expressed in terms of the means and the sizes of the component series.

(*vii*) It does nor require the arraying of data as the median does nor the grouping of data as the mode does.

Demerits. (*i*) It may not be represented in the actual data.

(*ii*) It gives greater importance to bigger items of a given series. Thus, in the case of extreme items, it gives a distorted picture of the distribution.

(*iii*) It cannot be located by inspection like median and mode.

(*iv*) It loses its accuracy if a single item of the series is ignored.

(*v*) It may lead to wrong conclusions if the details of the data from which it is computed are not given.

(*vi*) It cannot be calculated even if a single observation is missing.

(*vii*) It cannot be calculated if the extreme class is open, *e.g.*, below 5 or above 40.

(2) MEDIAN

Merits. (*i*) It is rigidly defined.

(*ii*) It is readily calculated and easily understood.

(*iii*) It is not affected by extreme values.

(*iv*) It can be determined by inspection in a frequency graph.

(*v*) It can be calculated for distribution with open-end classes.

Demerits. (*i*) It does not lend itself to algebraic treatment.

(*ii*) In the case of even number of observations, it cannot be calculated exactly. We simply estimate it by taking the mean of two middle items.

(*iii*) It is not based on all observations. It is insensitive, *i.e.,* we may replace the observations below it by some other observations below it and the observations above it by some other observations above it, without affecting the median itself.

(*iv*) As compared with mean, it is affected much by fluctuations of sampling.

(*v*) It can be determined only by arranging the data—an operation which involves considerable work.

(3) MODE

Merits. (*i*) It is readily comprehensible and easy to calculate.

(*ii*) It can be located, in some cases, by inspection.

(*iii*) It is not affected by extreme values.

(*iv*) It is based on all the values of the variate.

(*v*) It is capable of further algebraic treatment.

(*vi*) It can be calculated for distributions with open-end classes.

Demerits. (*i*) It is ill-defined. A clearly defined mode does not always exist.

(*ii*) It is not based on all the observations.

(*iii*) It is not capable of further mathematical treatment.

(4) GEOMETRIC MEAN

Merits. (*i*) It is rigidly defined.

(*ii*) It is based upon all the observations.

(*iii*) It is suitable for further mathematical treatment.

(*iv*) It is not much affected by fluctuations of sampling.

Demerits. (*i*) For a non-mathematical student, it is not easy to understand and calculate.

(*ii*) If any one of the observations is zero, G.M. is zero and if an odd number of observations are negative, it cannot be calculated.

(5) HARMONIC MEAN

Merits. (*i*) It is rigidly defined.

(*ii*) It is based on all the observations.

(*iii*) It is suitable for further mathematical treatment.

Demerits. (*i*) It is difficult to calculate.

(*ii*) It is difficult to understand.

3

Measures of Dispersion, Skewness and Kurtosis ; Moments of Frequency Distributions

3.1. DISPERSION

A measure of central tendency by itself can exhibit only one of the important characteristics of distribution. It can represent a series only 'as best as a single figure can'. It is inadequate to give us a complete idea of the distribution. It must be supported and supplemented by some other measures. One such measure is **Dispersion.**

Two or more frequency distributions may have exactly identical averages but even then they may differ markedly in several ways. Further analysis is, therefore, essential to account for these differences. Consider the following example :

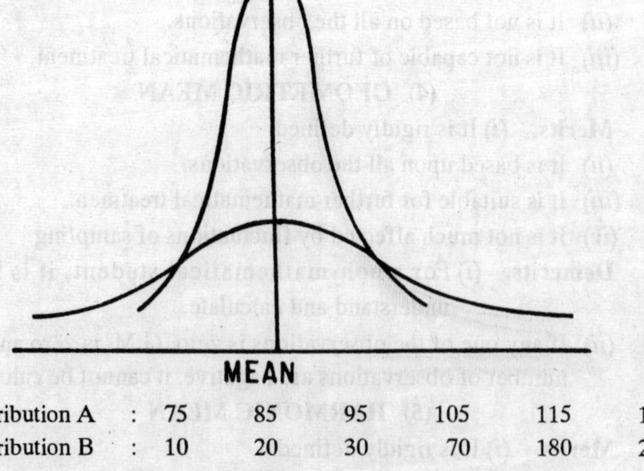

Distribution A	:	75	85	95	105	115	125
Distribution B	:	10	20	30	70	180	290

A.M. of each distribution is $\frac{600}{6} = 100$. In distribution A, the values of the variate differ from 100 but the difference is small. In distribution B, the items are widely scattered and lie far from the mean. Although the A.M. is the same, yet the two distributions widely differ from each other in their formation.

Therefore, while studying a distribution, it is equally important to know how the variates are clustered around or scattered away from the point of central tendency. Such variation is called *dispersion* or *spread* or *scatter* or *variability.* Thus, *dispersion is the extent to which the values are dispersed about the central value.*

3.2. REQUISITES FOR AN IDEAL MEASURE OF DISPERSION

(*i*) It should be rigidly defined.

(*ii*) It should be easy to calculate and easy to understand.

(*iii*) It should be based on all the observations.

(*iv*) It should be amenable to further mathematical treatment.

(*v*) It should be affected as little as possible by fluctuations of sampling.

3.3. MEASURES OF DISPERSION

The following are the measures of dispersion :

(*a*) Range.

(*b*) Quartile deviation or Semi-inter-quartile range.

(*c*) Average (or mean) deviation.

(*d*) Standard deviation.

(*a*) **Range.** Range is the difference between the extreme values of the variate.

Range = L – S, where

L = Largest and S = Smallest

$$\text{Co-efficient of Range} = \frac{L - S}{L + S}$$

It is easily understood and computed. But it suffers from the drawback that it depends exclusively on the two extreme values. It is not a reliable measure of dispersion.

(*b*) **Quartile Deviation.** The difference between the upper and lower quartiles *i.e.,* $Q_3 - Q_1$ is known as the *inter-quartile range* and half of it *i.e.,* $\frac{1}{2}(Q_3 - Q_1)$, is called the *semi-inter-quartile range* or the *quartile deviation.*

$$\text{Quartile Deviation} = \frac{1}{2}(Q_3 - Q_1).$$

It is definitely a better measure of dispersion than range as it makes use of 50% of the data. But since it ignores the other 50% of the data, it is also not a reliable measure of dispersion.

$$\text{Co-efficient of Quartile Deviation} = \frac{Q_3 - Q_1}{Q_3 + Q_1}$$

Example 1. *Find out the quartile deviation of the income of a certain person given in rupees for 12 months in a year.*

139, 150, 151, 151, 157, 158, 160, 161, 162, 162, 173, 175.

Sol.

S. No.	*Income (Rs.)*
1	139
2	150
3	151
4	151
5	157
6	158
7	160
8	161
9	162
10	162
11	173
12	175
N = 12	

$Q_1 = \text{size of}\left(\frac{N+1}{4}\text{ th} = \right) 3.25\text{th item}$

= size of 3rd item + .25 (size of 4th item – size of 3rd item)
= 151 + .25 (151 – 151) = Rs. 151

$Q_3 = \text{size of}\ \frac{3(N+1)}{4}\text{ th} = 9.75\text{th item}$

= size of 9th item + .75 (size of 10th item – size of 9th item)
= 162 + .75 (162 – 162) = Rs. 162

$\therefore$ Quartile Deviation $= \frac{1}{2}(Q_3 - Q_1) = \frac{11}{2}$ = Rs. 5.5.

Example 2. *Calculate the quartile deviation and co-efficient of quartile deviation from the following data :*

Height (in inches)	*Frequency*
50	*10*
51	*12*
52	*15*
53	*10*
54	*14*
55	*18*
56	*6*

Sol.

Height (in inches)	*f*	*c.f.*
50	10	10
51	12	22
52	15	37
53	10	47
54	14	61
55	18	79
56	6	85
		N = 85

Q_1 = size of item corresponding to *c.f.* just greater than $\frac{N+1}{4}$

= 21.5 *i.e.,* 22 is 51 inches

Q_3 = size of item corresponding to *c.f.* just greater than $\frac{3(N+1)}{4}$

= 64.5 *i.e.,* 79 is 55 inches

$$\therefore \quad \text{Q.D.} = \frac{Q_3 - Q_1}{2} = \frac{55 - 51}{2} = 2 \text{ inches}$$

$$\text{Co-eff. of Q.D.} = \frac{Q_3 - Q_1}{Q_3 + Q_1} = \frac{55 - 51}{55 + 51} = \frac{2}{53} = 0.038.$$

Example 3. *Calculate the semi-inter quartile range of the marks of 39 students in Statistics given below :*

Marks	:	*0—5*	*5—10*	*10—15*	*15—20*	*20—25*	*25—30*
No. of students	:	*4*	*6*	*8*	*12*	*7*	*2*

Sol. Cumulative frequency table is given below :

Marks	*No. of students (f)*	*c.f.*
0— 5	4	4
5—10	6	10
10—15	8	18
15—20	12	30
20—25	7	37
25—30	2	39

Here $N = \Sigma f = 39$

$\frac{N}{4} = 9.75 \quad \therefore \quad$ Class of Q_1 is 5—10

$$Q_1 = l + \frac{h}{f}\left(\frac{N}{4} - C\right) = 5 + \frac{5}{6}(9.75 - 4)$$

$$= 5 + \frac{5 \times 5.75}{6} = 5 + \frac{28.75}{6} = 5 + 4.79 = 9.79$$

$\frac{3N}{4} = 29.25 \quad \therefore \quad$ Class of Q_3 is 15—20

$$Q_3 = l + \frac{h}{f}\left(\frac{3N}{4} - C\right) = 15 + \frac{5}{12}(29.25 - 18)$$

$$= 15 + \frac{5 \times 11.25}{12} = 15 + \frac{56.25}{12} = 15 + 4.69 = 19.69$$

$$\text{Semi-inter quartile range} = \frac{1}{2}(Q_3 - Q_1) = \frac{1}{2}(19.69 - 9.79)$$

$$= \frac{1}{2} \times 9.90 = 4.95 \text{ marks.}$$

Example 4. *Calculate the quartile deviation of the marks of 63 students in Statistics given below :*

Marks	*No. of students*	*Marks*	*No. of students*
0—10	5	*50— 60*	7
10—20	7	*60— 70*	3
20—30	10	*70— 80*	2
30—40	16	*80— 90*	2
40—50	11	*90—100*	0

Sol. Please try yourself. [**Ans.** 12.32 marks]

(*c*) Average Deviation or Mean Deviation

If $x_1, x_2, x_3, \ldots, x_n$ occur $f_1, f_2, f_3, \ldots, f_n$ times respectively and $N = \sum_{i=1}^{n} f_i$, the mean deviation from the average A (usually mean or median) is given by

$$\textit{Mean Deviation} = \frac{1}{N} \sum_{i=1}^{n} f_i \,|\, x_i - A \,|,$$

where $|\, x_i - A \,|$ represents the modulus or the absolute value of the deviation $(x_i - A)$.

Since the mean deviation is based on all the values of the variate, it is a better measure of dispersion than range or quartile deviation. But some artificiality is created due to the ignoring of the signs of the deviations $(x_i - A)$. This renders it useless for further mathematical treatment.

Co-efficient of Mean Deviation

$$= \frac{\text{Mean Deviation}}{\text{Average from which it is calculated}}.$$

Example 1. *Calculate the mean deviation from the arithmetic mean of the following series of weekly wages (in Rs.) of workers :*

21, 23, 25, 28, 30, 32, 38, 39, 46, 48.

Sol. Mean $\bar{x} = \frac{1}{10}(21 + 23 + 25 + 28 + 30 + 32 + 38 + 39 + 46 + 48)$

$= 33$

x	$\|x-\bar{x}\|$
21	12
23	10
25	8
28	5
30	3
32	1
38	5
39	6
46	13
48	15
$n = 10$	$\Sigma\|x_i-\bar{x}\| = 78$

Mean Deviation from A.M. $= \frac{1}{n}\Sigma|x_i - \bar{x}| = \frac{78}{10}$ = Rs. 7.8

Note. Co-eff. of M.D. $= \frac{\text{M.D.}}{\text{A.M.}} = \frac{7.8}{33} = 0.24.$

Example 2. *Find the average deviation from the mean for the following distribution :*

x :	3	5	7	9	11	13
f :	2	7	10	9	5	1

Sol.

x	f	$d_x = x - a$ $= x - 8$	fd_x	$\|x-\bar{x}\|$	$f\|x-\bar{x}\|$
3	2	−5	−10	4.65	9.30
5	7	−3	−21	2.65	18.55
7	10	−1	−10	.65	6.50
9	9	1	9	1.35	12.15
11	5	3	15	3.35	16.75
13	1	5	5	5.35	5.35
	N = 34		−12		68.60

Mean $\bar{x} = a + \frac{\Sigma fd_x}{N} = 8 - \frac{12}{34} = 8 - .35 = 7.65$

Average Deviation $= \frac{1}{N}\Sigma f|x - \bar{x}| = \frac{68.60}{34} = 2.02.$

Example 3. *Compute the mean deviation from the mean of the following data :*

Marks	:	10	15	20	25	30
No. of students	:	2	4	6	8	5

Sol. Please try yourself. **[Ans.** 5.12]

Example 4. *Find the mean deviation about median for the following distribution :*

x :	5	7	9	11	13	15	17
f :	2	4	6	8	10	12	8

Sol.

x	f	c.f.	$\lvert x - M_d \rvert$	$f\lvert x - M_d \rvert$
5	2	2	8	16
7	4	6	6	24
9	6	12	4	24
11	8	20	2	16
13	10	30	0	0
15	12	42	2	24
17	8	50	4	32
	N = 50			136

$\frac{N+1}{2}$ = 25.5 Median = value of x corresponding to *c.f.* (just greater than $\frac{N+1}{2}$ *i.e.,*) 30 is 13

Mean deviation about median $= \frac{1}{N} \Sigma f \lvert x - M_d \rvert$

$= \frac{136}{50} = 2.72.$

Example 5. *Find the mean deviation about median for the following data :*

Height (in inches) :	58	60	62	64	66	59	61	63	65
No. of students :	1	32	33	20	8	20	35	22	10

(K.U. 1982 S)

Sol. Arranging the heights in ascending order

Height (in inches) x	*No. of students* f	c.f.	$\lvert x - M_d \rvert$	$f\lvert x - M_d \rvert$
58	1	1	4	4
59	20	21	3	60
60	32	53	2	64
61	35	88	1	35
62	33	121	0	0
63	22	143	1	22
64	20	163	2	40
65	10	173	3	30
66	8	181	4	32
	N = 181			287

Median = height corresponding to *c.f.* $\left(\text{just greater than } \frac{N+1}{2} \text{ i.e.}\right)$ 121 is 62

Mean Deviation about median $= \frac{1}{N} \Sigma f \mid x - M_d \mid$

$$= \frac{287}{62} = 4.63 \text{ inches.}$$

Example 6. *Calculate the mean deviation and its co-efficient from the median for the following data :*

Expenditure is Rs. per day	*No. of families*
3	*2*
4	*9*
5	*11*
6	*14*
7	*20*
8	*25*
9	*24*
10	*23*
11	*20*
12	*16*
13	*5*
14	*4*
15	*2*

Sol. Please try yourself. **[Ans.** Rs. 2.12, 0.24**]**

Example 7. *Calculate the mean deviation from the mean for the following distribution :*

Marks	:	*0—10*	*10—20*	*20—30*	*30—40*	*40—50*
No. of students	:	*5*	*8*	*15*	*16*	*6*

Sol.

Mid-value	*f*	$d_x = x - a$ $= x - 25$	fd_x	$\lvert x - \bar{x} \rvert$	$f \lvert x - \bar{x} \rvert$
5	5	– 20	– 100	22	110
15	8	– 10	– 80	12	96
25	15	0	0	2	30
35	16	10	160	8	128
45	6	20	120	18	108
	N = 50		100		472

Mean $\bar{x} = a + \frac{\Sigma fd_x}{N} = 25 + \frac{100}{50} = 27$

Mean deviation from the mean

$$= \frac{1}{N} \Sigma f \,|\, x - \bar{x} \,| = \frac{472}{50} = 9.44 \text{ marks.}$$

Example 8. *Find the mean deviation from the mean for the following distribution :*

Class	:	*0—6*	*6—12*	*12—18*	*18—24*	*24—30*
Frequency	:	*8*	*10*	*12*	*9*	*5*

Sol. Please try yourself. **[Ans.** 6.3]

Example 9. *Find the mean deviation from the median for the following frequency distribution :*

Marks	:	*0—10*	*10—20*	*20—30*	*30—40*	*40—50*
No. of students	:	*5*	*8*	*15*	*16*	*6*

Sol.

Mid-value	*f*	*c.f.*	$\|x - M_d\|$	$f\|x - M_d\|$
5	5	5	23	115
15	8	13	13	104
25	15	28	3	45
35	16	44	7	112
45	6	50	17	102
	N = 50			478

$\frac{N}{2} = 25 \quad \therefore$ Median class corresponds to *c.f.* 28

i.e., median class is 20—30

$$\text{Median } M_d = l + \frac{h}{f}\left(\frac{N}{2} - C\right)$$

$$= 20 + \frac{10}{15}(25 - 13) = 20 + 8 = 28$$

Mean deviation from median

$$= \frac{1}{N} \Sigma f \,|\, x - M_d \,| = \frac{478}{50} = 9.56 \text{ marks.}$$

Example 10. *Compute mean deviation about median from the following data :*

Marks	:	*0—10*	*10—20*	*20—30*	*30—40*	*40—50*
No. of students	:	*5*	*10*	*20*	*5*	*10*

Sol. Please try yourself. **[Ans.** Rs. 9]

Example 11. *Calculate the mean deviation about mean and also about median from the following data :*

Income per week in Rs.	:	*20—30*	*30—40*	*40—50*	*50—60*	*60—70*
No. of families	:	*120*	*201*	*150*	*75*	*25*

(K.U. 1981 S)

Sol. Please try yourself. **[Ans.** Rs. 9.2, Rs. 9.1]

(*d*) Standard Deviation

Root-mean Square Deviation. The root-mean square deviation denoted by *s*, is defined as the positive square root of the mean of the squares of the deviations from an arbitrary origin A. Thus

$$s = +\sqrt{\frac{1}{N}\Sigma f_i(x_i - A)^2}$$

When the deviations are taken from the mean $\bar{x}$, the root-mean square deviation is called the *standard deviation* and is denoted by the Greek letter σ. Thus

$$\sigma = \sqrt{\frac{1}{N}\Sigma f_i(x_i - \bar{x})^2}\,.$$

Note. The square of the standard deviation σ^2 is called *variance.*

3.4. SHORT-CUT METHODS FOR CALCULATING STANDARD DEVIATION (σ)

(*i*) Direct Method

$$\sigma = \sqrt{\frac{1}{N}\Sigma f_i(x_i - \bar{x})^2}$$

$$\Rightarrow \quad \sigma^2 = \frac{1}{N}\Sigma f_i\,(x_i^2 - 2\,x_i\bar{x} + \bar{x}^2)$$

$$= \frac{1}{N}\Sigma f_i x_i^2 - 2\bar{x}\,.\,\frac{1}{N}\Sigma f_i x_i + \bar{x}^2\,.\,\frac{1}{N}\Sigma f_i$$

(taking the constants $\bar{x}, \bar{x}^2$ outside the summation sign)

$$= \frac{1}{N}\Sigma f_i x_i^2 - 2\bar{x}\,.\,\bar{x} + \bar{x}^2\,.\,\frac{1}{N}\,.\,N$$

$$= \frac{1}{N}\Sigma f_i x_i^2 - \bar{x}^2$$

$$\therefore \quad \sigma = \sqrt{\frac{1}{N}\Sigma f_i x_i^2 - \bar{x}^2}$$

$$= \sqrt{\frac{1}{N}\Sigma f_i x_i^2 - \left(\frac{1}{N}\Sigma f_i x_i\right)^2}$$

(*ii*) Change of Origin

Let the origin be shifted to an arbitrary point '*a*'. Let $d = x - a$ denote the deviation of variate *x* from the new origin

$$d = x - a \quad \Rightarrow \quad \bar{d} = \bar{x} - a$$

$$\therefore \quad d - \bar{d} = x - \bar{x}$$

$$\sigma_x = \sqrt{\frac{1}{N}\Sigma f(x - \bar{x})^2} = \sqrt{\frac{1}{N}\Sigma f(d - \bar{d}\,)^2}$$

$$= \sigma_d$$

$\therefore$ S.D. remains unchanged by shift of origin.

$$\sigma_x = \sigma_d = \sqrt{\frac{1}{N}\Sigma fd^2 - \left(\frac{1}{N}\Sigma fd\right)^2}.$$

Note. In the case of series of individual observations, if the mean is a whole number, take $a = \bar{x}$. In the case of discrete series, when the values of x are not equidistant, take 'a' somewhere in the middle of the x-series.

(*iii*) **Shift of origin and change of scale** (Step Deviation Method) (*Kanpur 1989*)

Let the origin be shifted to an arbitrary point 'a'. Let the new scale be $\frac{1}{h}$ times the original scale.

Let $u = \frac{x-a}{h}$ then $hu = x - a$

$\Rightarrow$ $h\bar{u} = \bar{x} - a$ $\therefore$ $h(u - \bar{u}) = x - \bar{x}$

$$\sigma_x = \sqrt{\frac{1}{N}\Sigma f(x-\bar{x})^2} = \sqrt{\frac{1}{N}\Sigma f h^2 (u-\bar{u})^2}$$

$$= h\sqrt{\frac{1}{N}\Sigma f(u-\bar{u})^2} = h\,\sigma_u$$

which is independent of a but not h. Hence S.D. is independent of change of origin but not of change of scale.

$$\sigma_x = h\sigma_u = h\sqrt{\frac{1}{N}\Sigma fu^2 - \left(\frac{1}{N}\Sigma fu\right)^2}.$$

Note. In the case of Discrete Series, when the values of x are equidistant at intervals of h or in the case of continuous series having equal class intervals of width h, use Step Deviation Method.

☞3.5. RELATION BETWEEN σ AND s

By def., we have

$$s^2 = \frac{1}{N}\Sigma f_i (x_i - a)^2 = \frac{1}{N}\Sigma f_i (x_i - \bar{x} + \bar{x} - a)^2$$

$$= \frac{1}{N}\Sigma f_i (x_i - \bar{x} + d)^2, \qquad \text{where } d = \bar{x} - a$$

$$= \frac{1}{N}\Sigma f_i[(x_i - \bar{x})^2 + d^2 + 2d\,(x_i - \bar{x})]$$

$$= \frac{1}{N}\Sigma f_i (x_i - \bar{x})^2 + \frac{d^2}{N}\Sigma f_i + \frac{2d}{N}\Sigma f_i (x_i - \bar{x})$$

$$= \sigma^2 + \frac{d^2}{N}.N + \frac{2d}{N}(0) \quad [\because \ \Sigma f_i (x_i - \bar{x}) = \text{algebraic sum of the deviation from mean} = 0]$$

$$= \sigma^2 + d^2$$

Hence $s^2 = \sigma^2 + d^2$

$\because \quad d^2 \geq 0 \quad \therefore \quad s^2 \geq \sigma^2$

Hence variance is the minimum value of mean square deviation.

(D.U. 1987)

Clearly s^2 is least when $d = 0$, *i.e.*, $\bar{x} = a$

$\therefore$ Mean square deviation (s^2) and consequently root-mean square deviation (s) is least when the deviations are measured from the mean.

Hence *standard deviation is the least possible root-mean square deviation.* *(D.U. 1991, 92)*

3.6. RELATIONS BETWEEN MEASURES OF DISPERSION

Mean Deviation = $\frac{4}{5}$ (standard deviation) = $\frac{4}{5}\sigma$

Semi-interquartile range = $\frac{2}{3}$ (standard deviation) = $\frac{2}{3}\sigma$.

3.7. CO-EFFICIENT OF DISPERSION

Whenever we want to compare the variability of two series which differ widely in their averages or which are measured in different units, we calculate the co-efficients of dispersion, which being ratios, are numbers independent of the units of measurement. The co-efficients of dispersion (C.D.) based on different measures of dispersion are as follows :

(*a*) Based on range :

$$\text{C.D} = \frac{x_{max} - x_{min}}{x_{max} + x_{min}}$$

(*b*) Based on quartile deviation :

$$\text{C.D.} = \frac{Q_3 - Q_1}{Q_3 + Q_1}$$

(*c*) Based on mean deviation :

$$\text{C.D.} = \frac{\text{mean deviation}}{\text{average from which it is calculated}}$$

(*d*) Based on standard deviation :

$$\text{C.D.} = \frac{\text{S.D.}}{\text{Mean}} = \frac{\sigma}{\bar{x}}.$$

Co-efficient of variation. It is the percentage variation in the mean, standard deviation being considered as the total variation in the mean.

$$\text{C.V.} = \frac{\sigma}{\bar{x}} \times 100.$$

Example 1. *Compute standard deviation for the following data :*

5, 8, 7, 11, 9, 10, 8, 2, 4, 6.

Sol.

x	$d = x - 7$	d^2
5	-2	4
8	1	1
7	0	0
11	4	16
9	2	4
10	3	9
8	1	1
2	-5	25
4	-3	9
6	-1	1
70	0	70

Here $n = 10$, $\bar{x} = \frac{\Sigma x}{n} = \frac{70}{10} = 7$, a whole number.

∴ We measure deviations from 7.

$$\sigma_x = \sigma_d = \sqrt{\frac{1}{n}\Sigma d^2 - \left(\frac{1}{n}\Sigma d\right)^2}$$

$$= \sqrt{\frac{70}{10} - 0} = \sqrt{7} = 2.65.$$

Example 2. *Compute the standard deviation for the following data relating to marks obtained by 15 students :*

12, 21, 21, 23, 27, 28, 30, 34, 37, 39, 39, 39, 40, 49, 54.

Sol.

x	$d = x - 33$	d^2
12	-21	441
21	-12	144
21	-12	144
23	-10	100
27	-6	36
28	-5	25
30	-3	9
34	1	1
37	4	16
39	6	36
39	6	36
39	6	36
40	7	49
49	16	256
54	21	441
493	-2	1770

Here $n = 15$, Taking $a = 33$ (close to the mean $\bar{x}$)

$$\sigma_x = \sigma_d = \sqrt{\frac{1}{n}\Sigma d^2 - \left(\frac{1}{n}\Sigma d\right)^2}$$

$$= \sqrt{\frac{1770}{15} - \left(\frac{-2}{15}\right)^2} = 10.9 \text{ marks.}$$

Example 3. *Calculate the mean and standard deviation for the following distribution :*

x :	25	35	45	55	65	75	85
f :	3	61	132	153	140	51	2

Sol.

x	f	$u = \dfrac{x-55}{10}$	fu	fu^2
25	3	–3	–9	27
35	61	–2	–122	244
45	132	–1	–132	132
55	153	0	0	0
65	140	1	140	140
75	51	2	102	204
85	2	3	6	18
	N = 542		15	765

Mean $\bar{x} = a + h\dfrac{\Sigma fu}{N} = 55 + 10 \times \dfrac{-15}{542} = 55 - .28 = 54.72$

$$\sigma_x = h\sigma_u = h\sqrt{\frac{1}{N}\Sigma fu^2 - \left(\frac{\Sigma fu}{N}\right)^2}$$

$$= 10\sqrt{\frac{765}{542} - \left(\frac{-15}{542}\right)^2} = \frac{10}{542}\sqrt{414405} = 11.9.$$

Example 4. *Calculate mean and standard deviation for the following distribution :*

x :	56	63	70	77	84	91	98
f :	3	6	14	16	13	6	2

Sol. Please try yourself. **[Ans.** 76.53, 9.87]

Example 5. *For the following data, find the standard deviation and the co-efficient of variation.*

Marks	*No. of students*	*Marks*	*No. of students*
0—10	*5*	*40—50*	*30*
10—20	*10*	*50—60*	*20*
20—30	*20*	*60—70*	*10*
30—40	*40*	*70—80*	*4*

Sol.

Mid-values x	f	$u = \frac{x-35}{10}$	fu	fu^2
5	5	– 3	– 15	45
15	10	– 2	– 20	40
25	20	– 1	– 20	20
35	40	0	0	0
45	30	1	30	30
55	20	2	40	80
65	10	3	30	90
75	4	4	16	64
	N = 139		61	369

$$\sigma_x = h\sigma_u = h\sqrt{\frac{1}{N}\Sigma fu^2 - \left(\frac{\Sigma fu}{N}\right)^2}$$

$$= 10\sqrt{\frac{369}{139} - \left(\frac{61}{139}\right)^2} = 15.7$$

Also $\quad \bar{x} = a + h \cdot \frac{\Sigma fu}{N} = 35 + 10\left(\frac{61}{139}\right) = 35 + 4.4 = 39.4$

Co-efficient of variation $= \frac{\sigma}{\bar{x}} \times 100 = \frac{15.7}{39.4} \times 100 = 39.8.$

Example 6. *Find the standard deviation for the following data giving wages of 230 persons :*

Wages in Rs.	*No. of persons*	*Wages in Rs.*	*No. of persons*
70— 80	12	110—120	50
80— 90	18	120—130	45
90—100	35	130—140	20
100—110	42	140—150	8

Sol. Please try yourself. [**Ans.** Rs. 17.1]

Example 7. *Compute the standard deviation of the following data :*

Wages (Rs.) per day	:	*1—3*	*3—5*	*5—7*	*7—9*	*9—11*
No. of workers	:	*15*	*18*	*27*	*10*	*6*

(G.N.D.U. 1982)

Sol. Please try yourself. [**Ans.** Rs. 2.33]

Example 8. *Find the mean and standard deviation of the following :*

Series	*Frequency*
15—20	*2*
20—25	*5*
25—30	*8*
30—35	*11*
35—40	*15*
40—45	*20*
45—50	*20*
50—55	*17*
55—60	*16*
60—65	*13*
65—70	*11*
70—75	*5*

Sol.

Mid-values x	f	$u = \frac{x - 47.5}{5}$	fu	fu^2
17.5	2	– 6	– 12	72
22.5	5	– 5	– 25	125
27.5	8	– 4	– 32	128
32.5	11	– 3	– 33	99
37.5	15	– 2	– 30	60
42.5	20	– 1	– 20	20
47.5	20	0	0	0
52.5	17	1	17	17
57.5	16	2	32	64
62.5	13	3	39	117
67.5	11	4	44	176
72.5	5	5	25	125
	N = 143		5	1003

$$\bar{x} = a + h \cdot \frac{\Sigma fu}{N} = 47.5 + 5 \times \frac{5}{143} = 47.7$$

$$\sigma_x = h\sigma_u = h\sqrt{\frac{1}{N}\Sigma fu^2 - \left(\frac{\Sigma fu}{N}\right)^2}$$

$$= 5\sqrt{\frac{1003}{143} - \left(\frac{5}{143}\right)^2} = 5 \times 2.65 = 13.25.$$

Example 9. *Find the standard deviation and the semi-quartile range from the following data :*

Wages upto Rs.	*No. of persons*
10	*12*
20	*30*
30	*65*
40	*107*
50	*157*
60	*202*
70	*222*
80	*230*

(K.U. 1982)

Sol.

Wages	*Mid-values* x	f	$u = \frac{x-45}{10}$	fu	fu^2	*c.f.*
0—10	5	12	– 4	– 48	192	12
10—20	15	18	– 3	– 54	162	30
20—30	25	35	– 2	– 70	140	65
30—40	35	42	– 1	– 42	42	107
40—50	45	50	0	0	0	157
50—60	55	45	1	45	45	202
60—70	65	20	2	40	80	222
70—80	75	8	3	24	72	230
		N = 230		– 105	733	

$$\sigma_x = h\sigma_u = h\sqrt{\frac{1}{N}\Sigma fu^2 - \left(\frac{\Sigma fu}{N}\right)^2}$$

$$= 10\sqrt{\frac{733}{230} - \left(\frac{-105}{230}\right)^2} = \text{Rs. } 17.31$$

To find Q_1 $\frac{N}{4} = 57.5$ *c.f.* just greater than 57.5 is 65

∴ Class of Q_1 is 20—30. $l = 20$, $h = 10$, $f = 35$, $C = 30$

$$Q_1 = l + \frac{h}{f}\left(\frac{N}{4} - C\right) = 20 + \frac{10}{35}(57.5 - 30) = \text{Rs. } 27.86$$

To find Q_3 $\frac{3N}{4} = 172.5$ *c.f.* just greater than 172.5 is 202

∴ Class of Q_3 is 50—60, $l = 50$, $h = 10$, $f = 45$, $C = 157$

$$Q_3 = l + \frac{h}{f}\left(\frac{3N}{4} - C\right)$$

$$= 50 + \frac{10}{45}(172.5 - 157) = \text{Rs. } 53.44$$

Semi-quartile Range = $\frac{1}{2}$ ($Q_3 - Q_1$)

$= \frac{1}{2}$ (53.44 – 27.86) = Rs. 12.79.

Example 10. *Calculate the standard deviation and semi-quartile range for the following table giving the age distribution of 542 persons :*

Age (in years)	*No. of persons*
20—30	*3*
30—40	*61*
40—50	*132*
50—60	*153*
60—70	*140*
70—80	*51*
80—90	*2*

(K.U. 1981)

Sol. Please try yourself. **[Ans.** 11.9 years, 9.4 years]

Example 11. *A collar manufacturer is considering the production of a new type of collar to attract young men. The following statistics of neck circumferences are available based upon the measurements of typical group of college students :*

Mid-value (inches)	*No. of students*	*Mid-value (inches)*	*No. of students*
12.5	*4*	*15.0*	*29*
13.0	*19*	*15.5*	*18*
13.5	*30*	*16.0*	*1*
14.0	*63*	*16.5*	*1*
14.5	*66*		

Compute the mean, the standard deviation and variance.

Sol.

Mid-values x	f	$u = \frac{x - 14.5}{.5}$	fu	fu^2
12.5	4	– 4	– 16	64
13.0	19	– 3	– 57	171
13.5	30	– 2	– 60	120
14.0	63	– 1	– 63	63
14.5	66	0	0	0
15.0	29	1	29	29
15.5	18	2	36	72
16.0	1	3	3	9
16.5	1	4	4	16
	N = 231		– 124	544

$$\text{Mean } \overline{x} = a + \frac{h}{N}\Sigma fu = 14.5 + \frac{.5}{231}(-124) = 14.24$$

$$\text{S.D. } \sigma_x = h\,\sigma_u = h\sqrt{\frac{1}{N}\Sigma fu^2 - \left(\frac{\Sigma fu}{N}\right)^2}$$

$$= 5 \times \sqrt{\frac{544}{231} - \left(\frac{-124}{231}\right)^2} = .5 \times \sqrt{2.0668} = .5\,(1.44) = .72$$

$$\text{Variance} = \sigma_x^2 = (.5\sqrt{2.0668})^2 = .25\,(2.0668) = .5167.$$

Example 12. *Find the standard deviation of the following series :*

Expenditure	*No. of students*
Below Rs. 5	*6*
" Rs. 10	*16*
" Rs. 15	*28*
" Rs. 20	*38*
" Rs. 25	*46*

Sol.

Expenditure	*Mid-values*	*f*	$u = \frac{x - 12.5}{5}$	*fu*	fu^2
0— 5	2.5	6	− 2	− 12	24
5—10	7.5	10	− 1	− 10	10
10—15	12.5	12	0	0	0
15—20	17.5	10	1	10	10
20—25	22.5	8	2	16	32
		N = 46		4	76

$$\sigma_x = h\sigma_u = h\sqrt{\frac{1}{N}\Sigma fu^2 - \left(\frac{\Sigma fu}{N}\right)^2} = 5\sqrt{\frac{76}{46} - \left(\frac{4}{46}\right)^2} = \text{Rs. } 6.4.$$

Example 13. *Calculate the mean and the standard deviation of the following series :*

Marks	*No. of students*
More than 0	*100*
" " 10	*90*
" " 20	*75*
" " 30	*50*
" " 40	*25*
" " 50	*15*
" " 60	*5*
" " 70	*0*

Sol.

Marks	Mid-values	f	$u = \frac{x-35}{10}$	fu	fu^2
0—10	5	10	– 3	– 30	90
10—20	15	15	– 2	– 30	60
20—30	25	25	– 1	– 25	25
30—40	35	25	0	0	0
40—50	45	10	1	10	10
50—60	55	10	2	20	40
60—70	65	5	3	15	45
		N = 100		– 40	270

Mean $\quad \bar{x} = a + h\frac{\Sigma fu}{N} = 35 + 10\left(\frac{-40}{100}\right) = 31$

Standard deviation $\quad \sigma = h\sqrt{\frac{1}{N}\Sigma fu^2 - \left(\frac{\Sigma fu}{N}\right)^2}$

$$= 10\sqrt{\frac{270}{100} - \left(\frac{-40}{100}\right)^2} = 10\sqrt{2.7 - .16}$$

$$= 10 \times 1.59 = 15.9 \text{ marks.}$$

Example 14. *Classify the following scores in class-intervals 0—9, 10—19,......, 40—49 and calculate the variance of the distribution :*

24, 12, 37, 49, 30, 6, 28, 10, 13, 18, 12, 19, 2, 33, 4, 25, 42, 22, 16, 29, 0, 21, 23, 31, 21, 34, 20, 27, 23, 35.

Sol.

Class-intervals	Tally Marks	f	Mid-values (x)	$u = \frac{x-24.5}{10}$	fu	fu^2
0— 9	||||	4	4.5	– 2	– 8	16
10—19	~~||||~~ ||	7	14.5	– 1	– 7	7
20—29	~~||||~~ ~~||||~~ |	11	24.5	0	0	0
30—39	~~||||~~ |	6	34.5	1	6	6
40—49	||	2	44.5	2	4	8
		N = 30			– 5	37

Variance $\sigma^2 = h^2\left[\frac{1}{N}\Sigma fu^2 - \left(\frac{\Sigma fu}{N}\right)^2\right]$

$$= (10)^2 \left[\frac{37}{30} - \left(\frac{-5}{30} \right)^2 \right] = 120.5.$$

Example 15. *A student obtained the mean and standard deviation of 100 observations as 40 and 5 respectively. It was later discovered that he had wrongly copied down an observation as 50 instead of 40. Calculate the correct mean and standard deviation.*

Sol. Here $n = 100$ Incorrect mean $\overline{x} = 40$.

Using $\overline{x} = \frac{\Sigma x}{n}$ Incorrect $\Sigma x = n\overline{x} = 4000$

Corrected $\Sigma x = 4000 - 50 + 40 = 3990$

$\therefore$ Correct mean $= \frac{\text{Correct } \Sigma x}{n} = \frac{3990}{100} = 39.9$

Incorrect variance $\sigma^2 = (5)^2 = 25$

Using $\sigma^2 = \frac{1}{n} \Sigma x^2 - (\overline{x})^2$

$$25 = \frac{1}{100} \Sigma x^2 - (40)^2$$

$\Rightarrow$ $2500 = \Sigma x^2 - 160000 \Rightarrow \Sigma x^2 = 162500$

$\therefore$ Incorrect $\Sigma x^2 = 162500$

Corrected $\Sigma x^2 = 162500 - (50)^2 + (40)^2 = 161600$

$\therefore$ Corrected $\sigma^2 = \frac{1}{n}$ (correct Σx^2) $-$ (correct $\overline{x}$)^2

$$= \frac{161600}{100} - (39.9)^2 = 23.99$$

$\therefore$ Correct S.D. $= \sqrt{23.99} = 4.9$.

Example 16. *From a central frequency distribution consisting of 18 observations the mean and the standard deviation were found to be 7 and 4 respectively. But on comparing with the original data, it was found that a figure 12 was miscopied as 21 in calculations. Calculate the correct mean and standard deviation.*

Sol. Please try yourself. **[Ans.** 6.5, 2.5]

Example 17. *The mean of 5 observations is 4.4 and the variance is 8.24. If three of the five observations are 1, 2 and 6, find the other two.*

(D.U. 1988)

Sol. Here $n = 5, \ \overline{x} = 4.4$

Using $\overline{x} = \frac{\Sigma x}{n}, \ \Sigma x = n\overline{x} = 22$

If x_1, x_2 are the two missing observations, then

$$\Sigma x = x_1 + x_2 + 1 + 2 + 6 = 22 \Rightarrow x_1 + x_2 = 13 \quad ...(i)$$

Also $\sigma^2 = \frac{1}{n} \Sigma x^2 - (\overline{x})^2$

$$\Rightarrow \qquad 8.24 = \frac{1}{5}\Sigma x^2 - (4.4)^2$$

$$\Rightarrow \qquad 41.20 = \Sigma x^2 - 5 \times 19.36 \qquad \Rightarrow \qquad \Sigma x^2 = 138$$

$$\Rightarrow \qquad x_1^2 + x_2^2 + 1^2 + 2^2 + 6^2 = 138 \qquad \Rightarrow \qquad x_1^2 + x_2^2 = 97$$

$$\Rightarrow \qquad (x_1 + x_2)^2 - 2x_1 x_2 = 97 \qquad \Rightarrow \quad 169 - 2x_1x_2 = 97$$

[using (*i*)]

$$\therefore \qquad x_1 x_2 = 36$$

Now $\qquad (x_1 - x_2)^2 = (x_1 + x_2)^2 - 4x_1x_2 = (13)^2 - 4(36) = 25$

$$\therefore \qquad x_1 - x_2 = 5 \qquad \text{...(ii)}$$

From (*i*) and (*ii*) $\qquad x_1 = 9, \qquad x_2 = 4.$

Example 18. *An original frequency table with mean 11 and variance 9.9 was lost but the following table derived from it was found. Construct the original table.*

Step deviation	:	*– 2*	*– 1*	*0*	*1*	*2*
Frequency	:	*1*	*6*	*7*	*4*	*2*

Sol.

Step-deviation (*u*)	*f*	*fu*	*fu*^{2}
– 2	1	– 2	4
– 1	6	– 6	6
0	7	0	0
1	4	4	4
2	2	4	8
	N = 20	0	22

$$\sigma = h\sqrt{\frac{1}{N}\Sigma fu^2 - \left(\frac{\Sigma fu}{N}\right)^2}$$

$$\Rightarrow \qquad \text{Variance } \sigma^2 = h^2\left[\frac{1}{N}\Sigma fu^2 - \left(\frac{\Sigma fu}{N}\right)^2\right]$$

$$\Rightarrow \qquad 9.9 = h^2\left[\frac{22}{20} - 0\right] \quad \Rightarrow \quad h^2 = 9 \quad \therefore \quad h = 3$$

Now mean $\qquad \bar{x} = a + h\frac{\Sigma fu}{N}$

$$\Rightarrow \qquad 11 = a + 3 \times (0) \qquad \Rightarrow \qquad a = 11$$

Also $\qquad u = \frac{x - a}{h} \qquad$ Taking $u = -2$

$$-2 = \frac{x - 11}{3} \qquad \Rightarrow \qquad x = 5$$

∴ Mid-value of first class-interval is 5. Since the size of each class-interval is $h = 3$, the class-intervals are

3.5—6.5, 6.5—9.5, 9.5—12.5, 12.5—15.5, 15.5—18.5.

Hence the original table is

Class-intervals	: 3.5—6.5	6.5—9.5	9.5—12.5	12.5—15.5	15.5—18.5
f	: 1	6	7	4	2

Example 19. *Goals scored by two teams A and B in a football season were as follows :*

No. of goals scored in a match	*No. of matches A*	*B*
0	27	17
1	9	9
2	8	6
3	5	5
4	4	3

Find out which team is more consistent.

(Bundelkhand 1985 ; K.U. 1982 S)

Sol. *Calculation of co-efficient of variation for the team A :*

No. of goals scored (x)	*No. of matches* (f)	$d_x = x - 2$	fd_x	fd_x^2
0	27	−2	−54	108
1	9	−1	−9	9
2	8	0	0	0
3	5	1	5	1
4	4	2	8	5 6
	N = 53		−50	138

$$\bar{x} = a + \frac{\Sigma fd_x}{N} = 2 + \frac{-50}{53} = 2 - 94 = 1.06$$

$$\sigma = \sqrt{\frac{1}{N}\Sigma fd_x^{\,2} - \left(\frac{\Sigma fd_x}{N}\right)^2}$$

$$= \sqrt{\frac{138}{53} - \left(\frac{-50}{53}\right)^2} = 1.31$$

Co-efficient of variation for the team A

$$= \frac{\sigma}{\bar{x}} \times 100 = \frac{1.31 \times 100}{1.06} = 123.6.$$

Calculation of co-efficient of variation for the team B :

No. of goals scored (*x*)	*No. of matches* (*f*)	$d_x = x - 2$	fd_x	fd_x^2
0	17	− 2	− 34	68
1	9	− 1	− 9	9
2	6	0	0	0
3	5	1	5	5
4	3	2	6	12
	N = 40		− 32	94

$$\bar{x} = a + \frac{\Sigma fd_x}{N} = 2 - \frac{32}{40} = 2 - .8 = 1.2$$

$$\sigma = \sqrt{\frac{1}{N}\Sigma fd_x^{\ 2} - \left(\frac{\Sigma fd_x}{N}\right)^2}$$

$$= \sqrt{\frac{94}{40} - \left(\frac{-32}{40}\right)^2} = 1.3$$

Co-efficient of variation for the team B

$$= \frac{\sigma}{\bar{x}} \times 100 = \frac{1.3 \times 100}{1.2} = 108.3$$

Since the co-efficient of variation is less for the team B, hence team B is more consistent.

Example 20. *The scores of two golfers for 10 rounds each are :*

A : 58, 59, 60, 54, 65, 66, 52, 75, 69, 52

B : 84, 56, 92, 65, 86, 78, 44, 54, 78, 68

Which may be regarded as the more consistent player ?

Sol. Please try yourself. **[Ans.** A]

Example 21. *During the first 10 weeks of a session, the marks of two students, X and Y, taking the course were :*

X : 58, 59, 60, 54, 65, 66, 52, 75, 69, 52

Y : 56, 87, 89, 78, 71, 73, 84, 65, 66, 46

Which of the two you would consider to be more consistent ?

Sol. Please try yourself. **[Ans.** X]

Example 22. *In two samples, where the variates x_1 and x_2 are measured in same units from their respective means*

$n_1 = 36$, $\Sigma x_1^2 = 49428$

$n_2 = 49$, $\Sigma x_2^2 = 71258$

Compute the value of the standard deviation of the two samples. What additional information is required to calculate the co-efficient of variation of the above two samples ?

Sol. Standard deviation of the first sample

$$= \sqrt{\frac{\Sigma x_1^2}{n_1}} = \sqrt{\frac{49428}{36}} = \frac{222.06}{6} = 37.01$$

Standard deviation of the second sample

$$= \sqrt{\frac{\Sigma x_2^2}{n_2}} = \sqrt{\frac{71258}{49}} = \frac{266.56}{7} = 38.08$$

Since co-efficient of variation $= \frac{\sigma}{\bar{x}} \times 100$

we require arithmetic means $\bar{x}_1$ and $\bar{x}_2$ to calculate co-efficient of variation.

Example 23. *In any two series, where ξ_1 and ξ_2 represent the deviations from an assumed average 100,*

$$n_1 = 150 \qquad \Sigma\xi_1 = 100 \qquad \Sigma\xi_1^2 = 245320$$

$$n_2 = 200 \qquad \Sigma\xi_2 = 250 \qquad \Sigma\xi_2^2 = 43850$$

Calculate the co-efficient of variation for the two series.

Sol. (*i*) Assumed means 'a' = 100, $\Sigma d_x = \Sigma\xi_1 = 100$

$$\Sigma d_x^2 = \Sigma\xi_1^2 = 245320$$

$$\therefore \qquad \bar{x} = a + \frac{\Sigma d_x}{n} = 100 + \frac{100}{150} = 100 + .67 = 100.67$$

$$\sigma = \sqrt{\frac{1}{n}\Sigma d_x^2 - \left(\frac{\Sigma d_x}{n}\right)^2} = \sqrt{\frac{245320}{150} - \left(\frac{100}{150}\right)^2}$$

$$= 40.44$$

Co-efficient of variation $= \frac{\sigma}{\bar{x}} \times 100 = \frac{40.44}{100.67} \times 100 = 40.17.$

(*ii*) Please try yourself. [**Ans.** 14.6]

MISCELLANEOUS PROBLEMS

Example 1. *Find the mean and the variance of first n natural numbers.*

Sol. $\bar{x} = \frac{1 + 2 + \ldots\ldots + n}{n}$ $\qquad \left| \bar{x} = \frac{\Sigma x_i}{n} \right.$

$$= \frac{\frac{n}{2}(n+1)}{n} = \frac{1}{2}(n+1)$$

Mean square deviation from $x = 0$ is given by

$$s^2 = \frac{1}{n}\Sigma(x_i - a)^2 = \frac{1}{n}\Sigma x_i^2 \qquad |\because \quad a = 0$$

$$= \frac{1}{n}(1^2 + 2^2 + \ldots + n^2)$$

$$= \frac{1}{n} \cdot \frac{n(n+1)(2n+1)}{6} = \frac{1}{6}(n+1)(2n+1)$$

Variance $\sigma^2 = s^2 - d^2$, where $d = \overline{x} - a = \overline{x}$ $\qquad (\because \quad a = 0)$

$$= \frac{1}{6}(n+1)(2n+1) - \frac{1}{4}(n+1)^2$$

$$= \frac{1}{12}(n+1)[4n + 2 - 3(n+1)]$$

$$= \frac{1}{12}(n+1)(n-1) = \frac{1}{12}(n^2 - 1).$$

Example 2. *Find the mean deviation from the mean and standard deviation of the series a, a + d, a + 2d, ..., a + 2nd, and prove that the latter is greater than the former.* (D.U. 1985 ; Kanpur 1990)

Sol. Mean $\overline{x} = \dfrac{a + (a+d) + (a+2d) + \ldots.. + (a+2nd)}{2n+1}$ $\quad \Big| \; \overline{x} = \dfrac{\Sigma x_i}{n}$

$$= \frac{\frac{2n+1}{2}(a + a + 2nd)}{2n+1} \quad \Big| \text{ For an A.P., } S_n = \frac{n}{2}(a + l)$$

$$= a + nd$$

The series being

$a, a + d, a + 2d, \ldots\ldots, a + (n-1)d, a + nd, a + (n+1)d, \ldots\ldots, a + 2nd$

Mean deviation from mean

$$= \frac{1}{N}\Sigma f_i \,|\, x_i - \overline{x} \,|$$

$$= \frac{1}{2n+1}\Sigma \,|\, x_i - a - nd \,|$$

$$= \frac{1}{2n+1}[nd + (n-1)d + (n-2)d + \ldots + d + 0 + d + \ldots + nd]$$

$$= \frac{2d}{2n+1}[n + (n-1) + (n-2) \ldots + 1]$$

$$= \frac{2d}{2n+1} \cdot \frac{n(n+1)}{2} = \frac{n(n+1)d}{2n+1}$$

Now $\quad \sigma^2 = \dfrac{1}{N}\Sigma f_i (x_i - \overline{x})^2$

$$= \frac{1}{2n+1}\Sigma (x_i - \overline{x})^2$$

$$= \frac{1}{2n+1}[n^2d^2 + (n-1)^2 d^2 + (n-2)^2 d^2 + \ldots\ldots + d^2 + 0 + d^2 + \ldots\ldots + n^2d^2]$$

$$= \frac{2d^2}{2n+1}[n^2 + (n-1)^2 + (n-2)^2 + \ldots + 1^2]$$

$$= \frac{2d^2}{2n+1} \cdot \frac{n(n+1)(2n+1)}{6} = \frac{n(n+1)d^2}{3}$$

$\therefore$ Standard deviation $= \sigma = d\sqrt{\dfrac{n(n+1)}{3}}$

Now, standard deviation > mean deviation from the mean

if $$d\sqrt{\frac{n(n+1)}{3}} > \left[\frac{n(n+1)d}{2n+1}\right]$$

or if $$\frac{n(n+1)}{3} > \frac{n^2(n+1)^2}{(2n+1)^2}$$

or if $$(2n+1)^2 > 3n(n+1)$$

or if $$4n^2 + 4n + 1 - 3n^2 - 3n > 0$$

or if $$n^2 + n + 1 > 0$$

or if $$(n + \tfrac{1}{2})^2 + \tfrac{3}{4} > 0 \quad \text{which is true.}$$

Hence the result.

Example 3. *Show that if the variable takes the values 0, 1, 2, ..., n with frequencies proportional to the binomial co-efficients* $^nC_0, {}^nC_1, {}^nC_2, ..., {}^nC_n$ *respectively, then the mean of distribution is* $\left(\dfrac{n}{2}\right)$; *the mean square deviation about* $x = 0$ *is* $\dfrac{n(n+1)}{4}$ *and the variance is* $\left(\dfrac{n}{4}\right)$. (D.U. 1981, 86, 89, 90)

Sol. $N = \Sigma f = k\,.\,{}^nC_0 + k\,.\,{}^nC_1 + k\,.\,{}^nC_2 + ... + k\,.\,{}^nC_n$

$$= k(1+1)^n = k\,.\,2^n$$

$$\Sigma fx = k\,{}^nC_1 + 2k\,.\,{}^nC_2 + 3k\,.\,{}^nC_3 + ... + nk\,.\,{}^nC_n$$

$$= k\left[n + n(n-1) + \frac{n(n-1)(n-2)}{2\,!} + ... + n\right]$$

$$= kn\left[1 + (n-1) + \frac{(n-1)(n-2)}{2\,!} + ... + 1\right]$$

$$= kn\left[{}^{n-1}C_0 + {}^{n-1}C_1 + {}^{n-1}C_2 + ... + {}^{n-1}C_{n-1}\right]$$

$$= kn(1+1)^{n-1} = kn\,.\,2^{n-1}$$

$\therefore$ Mean $\bar{x} = \dfrac{\Sigma fx}{\Sigma f} = \dfrac{kn\,.\,2^{n-1}}{k\,.\,2^n} = \dfrac{n}{2}.$

The mean square deviation s^2 about the point $x = 0$ is given by

$$s^2 = \frac{1}{N}\Sigma fx^2 = \frac{1}{k\,.\,2^n}\left[1^2\,.\,k\,{}^nC_1 + 2^2\,.\,k\,{}^nC_2 + 3^2\,.\,k\,{}^nC_3 + ... \right.$$
$$\left. ... + n^2\,.\,k\,{}^nC_n\right]$$

$$= \frac{1}{2^n}\left[n + 2n(n-1) + \frac{3n(n-1)(n-2)}{2\,!} + \ldots\ldots + n^2 \,.\, 1 \right]$$

$$= \frac{n}{2^n}\left[1 + 2(n-1) + \frac{3}{2\,!}(n-1)(n-2) + \ldots\ldots + n \right]$$

$$= \frac{n}{2^n}\left[\left\{ 1 + (n-1) + \frac{(n-1)(n-2)}{2\,!} + \ldots\ldots + 1 \right\} \right.$$

$$\left. + \{(n-1) + (n-1)(n-2) + \ldots\ldots + (n-1)\} \right]$$

$$= \frac{n}{2^n}\,[({}^{n-1}C_0 + {}^{n-1}C_1 + {}^{n-1}C_2 + \ldots\ldots + {}^{n-1}C_{n-1})$$

$$+ (n-1)({}^{n-2}C_0 + {}^{n-2}C_1 + \ldots\ldots + {}^{n-2}C_{n-2})]$$

$$= \frac{n}{2^n}\,[(1+1)^{n-1} + (n-1)(1+1)^{n-2}]$$

$$= \frac{n}{2^n}\,[2^{n-1} + (n-1)\,.\,2^{n-2}]$$

$$= n\left(\frac{1}{2} + \frac{n-1}{4} \right) = \frac{n(n+1)}{4}$$

Variance $\sigma^2 = s^2 - d^2$, where $d = \bar{x} - a = \bar{x} - 0 = \bar{x}$

$$= \frac{n(n+1)}{4} - \frac{n^2}{4} = \frac{n}{4}.$$

Example 4. *Show that, if the variable takes the values 0, 1, 2,......, n with frequencies given by the terms of the binomial series q^n, ${}^nC_1 q^{n-1}p$, ${}^nC_2 q^{n-2}p^2$,......., p^n, where $p + q = 1$, then the mean square deviation is $n^2p^2 + npq$ and the variance is npq.*

Sol. $N = \Sigma f = q^n + {}^nC_1 q^{n-1} p + {}^nC_2 q^{n-2} p^2 + \ldots\ldots + p^n$

$$= (q+p)^n = 1 \qquad (\because \quad p + q = 1)$$

Mean $\bar{x} = \frac{1}{N}\Sigma fx = \Sigma fx$

$$= {}^nC_1 q^{n-1} p + 2\,{}^nC_2 q^{n-2} p^2 + \ldots\ldots + np^n$$

$$= nq^{n-1} p + 2\,\frac{n(n-1)}{2\,!}\,q^{n-2} p^2 + \ldots\ldots + np^n$$

$$= np[q^{n-1} + (n-1)q^{n-2} p + \ldots\ldots + p^{n-1}]$$

$$= np[q^{n-1} + {}^{n-1}C_1 q^{n-2} p + \ldots\ldots + {}^{n-1}C_{n-1} p^{n-1}]$$

$$= np\,(q+p)^{n-1} = np \qquad (\because \quad p + q = 1)$$

Mean square deviation s^2 about $x = 0$ (say) is given by

$$s^2 = \frac{1}{N}\Sigma fx^2 = \Sigma fx^2$$

$$= 1^2 \,.\, {}^nC_1 \, q^{n-1}p + 2^2 \,.\, {}^nC_2 \, q^{n-2}p^2 + 3^2 \,.\, {}^nC_3 \, q^{n-3}p^3 + \ldots\ldots + n^2 \,.\, p^n$$

$$= nq^{n-1}p + 2n(n-1)q^{n-2}p^2 + \frac{3n(n-1)(n-2)}{2\,!} q^{n-3}p^3 + \ldots + n^2p^n$$

$$= np\left[q^{n-1} + 2(n-1)q^{n-2}p + \frac{3(n-1)(n-2)}{2\,!} q^{n-3}p^2 + \ldots + np^{n-1} \right]$$

$$= np\left[\left\{ q^{n-1} + (n-1)q^{n-2}p + \frac{(n-1)(n-2)}{2\,!} q^{n-3}p^2 + \ldots + p^{n-1} \right\}\right.$$

$$\left. + \{(n-1)q^{n-2}p + (n-1)(n-2)q^{n-3}p^2 + \ldots\ldots + (n-1)p^{n-1}\}\right]$$

$$= np[(q+p)^{n-1} + (n-1)p\{q^{n-2} + (n-2)q^{n-3}p + \ldots\ldots + p^{n-2}\}]$$

$$= np[1 + (n-1)p(q+p)^{n-2}] = np[1 + (n-1)p]$$

$$= np[1 + np - p] = np(np + q) \qquad |\because \; p + q = 1 \quad \therefore \quad 1 - p = q$$

$$= n^2p^2 + npq$$

Variance $\sigma^2 = s^2 - d^2$ where $d = \bar{x} - a = \bar{x} - 0 = \bar{x}$

$$= n^2p^2 + npq - n^2p^2 = npq.$$

Example 5. *Show that in a discrete series if deviations are small compared with mean M so that* $\left(\frac{x}{M}\right)^3$ *and higher powers of* $\left(\frac{x}{M}\right)$ *are neglected, prove that*

(*i*) $G = M\left(1 - \frac{\sigma^2}{2M^2}\right)$ (D.U. 1982)

(*ii*) $M^2 - G^2 = \sigma^2$ (M.D.U. 1981 S)

(*iii*) $H = M\left(1 - \frac{\sigma^2}{M^2}\right)$ (Kanpur 1987 ; M.D.U. 82 ; D.U. 82)

(*iv*) $M = H\left(1 + \frac{\sigma^2}{M^2}\right)$ (M.D.U. 1982 S)

(*v*) $MH = G^2$

(*vi*) $M - 2G + H = 0$

where G is the geometric mean and H is harmonic mean.

Sol. If X is the variable, then we are given that

$$x = \mathrm{X} - \mathrm{M} \qquad \text{or} \qquad \mathrm{X} = x + \mathrm{M} = \mathrm{M}\left(1 + \frac{x}{\mathrm{M}}\right)$$

(*i*) By def. $\log \mathrm{G} = \frac{1}{\mathrm{N}} \Sigma f \log \mathrm{X} = \frac{1}{\mathrm{N}} \Sigma f \log \mathrm{M}\left(1 + \frac{x}{\mathrm{M}}\right)$

$$= \frac{1}{\mathrm{N}} \Sigma f \log \mathrm{M} + \frac{1}{\mathrm{N}} \Sigma f \log\left(1 + \frac{x}{\mathrm{M}}\right)$$

$$= \log M \cdot \frac{1}{N} \Sigma f + \frac{1}{N} \Sigma f \left(\frac{x}{M} - \frac{x^2}{2 M^2} \right)$$

$\left[\because \log (1 + x) = x - \frac{x^2}{2} + \frac{x^3}{3} \right.$,......., neglecting higher powers of $\left. \frac{x}{M} \right]$

$$= \log M \cdot \frac{1}{N} \cdot N + \frac{1}{MN} \Sigma fx - \frac{1}{2M^2 N} \Sigma fx^2$$

$$= \log M + \frac{1}{MN} (0) - \frac{1}{2M^2} \cdot \sigma^2$$

$\left[\because \Sigma fx \right.$ = alg. sum of the deviations of all the variates from mean = 0 and $\left. \frac{1}{N} \Sigma fx^2 = \sigma^2 \right]$

$$= \log M - \frac{\sigma^2}{2M^2}$$

or $\quad \log G - \log M = - \frac{\sigma^2}{2M^2}$

or $\quad \log \frac{G}{M} = - \frac{\sigma^2}{2M^2} \quad$ or $\quad \frac{G}{M} = e^{-\sigma^2/2M^2}$

or $\quad G = M e^{-\sigma^2/2M^2}$

$$= M \left(1 - \frac{\sigma^2}{2M^2} \right) \quad \Bigg| \because e^x = 1 + x + \frac{x^2}{2!} + \ldots\ldots$$

(ii) $\quad G^2 = M^2 e^{-\sigma^2/M^2} = M^2 \left(1 - \frac{\sigma^2}{M^2} \right) = M^2 - \sigma^2$

$\therefore \quad M^2 - G^2 = \sigma^2.$

(iii) $\quad \frac{1}{H} = \frac{1}{N} \Sigma f \left(\frac{1}{x} \right)$

$$= \frac{1}{N} \Sigma f \frac{1}{M} \left(1 + \frac{x}{M} \right)^{-1}$$

$$= \frac{1}{N} \cdot \frac{1}{M} \Sigma f \left(1 - \frac{x}{M} + \frac{x^2}{M^2} \right)$$

$$= \frac{1}{MN} \Sigma f - \frac{1}{MN} \cdot \frac{1}{M} \Sigma fx + \frac{1}{MN} \cdot \frac{1}{M^2} \Sigma fx^2$$

$$= \frac{1}{MN} . N - \frac{1}{M^2N}(0) + \frac{1}{M^2}\sigma^2$$

$$\left| \because \ \Sigma fx = 0, \ \frac{1}{N}\Sigma fx^2 = \sigma^2 \right.$$

$$= \frac{1}{M}\left(1 + \frac{\sigma^2}{M^2}\right)$$

$$\Rightarrow \qquad H = M\left(1 + \frac{\sigma^2}{M^2}\right)^{-1} = M\left(1 - \frac{\sigma^2}{M^2}\right).$$

(iv) $$\frac{1}{H} = \frac{1}{M}\left(1 + \frac{\sigma^2}{M^2}\right) \Rightarrow M = H\left(1 + \frac{\sigma^2}{M^2}\right)$$

(v) $$MH = M . M\left(1 - \frac{\sigma^2}{M^2}\right)$$

$$= M^2 - \sigma^2 = G^2 \qquad \text{[using (ii)]}$$

(vi) $$M - 2G + H = M - 2M\left(1 - \frac{\sigma^2}{2M^2}\right) + M\left(1 - \frac{\sigma^2}{M^2}\right)$$

$$= M - 2M + \frac{\sigma^2}{M} + M - \frac{\sigma^2}{M} = 0.$$

Example 6. *If the deviation $X_i = x_i - M$ is very small in comparison with mean M and $\left(\frac{X_i}{M}\right)^3$ and higher powers of $\left(\frac{X_i}{M}\right)$ are neglected, prove that :*

$$V = \sqrt{\frac{2(M - G)}{M}}$$

where G is the geometric mean of the values $x_1, x_2, ..., x_n$ and V is the co-efficient of dispersion.

Sol. $$X_i = x_i - M$$

or $$x_i = X_i + M = M\left(1 + \frac{X_i}{M}\right)$$

$$\log G = \frac{1}{N}\Sigma f_i \log x_i$$

$$= \frac{1}{N}\Sigma f_i \log M\left(1 + \frac{X_i}{M}\right)$$

$$= \frac{1}{N}\Sigma f_i \log M + \frac{1}{N}\Sigma f_i \log\left(1 + \frac{X_i}{M}\right)$$

$$= \log M . \frac{1}{N}\Sigma f_i + \frac{1}{N}\Sigma f_i\left(\frac{X_i}{M} - \frac{X_i{}^2}{2M^2}\right)$$

$$= \log M . \frac{1}{N} . N + \frac{1}{N} . \frac{1}{M} \Sigma f_i X_i - \frac{1}{N} . \frac{1}{2M^2} \Sigma f_i X_i^2$$

$$= \log M + \frac{1}{MN} (0) - \frac{1}{2M^2} . \sigma^2 = \log M - \frac{\sigma^2}{2M^2}$$

or $$\log G - \log M = -\frac{\sigma^2}{2M^2} \quad \text{or} \quad \log \frac{G}{M} = -\frac{\sigma^2}{2M^2}$$

or $$\frac{G}{M} = e^{-\sigma^2/2M^2}$$

or $$G = M\left(1 - \frac{\sigma^2}{2M^2}\right) \quad \text{or} \quad \frac{\sigma^2}{2M} = M - G.$$

$\therefore$ $$\sigma = \sqrt{2M(M - G)}$$

$$V = \frac{\sigma}{M} = \frac{1}{M} \sqrt{2M(M-G)} = \sqrt{\frac{2(M-G)}{M}}.$$

Example 7. *The mean and standard deviation of the variable x are M and σ respectively. If deviations are small compared with the value of the mean, show that :*

$$\textit{Mean}\ (\sqrt{x}) = \sqrt{M}\left(1 - \frac{\sigma^2}{8M^2}\right) \textit{ approx.}$$

(M.D.U. 1981 ; D.U. 1987, 93 ; Meerut 1991)

Sol. If x is the variate, then

$$x - M = X$$

so that $$x = M + X = M\left(1 + \frac{X}{M}\right)$$

$$\text{Mean}\ (\sqrt{x}) = \frac{1}{N} \Sigma f \sqrt{x} = \frac{1}{N} \Sigma f \sqrt{M} \left(1 + \frac{X}{M}\right)^{1/2}$$

$$= \sqrt{M} . \frac{1}{N} \Sigma f \left(1 + \frac{X}{2M} - \frac{X^2}{8M^2}\right) \text{approx.}$$

(by Binomial Theorem)

$$= \sqrt{M} . \frac{1}{N} \left[\Sigma f + \frac{1}{2M} \Sigma fX - \frac{1}{8M^2} \Sigma f X^2\right]$$

$$= \sqrt{M} . \frac{1}{N} \left[N + \frac{1}{2M} (0) - \frac{1}{8M^2} . N \sigma^2\right]$$

$$\left[\because \quad \Sigma fX = 0, \frac{1}{N} \Sigma f X^2 = \sigma^2\right]$$

$$= \sqrt{M} . \left(1 - \frac{\sigma^2}{8M^2}\right).$$

Example 8. *Prove that the mean deviation about the mean $\overline{x}$ of the variate x, the frequency of whose ith size x_i is f_i is given by*

$$\frac{2}{N}[\overline{x} \sum_{x_i < \overline{x}} f_i - \sum_{x_i < \overline{x}} f_i x_i] \qquad \text{(D.U. 1987, 93)}$$

Sol. Mean deviation about mean $\overline{x}$

$$= \frac{1}{N} \Sigma f_i \,|\, x_i - \overline{x} \,|$$

$$= \frac{1}{N}[\sum_{x_i < \overline{x}} f_i (\overline{x} - x_i) + \sum_{x_i > \overline{x}} f_i (x_i - \overline{x})]$$

$$= \frac{1}{N}[- \sum_{x_i < \overline{x}} f_i (x_i - \overline{x}) + \sum_{x_i > \overline{x}} f_i (x_i - \overline{x})]$$

$$= \frac{1}{N}[- \sum_{x_i < \overline{x}} f_i (x_i - \overline{x}) - \sum_{x_i < \overline{x}} f_i (x_i - \overline{x})]$$

$[\because \ \Sigma f_i (x_i - \overline{x}) = 0$

$$\therefore \sum_{x_i < \overline{x}} f_i (x_i - \overline{x}) + \sum_{x_i > \overline{x}} f_i (x_i - \overline{x}) = 0$$

$$\Rightarrow \sum_{x_i > \overline{x}} f_i (x_i - \overline{x}) = - \sum_{x_i < \overline{x}} f_i (x_i - \overline{x})]$$

$$= -\frac{2}{N} \sum_{x_i < \overline{x}} f_i (x_i - \overline{x})$$

$$= -\frac{2}{N}[\sum_{x_i < \overline{x}} f_i x_i - \sum_{x_i < \overline{x}} f_i \overline{x}]$$

$$= \frac{2}{N}[\overline{x} \sum_{x_i < \overline{x}} f_i - \sum_{x_i < \overline{x}} f_i x_i].$$

Example 9. *From a sample of n observations, the arithmetic mean and variance are calculated. It is then found that one of the values x_1 is in error and should be replaced by x_1'. Show that the adjustment to the variance to correct this error is*

$$\frac{1}{n}(x_1' - x_1)\left(x_1' + x_1 - \frac{x_1' - x_1 + 2T}{n}\right)$$

where T is the total of the original results.

Sol. Let the original observations be

$$x_1, x_2, x_3, \ldots\ldots, x_n.$$

Then $\qquad T = x_1 + x_2 + x_3 + \ldots\ldots + x_n$

$\therefore \qquad T - x_1 = x_2 + x_3 + \ldots\ldots + x_n$

$$T - x_1 + x_1' = x_1' + x_2 + x_3 + \ldots\ldots + x_n$$

Original arithmetic mean

$$= \frac{x_1 + x_2 + \ldots\ldots + x_n}{n} = \frac{T}{n}$$

New arithmetic mean

$$= \frac{x_1' + x_2 + \ldots\ldots + x_n}{n} = \frac{T - x_1 + x_1'}{n}$$

Original variance

$$= \frac{x_1^2 + x_2^2 + \ldots\ldots + x_n^2}{n} - \left(\frac{x_1 + x_2 + \ldots\ldots + x_n}{n} \right)^2$$

$$\left| \sigma^2 = \frac{1}{n} \Sigma x^2 - (\bar{x})^2 \right.$$

$$= \frac{x_1^2 + x_2^2 + \ldots\ldots + x_n^2}{n} - \left(\frac{T}{n} \right)^2$$

$$\text{New variance} = \frac{x_1'^2 + x_2^2 + \ldots + x_n^2}{n} - \left(\frac{T - x_1 + x_1'}{n} \right)^2$$

Adjustment = New variance – Original variance

$$= \frac{x_1'^2 - x_1^2}{n} + \left(\frac{T}{n} \right)^2 - \left(\frac{T - x_1 + x_1'}{n} \right)^2$$

$$= \frac{(x_1' + x_1)(x_1' - x_1)}{n} + \left(\frac{T}{n} + \frac{T - x_1 + x_1'}{n} \right) \left(\frac{T}{n} - \frac{T - x_1 + x_1'}{n} \right)$$

$$= \frac{(x_1' + x_1)(x_1' - x_1)}{n} + \left(\frac{2T - x_1 + x_1'}{n} \right) \left(\frac{x_1 - x_1'}{n} \right)$$

$$= \frac{(x_1' - x_1)}{n} \left[x_1' + x_1 - \frac{2T - x_1 + x_1'}{n} \right]$$

$$= \frac{1}{n} (x_1' - x_1) \left(x_1' + x_1 - \frac{x_1' - x_1 + 2T}{n} \right).$$

Example 10. *Show that the mean deviation from the median is less than that measured from any other value.* (M.U. 1984)

Sol. Let the values be equidistant, the distance between any two consecutive values being h. Let x_0 be the median.

Case I. *When the number of values is odd and = 2n + 1 (say).*

The $(2n + 1)$ values are

$$x_0 - nh, x_0 - (n - 1)\, h, \ldots\ldots, x_0 - h, x_0, x_0 + h, \ldots + x_0 + (n - 1)\, h, x_0 + nh$$

Let S_m be the sum of the absolute values of the deviations from x_m *i.e.,*

$$S_m = \Sigma \,|\, x_m - x \,|$$

where x assumes all the values from $x_0 - nh$ to $x_0 + nh$.

$$S_0 = \Sigma\,|\,x_0 - x\,|$$
$$= nh + (n-1)\,h + \ldots\ldots + h + 0 + h + \ldots\ldots + (n-1)\,h + h$$
$$= 2h\,[1 + 2 + \ldots\ldots + (n-1) + n]$$

$$S_1 = \Sigma\,|\,x_1 - x\,|$$

where $x_1 = x_0 + h$

$$= (n+1)h + nh + (n-1)h + \ldots\ldots + 2h + h + 0 + h + h + \ldots\ldots + (n-1)h$$
$$= 2h[1 + 2 + \ldots\ldots + (n-1)] + nh + (n+1)h$$

$S_2 = \Sigma\,|\,x_2 - x\,|$ where $x_2 = x_0 + 2h$

$$= (n+2)h + (n+1)h + nh + (n-1)h + \ldots\ldots + 2h + h + 0 + h + 2h + \ldots\ldots + (n-2)h$$
$$= 2h\,[1 + 2 + \ldots\ldots + (n-2)] + (n-1)h + nh + (n+1)h + (n+2)h$$

$\therefore$ $S_1 - S_0 = h > 0$

and $S_2 - S_1 = 3h > 0$

$\Rightarrow$ $S_0 < S_1 < S_2 \ldots\ldots$

Similarly it can be shown that

$$S_0 < S_{-1} < S_{-2} \ldots\ldots$$

Thus we notice that S_m increases as the absolute value of m increases. Therefore, S_m is least when the absolute values of m is least, *i.e.*, $m = 0$.

Hence S_0 is least.

Case II. *When the number of values is even and* = $2n$ (*say*).

The $2n$ values are

$$\left(x_0 - \frac{2n-1}{2}h\right), \left(x_0 - \frac{2n-3}{2}h\right), \ldots\ldots, x_0 - \frac{h}{2},$$
$$x_0 + \frac{h}{2}, \ldots\ldots, \left(x_0 + \frac{2n-3}{2}h\right), \left(x_0 + \frac{2n-1}{2}h\right)$$

$$S_0 = \Sigma\,|\,x_0 - x\,|$$
$$= \frac{2n-1}{2}h + \frac{2n-3}{2}h + \ldots\ldots + \frac{h}{2} + \frac{h}{2} + \ldots\ldots + \frac{2n-3}{2}h + \frac{2n-1}{2}h$$
$$= 2\left[\tfrac{1}{2}h + \tfrac{3}{2}h + \ldots\ldots + \frac{2n-1}{2}h\right]$$

$S_1 = \Sigma\,|\,x_1 - x\,|$ where $x_1 = x_0 + h$

$$= \frac{2n+1}{2}h + \frac{2n-1}{2}h + \ldots\ldots + \frac{3h}{2} + \frac{h}{2} + \frac{h}{2} + \frac{3h}{2} + \ldots\ldots + \frac{2n-3}{2}h$$
$$= 2\left[\tfrac{1}{2}h + \tfrac{3}{2}h + \ldots + \frac{2n-3}{2}h\right] + \frac{2n-1}{2}h + \frac{2n+1}{2}h$$

$$\therefore \quad S_1 - S_0 = \frac{2n+1}{2}h - \frac{2n-1}{2}h = h > 0$$

Similarly,

$$S_2 - S_1 = 3h > 0$$

$$\therefore \quad S_0 < S_1 < S_2$$

Similarly, $S_0 < S_{-1} < S_{-2}$ ......

Hence as in Case I above, S_0 is least.

Example 11. *Show that for any discrete distribution, the standard deviation is not less than the mean deviation from the mean.*

(D.U. 1983, 85, 86, 89)

Sol. We have to show that

S.D. ≥ Mean deviation from mean

⇒ (S.D.)^2 ≥ (Mean deviation from mean)^2

$$\Rightarrow \quad \frac{1}{N}\sum_{i=1}^{n} f_i (x_i - \bar{x})^2 \geq \left(\frac{1}{N}\sum_{i=1}^{n} f_i \,|x_i - \bar{x}|\right)^2$$

$$\Rightarrow \quad \frac{1}{N}\sum_{i=1}^{n} f_i z_i^2 \geq \left(\frac{1}{N}\sum_{i=1}^{n} f_i z_i\right)^2 \quad \text{where} \quad z_i = |x_i - \bar{x}|$$

$$\Rightarrow \quad N\sum_{i=1}^{n} f_i z_i^2 \geq \left(\sum_{i=1}^{n} f_i z_i\right)^2$$

$$\Rightarrow \quad (f_1 + f_2 + \ldots + f_n)(f_1 z_1^2 + f_2 z_2^2 + \ldots\ldots + f_n z_n^2)$$

$$\geq (f_1 z_1 + f_2 z_2 + \ldots\ldots + f_n z_n)^2$$

$$\Rightarrow \quad f_1 f_2 (z_1^2 + z_2^2) + \ldots\ldots \geq 2 f_1 f_2 z_1 z_2 + \ldots\ldots$$

$$\Rightarrow \quad f_1 f_2 (z_1 - z_2)^2 + \ldots\ldots \geq 0$$

which is essentially true as $f_1, f_2, \ldots\ldots$ are not negative. The equality holds when all the variate values are equal.

Example 12. *For a series, the value of mean deviation is 15. Find the most likely value of its quartile deviation.*

Sol. $\because \quad$ M.D. $= \frac{4}{5}\sigma \quad \therefore \quad 15 = \frac{4}{5}\sigma \quad$ or $\quad \sigma = \frac{75}{4}$

$$\text{Q.D.} = \frac{2}{3}\sigma = \frac{2}{3} \times \frac{75}{4} = 12.5.$$

Example 13. *If the mean and standard deviation of a variate x are m and σ respectively, obtain the mean and standard deviation of variate y, where* $y = \frac{ax+b}{c}$, *a, b, c being constants.* (Kanpur 1989)

Sol. $m = \frac{1}{N} \Sigma x$

$$\bar{y} = \frac{1}{N} \Sigma y = \frac{1}{N} \Sigma \frac{ax+b}{c} = \frac{a}{c} \cdot \frac{1}{N} \Sigma x + \frac{b}{c} \cdot \frac{1}{N} \cdot N$$

$$= \frac{a}{c} m + \frac{b}{c} = \frac{am+b}{c}$$

$$\sigma = \sqrt{\frac{1}{N} \Sigma (x-\bar{x})^2} = \sqrt{\frac{1}{N} \Sigma (x-m)^2}$$

$$\sigma_y = \sqrt{\frac{1}{N} \Sigma (y-\bar{y})^2} = \sqrt{\frac{1}{N} \sum \left(\frac{ax+b}{c} - \frac{am+b}{c} \right)^2}$$

$$= \sqrt{\frac{1}{N} \sum \frac{a^2}{c^2} (x-m)^2} = \sqrt{\frac{a^2}{c^2}} \cdot \sqrt{\frac{1}{N} \sum (x-m)^2}$$

$$= \left| \frac{a}{c} \right| \sigma \qquad \left| \because \sqrt{x^2} = |x| \right.$$

3.8. THEOREM

The standard deviations of two series containing n_1 and n_2 members are σ_1 and σ_2 respectively, being measured from their respective means $\bar{x}_1$ and $\bar{x}_2$. If the two series are grouped together as one series of $(n_1 + n_2)$ members, show that the standard deviation, σ, of this series, measured from its mean $\bar{x}$ is given by

$$\sigma^2 = \frac{n_1 \sigma_1^2 + n_2 \sigma_2^2}{n_1 + n_2} + \frac{n_1 n_2}{(n_1+n_2)^2} (\bar{x}_1 - \bar{x}_2)^2$$

(D.U. 1982, 84, 88)

Proof. Let S_1^2 and S_2^2 be the mean square deviations of the two series respectively and S^2 be the mean square deviation of the two series taken together.

Then if 'a' be the assumed mean, we have

$$S^2 = \frac{1}{n_1+n_2} \sum_{1}^{n_1+n_2} f(x-a)^2$$

$$= \frac{1}{n_1+n_2} \left[\sum_{1}^{n_1} f(x-a)^2 + \sum_{n_1+1}^{n_1+n_2} f(x-a)^2 \right]$$

$$= \frac{n_1 S_1^2 + n_2 S_2^2}{n_1+n_2} \left[\because S_1^2 = \frac{1}{n_1} \sum_{1}^{n_1} f(x-a)^2 \text{ etc.} \right]$$

$$= \frac{n_1(\sigma_1^2 + d_1^2) + n_2(\sigma_2^2 + d_2^2)}{n_1 + n_2}$$

$$\Big|\because \quad S^2 = \sigma^2 + d^2, \text{where } d = \bar{x} - a$$

$$= \frac{n_1\sigma_1^2 + n_2\sigma_2^2}{n_1 + n_2} + \frac{n_1 d_1^2 + n_2 d_2^2}{n_1 + n_2} \quad \text{...(1)}$$

Now $d_1 = \bar{x}_1 - a, \qquad d_2 = \bar{x}_2 - a$

If 'a' be the mean of the two combined series *i.e.*, if $a = \bar{x}$, then

$$S^2 = \sigma^2$$

Also $\bar{x} = \dfrac{n_1\bar{x}_1 + n_2\bar{x}_2}{n_1 + n_2}$

$$\therefore \quad d_1 = \bar{x}_1 - \bar{x} = \bar{x}_1 - \frac{n_1\bar{x}_1 + n_2\bar{x}_2}{n_1 + n_2} = \frac{n_2(\bar{x}_1 - \bar{x}_2)}{n_1 + n_2}$$

$$d_2 = \bar{x}_2 - \bar{x} = \bar{x}_2 - \frac{n_1\bar{x}_1 + n_2\bar{x}_2}{n_1 + n_2} = \frac{n_1(\bar{x}_2 - \bar{x}_1)}{n_1 + n_2}$$

$\therefore \quad n_1 d_1^2 + n_2 d_2^2$

$$= \frac{n_1 n_2^2(\bar{x}_1 - \bar{x}_2)^2}{(n_1 + n_2)^2} + \frac{n_2 n_1^2(\bar{x}_2 - \bar{x}_1)^2}{(n_1 + n_2)^2}$$

$$= \frac{n_1 n_2(\bar{x}_1 - \bar{x}_2)^2}{(n_1 + n_2)^2} \cdot (n_2 + n_1) = \frac{n_1 n_2}{n_1 + n_2}(\bar{x}_1 - \bar{x}_2)^2$$

$\therefore$ From (1) $\quad (\because \quad S^2 = \sigma^2)$

$$\sigma^2 = \frac{n_1\sigma_1^2 + n_2\sigma_2^2}{n_1 + n_2} + \frac{n_1 n_2}{(n_1 + n_2)^2}(\bar{x}_1 - \bar{x}_2)^2.$$

Example 1. *The means of two samples of sizes 50 and 100 respectively are 54.1 and 50.3 and the standard deviations are 8 and 7. Obtain the mean and standard deviation of the sample of size 150 obtained by combining the two samples.*

Sol. Here $n_1 = 50, \quad \bar{x}_1 = 54.1, \quad \sigma_1 = 8$

$n_2 = 100, \quad x_2 = 50.3, \quad \sigma_2 = 7$

If $\bar{x}$ is the mean of the combined series, then

$$\bar{x} = \frac{n_1\bar{x}_1 + n_2\bar{x}_2}{n_1 + n_2} = \frac{50(54.1) + 100(50.3)}{50 + 100}$$

$$= \frac{2705 + 5030}{150} = \frac{7735}{150} = 51.57$$

If σ be the standard deviation of the combined series, then

$$\sigma^2 = \frac{n_1\sigma_1^2 + n_2\sigma_2^2}{n_1 + n_2} + \frac{n_1 d_1^2 + n_2 d_2^2}{n_1 + n_2}$$

where $d_1 = \bar{x}_1 - \bar{x} = 54.1 - 51.57 = 2.53$

$d_2 = \bar{x}_2 - \bar{x} = 50.3 - 51.57 = -1.27$

$$\therefore \quad \sigma^2 = \frac{50(64) + 100(49)}{50 + 100} + \frac{50(6.4209) + 100(1.6129)}{50 + 100}$$

$$= \frac{320 + 4900 + 321.045 + 161.29}{150} = \frac{8582.335}{150}$$

$$= 57.2156 \; ; \qquad \sigma = 7.56.$$

Example 2. *The first of the two samples has 100 items with mean 15 and standard deviation 3. If the whole group has 250 items with mean 15.6 and standard deviation $\sqrt{13.44}$, find the standard deviation of the second group.*

(D.U. 1993)

Sol. Here $n_1 = 100, \quad \bar{x}_1 = 15, \quad \sigma_1 = 3$

$n = n_1 + n_2 = 250, \quad \bar{x} = 15.6, \quad \sigma = \sqrt{13.44}$

$\therefore \quad n_2 = 250 - 100 = 150$

Using $\bar{x} = \dfrac{n_1\bar{x}_1 + n_2\bar{x}_2}{n_1 + n_2}$, we have

$$15.6 = \frac{100(15) + 150(\bar{x}_2)}{250}$$

or $\quad 150\,\bar{x}_2 = 250 \times 15.6 - 1500 = 2400$

$\therefore \quad \bar{x}_2 = 16$

$\therefore \quad d_1 = \bar{x}_1 - \bar{x} = 15 - 15.6 = -0.6$

$d_2 = \bar{x}_2 - \bar{x} = 16 - 15.6 = 0.4$

The variance of the combined group σ^2 is given by the formula

$$\sigma^2 = \frac{n_1\sigma_1^2 + n_2\sigma_2^2}{n_1 + n_2} + \frac{n_1 d_1^2 + n_2 d_2^2}{n_1 + n_2}$$

or $\quad (n_1 + n_2)\,\sigma^2 = n_1(\sigma_1^2 + d_1^2) + n_2\,(\sigma_2^2 + d_2^2)$

$\therefore \quad 250 \times 13.44 = 100\,(9 + 0.36) + 150\,(\sigma_2^2 + 0.16)$

or $\quad 150\,\sigma_2^2 = 250 \times 13.44 - 100 \times 9.36 - 150 \times 0.16$

$= 3360 - 936 - 24 = 2400$

$\therefore \quad \sigma_2^2 = 16. \qquad$ Hence $\quad \sigma_2 = 4.$

Example 3. *An analysis of monthly wages paid to the workers in two firms A and B belonging to the same industry, gave the following results :*

	Firm A	*Firm B*
No. of wage earners	*586*	*648*
Average monthly wages	*Rs. 52.5*	*Rs. 47.5*
Variance of distribution of wages	*100*	*121*

(a) Which firm, A or B, pays out larger amount as monthly wages ?

(b) In which firm, A or B, is there greater variability in individual wages ?

(c) What are the measures of (i) average monthly wages, and (ii) the variability in individual wages, of all the workers in the two firms, A and B, taken together ?

Sol. (*i*) **Firm A :**

No. of wage earners $(n_1) = 586$

Average monthly wage $(\overline{x}_1) = \text{Rs. } 52.5$

$\therefore$ Total wages paid to the workers

= No. of workers × average monthly wage

$= n_1 \overline{x}_1 = 586 \times 52.5$

= Rs. 30765.

Firm B :

No. of wage earners (n_2) = 648

Average monthly wage $(\overline{x}_2)$ = Rs. 47.5

$\therefore$ Total wages paid to the workers $= n_2 \overline{x}_2 = 648 \times 47.5$

= Rs. 30780.

Hence the firm B pays larger amount as monthly wages.

(*ii*) Variance of distribution of wages in firm A $(\sigma_1^2) = 100$

" " " " " " " B $(\sigma_2^2) = 121$

Co-efficient of variation of distribution of wages for firm A

$$= \frac{\sigma_1}{\overline{x}_1} \times 100 = \frac{10}{52.5} \times 100 = \frac{10000}{525} = 19$$

Co-efficient of variation of distribution of wages for firm B

$$= \frac{\sigma_2}{\overline{x}_2} \times 100 = \frac{11}{47.5} \times 100 = \frac{11000}{475} = 23.2$$

Since C.V. for firm B is greater than C.V. for firm A

$\therefore$ *Firm B has greater variability in individual wages.*

(*iii*) Let $\overline{x}$ be the average wage of all the workers in the two firms A and B taken together.

Using $\overline{x} = \dfrac{n_1\overline{x}_1 + n_2\overline{x}_2}{n_1 + n_2}$, we have

$$\overline{x} = \frac{586(52.5) + 648(47.5)}{586 + 648} = \frac{61545}{1234}$$

= Rs. 49.87.

The combined variance σ^2 is given by the formula

$$\sigma^2 = \frac{n_1(\sigma_1^2 + d_1^2) + n_2(\sigma_2^2 + d_2^2)}{n_1 + n_2}$$

$$= \frac{586(100 + 6.92) + 648(121 + 5.62)}{586 + 648}$$

$$\left[\because \quad d_1 = \overline{x}_1 - \overline{x} = 52.5 - 49.87 = 2.63 \right.$$
$$\left. d_2 = \overline{x}_2 - \overline{x} = 47.5 - 49.87 = -2.37 \right]$$

$$= \frac{586 \times 106.92 + 648 \times 126.62}{1234}$$

$$= \frac{144704.88}{1234} = 117.26.$$

Example 4. *An analysis of monthly wages paid to the workers in two firms A and B belonging to the same industry gives the following results :*

	Firm A	*Firm B*
No. of workers	*500*	*600*
Average monthly wage	*Rs. 186*	*Rs. 175*
Variance of distribution of wages	*81*	*100*

(i) Which firm, A or B, has a larger wage bill ?

(ii) In which firm, A or B, is there greater variability in individual wages ?

(iii) Calculate (a) the average monthly wage and (b) the variance of the distribution of wages, of all workers in firms A and B taken together.

Sol. Please try yourself.

[**Ans.** (*i*) B (*ii*) B (*iii*) $\overline{x}$ = Rs. 180, $\sigma^2 = 121.36$]

Example 6. *Means and standard deviations of the scores of an intelligence test of two classes of different sizes of 25 and 75 are :*

$M_1 = 80$ *Marks* $M_2 = 85$ *Marks*

S.D. = 15 Marks $M_2 = 20$ *Marks*

Calculate the combined mean and standard deviation of the two classes.

Sol. Please try yourself. [**Ans.** Mean = 83.75 ; S.D. = 19]

3.9. SKEWNESS

For a symmetrical distribution, the frequencies are symmetrically distributed about the mean *i.e.*, variates equidistant from the mean have equal frequencies. Also, in the case of such a distribution, the mean, mode and median coincide and median lies half-way between the two quartiles.

Thus $M = M_0 = M_d$ and $Q_3 - M = M - Q_1$.

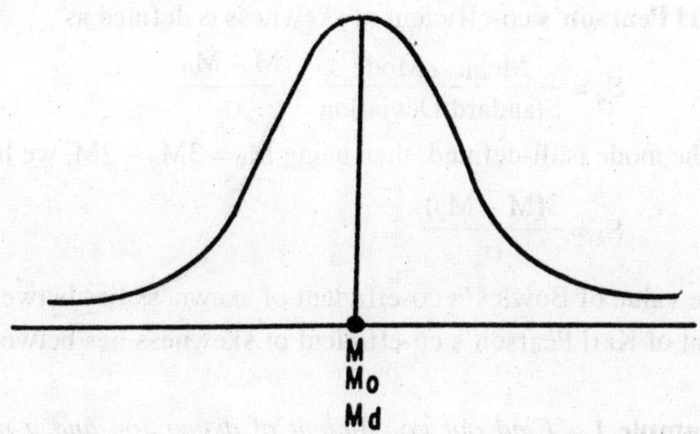

Skewness means lack of symmetry or lopsidedness in a frequency distribution. The object of measuring skewness is to estimate the extent to which a distribution is distorted from a perfectly symmetrical distribution. Skewness indicates whether the curve is turned more to one side than to other *i.e.,* whether the curve has a longer tail on one side.

Skewness can be positive as well as negative. Skewness is positive if the longer tail of the distribution lies towards the right and negative if it lies towards the left.

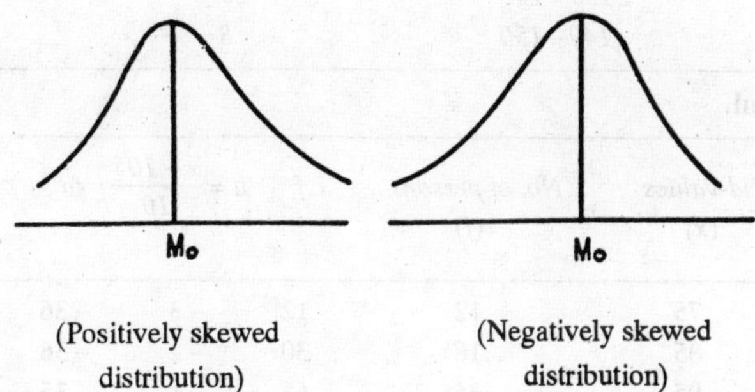

(Positively skewed distribution) (Negatively skewed distribution)

3.10. MEASURES OF SKEWNESS

Measures of skewness give us an idea about the extent of 'lopsidedness' in a series. Such measures should be

(*i*) Pure numbers so as to be independent of the units in which the variable is measured.

(*ii*) Zero when the distribution is symmetrical.

Relative measures of skewness are called the **co-efficient of skewness.** They are independent of the units of measurement and as such, they are pure numbers.

Bowley's co-efficient of skewness based on Quartiles is defined as

$$S_k = \frac{(Q_3 - M_d) - (M_d - Q_1)}{(Q_3 - M_d) + (M_d - Q_1)} = \frac{Q_3 + Q_1 - 2M_d}{Q_3 - Q_1}$$

Karl Pearson's co-efficient of skewness is defined as

$$S_k = \frac{\text{Mean} - \text{Mode}}{\text{Standard Deviation}} = \frac{M - M_0}{\sigma}$$

If the mode is ill-defined, then using $M_0 = 3M_d - 2M$, we have

$$S_k = \frac{3(M - M_d)}{\sigma}.$$

The value of Bowley's co-efficient of skewness lies between – 1 and + 1 and that of Karl Pearson's co-efficient of skewness lies between – 3 and + 3.

Example 1. *Find out co-efficient of dispersion and a measure of skewness from the following table giving the wages of 230 persons :*

Wages (in Rs.)	*No. of persons*
70— 80	*12*
80— 90	*18*
90—100	*35*
100—110	*42*
110—120	*50*
120—130	*45*
130—140	*20*
140—150	*8*

Sol.

Mid-values (*x*)	*No. of persons* (*f*)	*c.f.*	$u = \frac{x - 105}{10}$	*fu*	fu^2
75	12	12	– 3	– 36	108
85	18	30	– 2	– 36	72
95	35	65	– 1	– 35	35
105	42	107	0	0	0
115	50	157	1	50	50
125	45	202	2	90	180
135	20	222	3	60	180
145	8	230	4	32	128
	N = 230			125	753

Mean $M = a + h\frac{\Sigma fu}{N} = 105 + 10 \times \frac{125}{230} = 105 + 5.4 =$ Rs. 110.4.

The greatest frequency 50 lies in the class 110—120. Hence this is the modal class.

$$f_m = 50, \quad f_1 = 42, \quad f_2 = 45, \quad l = 110, \quad h = 10$$

$$\therefore \quad \text{Mode } M_0 = l + \frac{f_m - f_1}{2f_m - f_1 - f_2} \times h$$

$$= 110 + \frac{50 - 42}{100 - 42 - 45} \times 10$$

$$= 110 + \frac{80}{13} = 110 + 6.2 = \text{Rs. } 116.2.$$

$$\text{Standard deviation } \sigma = h\sqrt{\frac{1}{N}\Sigma fu^2 - \left(\frac{1}{N}\Sigma fu\right)^2}$$

$$= 10\sqrt{\frac{753}{230} - \left(\frac{125}{230}\right)^2} = \text{Rs. } 17.3$$

$$\therefore \quad \text{Co-efficient of dispersion} = \frac{\sigma}{M} = \frac{17.3}{110.4} = 0.16$$

$$\text{Measure of skewness } S_k = \frac{M - M_0}{\sigma} = \frac{110.4 - 116.2}{17.3}$$

$$= -0.33.$$

Example 2. *Find the mean, mode, standard deviation and co-efficient of skewness for the following :*

Years under	:	*10*	*20*	*30*	*40*	*50*	*60*
No. of persons	:	*15*	*32*	*51*	*78*	*97*	*109*

Sol. Arranging in groups, we have

Years	*Mid-values* (x)	*f*	$u = \frac{x - 35}{10}$	*fu*	fu^2
0—10	5	15	− 3	− 45	135
10—20	15	17	− 2	− 34	68
20—30	25	19	− 1	− 19	19
30—40	35	27	0	0	0
40—50	45	19	1	19	19
50—60	55	12	2	24	48
		N = 109		− 55	289

$$\text{Mean} \quad M = a + h\frac{\Sigma fu}{N} = 35 + 10 \times \frac{-55}{109} = 29.95$$

The greatest frequency 27 lies in the class 30—40. Hence this is the modal class.

$$\text{Mode} \quad M_0 = l + \frac{f_m - f_1}{2f_m - f_1 - f_2} \times h$$

$$= 30 + \frac{27 - 19}{54 - 19 - 19} \times 10 = 30 + \frac{80}{16} = 35$$

$$\text{Standard deviation } \sigma = h\sqrt{\frac{1}{N}\Sigma fu^2 - \left(\frac{1}{N}\Sigma fu\right)^2}$$

$$= 10\sqrt{\frac{289}{109} - \left(\frac{-55}{109}\right)^2} = 15.49$$

$$\therefore \quad \text{Co-efficient of skewness} = \frac{M - M_0}{\sigma} = \frac{29.95 - 35}{15.49}$$

$$= \frac{-5.05}{15.49} = -.32.$$

Example 3. *Calculate co-efficient of skewness from the following data :*

Marks above	:	*0*	*10*	*20*	*30*	*40*	*50*	*60*	*70*	*80*
Frequency	:	*150*	*140*	*100*	*80*	*80.*	*70*	*30*	*14*	*0*

Sol. Arranging in groups, we have

Marks	*Mid-values* (x)	f	$u = \frac{x - 45}{10}$	fu	fu^2
0—10	5	10	− 4	− 40	160
10—20	15	40	− 3	− 120	360
20—30	25	20	− 2	− 40	80
30—40	35	0	− 1	0	0
40—50	45	10	0	0	0
50—60	55	40	1	40	40
60—70	65	16	2	32	64
70—80	75	14	3	42	126
		N = 150		− 86	830

$$\text{Mean} \quad M = a + h\frac{\Sigma fu}{N} = 45 - \frac{86}{150} = 45 - 0.77 = 44.23$$

The greatest frequency 40 occurs in two groups 10—20 and 50—60. Instead of grouping to find mode, we find the median.

$$\frac{N}{2} = 75 \quad \therefore \quad \text{Median class is 40—50}$$

$$\text{Using} \quad M_d = l + \frac{h}{f}\left(\frac{N}{2} - C\right)$$

$$= 40 + \frac{10}{10}(75 - 70) = 40 + 5 = 45$$

Standard deviation $\sigma = h\sqrt{\frac{1}{N}\Sigma fu^2 - \left(\frac{1}{N}\Sigma fu\right)^2}$

$$= 10\sqrt{\frac{830}{150} - \left(\frac{-86}{150}\right)^2} = 22.81$$

$\therefore$ Co-efficient of skewness

$$S_k = \frac{3(M - M_d)}{\sigma} = \frac{3(39.27 - 45)}{22.8} = \frac{-17.19}{22.8} = -.8.$$

Example 4. *Compute the co-efficient of skewness from the following figures :*

25, 15, 23, 40, 27, 25, 23, 25, 20.

Sol.

Size of item (x)	$d_x = x - 20$	d_x^2
25	5	25
15	–5	25
23	3	9
40	20	400
27	7	49
25	5	25
23	3	9
25	5	25
20	0	0
	= 43	567

No. of items $n = 9$

Mean $M = 20 + \frac{\Sigma d_x}{n} = 20 + \frac{43}{9} = 20 + 4.78 = 24.78.$

Since 25 occurs 3 times and no other item occurs 3 or more than 3 times.

$\therefore$ Mode $M_0 = 25$

Standard deviation $\sigma = \sqrt{\frac{1}{n}\Sigma d_x^2 - \left(\frac{\Sigma d_x}{n}\right)^2}$

$$= \sqrt{\frac{567}{9} - \left(\frac{43}{9}\right)^2} = 6.33.$$

$\therefore$ Co-efficient of skewness

$$= \frac{M - M_0}{\sigma} = \frac{24.78 - 25}{6.33} = -.03.$$

Example 5. *Find out the co-efficient of skewness for the following distribution :*

Variable	*Frequency*	*Variable*	*Frequency*
0— 5	2	20—25	21
5—10	5	25—30	16
10—15	7	30—35	8
15—20	13	35—40	3

Sol. Please try yourself. **[Ans.** – 1]

Example 6. *Calculate the quartile co-efficient of skewness of the following distribution :*

Variate (x)	*: 1—5*	*6—10*	*11—15*	*16—20*	*21—25*	*26—30*	*31—35*
Frequency (f) :	*3*	*4*	*68*	*30*	*10*	*6*	*2*

Sol.

Class	*f*	*c.f.*
1— 5	3	3
6—10	4	7
11—15	68	75
16—20	30	105
21—25	10	115
26—30	6	121
31—35	2	123
		N = 123

$\frac{N}{2} = 61.5$ Median class is 11—15 *i.e.,* 10.5—15.5

$$\therefore \quad \text{Median } M_d = l + \frac{h}{f}\left(\frac{N}{2} - C\right)$$

$$= 10.5 + \frac{5}{68}(61.5 - 7) = 10.5 + \frac{272.5}{68} = 14.5$$

$$\frac{N}{4} = 30.75$$

$\therefore$ Class of lower quartile Q_1 is 11—15 *i.e.,* 10.5—15.5

$$Q_1 = l + \frac{h}{f}\left(\frac{N}{4} - C\right) = 10.5 + \frac{5}{68}(30.75 - 7)$$

$$= 10.5 + \frac{118.75}{68} = 12.25$$

$$\frac{3N}{4} = 92.25$$

∴ Class of upper quartile Q_3 is 16—20 *i.e.,* 15.5—20.5

$$Q_3 = l + \frac{h}{f}\left(\frac{3N}{4} - C\right) = 15.5 + \frac{5}{30}(92.25 - 75)$$

$$= 15.5 + \frac{16.50}{6} = 18.25$$

Quartile co-efficient of skewness

$$= \frac{Q_3 + Q_1 - 2M_d}{Q_3 - Q_1} = \frac{30.50 - 29}{6} = \frac{1.50}{6} = 0.25.$$

Example 7. *Compute quartile co-efficient of dispersion and skewness of the following data :*

Central size :	*1*	*2*	*3*	*4*	*5*	*6*	*7*	*8*	*9*	*10*
Frequency :	*2*	*9*	*11*	*14*	*20*	*24*	*20*	*16*	*5*	*2*

Sol. Arranging in groups, we have

Class	*f*	*c.f.*
0.5— 1.5	2	2
1.5— 2.5	9	11
2.5— 3.5	11	22
3.5— 4.5	14	36
4.5— 5.5	20	56
5.5— 6.5	24	80
6.5— 7.5	20	100
7.5— 8.5	16	116
8.5— 9.5	5	121
9.5—10.5	2	123

$$N = 123 \quad \Rightarrow \quad \frac{N}{2} = 61.5$$

∴ Median class is 5.5—6.5

Median $\quad M_d = 5.5 + \frac{1}{24}(61.5 - 56) = 5.5 + .23 = 5.73$

$$\frac{N}{4} = 30.75$$

∴ Class of lower quartile Q_1 is 2.5—3.5

$$Q_1 = 2.5 + \frac{1}{11}(30.75 - 22) = 2.5 + .8 = 3.3$$

Also $\quad \frac{3N}{4} = 92.25$

∴ Class of upper quartile Q_3 is 6.5—7.5

$$Q_3 = 6.5 + \frac{1}{20}(92.25 - 80) = 6.5 + .61 = 7.11$$

Quartile co-efficient of dispersion

$$= \frac{Q_3 - Q_1}{Q_3 + Q_1} = \frac{7.11 - 3.3}{7.11 + 3.3} = \frac{3.81}{10.41} = 0.37$$

Quartile co-efficient of skewness

$$= \frac{Q_3 + Q_1 - 2M_d}{Q_3 - Q_1}$$

$$= \frac{10.41 - 11.46}{3.81} = \frac{-1.05}{3.81} = -0.28.$$

Example 8. *Calculate Karl Pearson's co-efficient of skewness for the following distribution :*

Class Interval	*Frequency*
0— 4	*5*
4— 8	*7*
8—12	*10*
12—16	*15*
16—20	*8*
20—24	*4*

Sol. Please try yourself. **[Ans. – .26]**

Example 9. *From the following table, compute quartile deviation as well as co-efficient of skewness :*

Size	*Frequency*	*Size*	*Frequency*
4— 8	*6*	*24—28*	*12*
8—12	*10*	*28—32*	*10*
12—16	*18*	*32—36*	*6*
16—20	*30*	*36—40*	*2*
20—24	*15*		

Sol.

Size	*f*	*c.f.*
4— 8	6	6
8—12	10	16
12—16	18	34
16—20	30	64
20—24	15	79
24—28	12	91
28—32	10	101
32—36	6	107
36—40	2	109

$N = 109 \quad \Rightarrow \quad \frac{N}{4} = 27.25$

$\therefore$ Class of lower quartile Q_1 is 12—16

$$Q_1 = l + \frac{h}{f}\left(\frac{N}{4} - C\right) = 12 + \frac{4}{18}(27.25 - 16) = 14.5$$

$$\frac{3N}{4} = 81.75 \quad \therefore \quad \text{Class of upper quartile } Q_3 \text{ is } 24\text{—}28$$

$$Q_3 = l + \frac{h}{f}\left(\frac{3N}{4} - C\right) = 24 + \frac{4}{12}(81.75 - 79) = 24.92$$

$$\therefore \quad Q.D = \tfrac{1}{2}(Q_3 - Q_1) = \tfrac{1}{2}(24.92 - 14.5) = 5.21$$

$$\frac{N}{2} = 54.5 \quad \therefore \quad \text{Median class is } 16\text{—}20$$

$$M_d = l + \frac{h}{f}\left(\frac{N}{2} - C\right) = 16 + \frac{4}{30}(54.5 - 34) = 18.73$$

$$\text{Quartile co-efficient of skewness} = \frac{Q_3 + Q_1 - 2M_d}{Q_3 - Q_1}$$

$$= \frac{24.92 + 14.5 - 2(18.73)}{24.92 - 14.5} = \frac{1.96}{10.42} = .19.$$

Example 10. *In a frequency distribution, the co-efficient of skewness based upon the quartiles is 0.6. If the sum of the upper and lower quartiles is 100 and median is 38, find the value of the upper and lower quartiles.*

Sol. Co-efficient of skewness based on quartiles

$$= \frac{Q_3 + Q_1 - 2M_d}{Q_3 - Q_1}$$

Here $Q_3 + Q_1 = 100$ ...(1)

$M_d = 38$

$$\therefore \quad 0.6 = \frac{100 - 2 \times 38}{Q_3 - Q_1}$$

or $Q_3 - Q_1 = \dfrac{24}{0.6} = 40$ ...(2)

From (1) and (2), $Q_3 = 70$, $Q_1 = 30$.

Example 11. *A frequency distribution gives the following results :*
Co-efficient of variation = 5
Karl Pearson's co-efficient of skewness = 0.5
Standard deviation = 2
Find the mean and mode of the distribution.

Sol. Let M, M_0 denote the mean and mode of the distribution.

Co-efficient of variation = 5

$$\Rightarrow \quad \frac{\sigma}{M} \times 100 = 5$$

or $$M = 20\sigma = 20 \times 2 = 40$$

Karl Pearson's co-efficient of skewness

$$= \frac{M - M_0}{\sigma}$$

$$\therefore \quad 0.5 = \frac{40 - M_0}{2}$$

or $$1 = 40 - M_0$$

$$\therefore \quad M_0 = 39.$$

Example 12. *Find the co-efficient of variation of a frequency distribution given that its mean is 120, mode is 123 and Karl Pearson's co-efficient of skewness is – 0.3.*

Sol. Mean $M = 120$

Mode $M_0 = 123$

Karl Pearson's co-efficient of skewness = – 0.3

$$\therefore \quad \frac{M - M_0}{\sigma} = -0.3$$

$$\frac{-3}{\sigma} = -0.3 \quad \therefore \quad \sigma = 10$$

Co-efficient of variation

$$= \frac{\sigma}{M} \times 100 = \frac{10}{120} \times 100$$

$$= 8.33.$$

3.11. MOMENTS

The *rth moment* of a variable *x about any point A* is denoted by μ_r' and is defined as

$$\mu_r' = \frac{1}{N} \Sigma f(x - A)^r \qquad \text{where } N = \Sigma f$$

The *r*th moment of a variable *x* about the mean M is denoted by μ_r and is defined as

$$\mu_r = \frac{1}{N} \Sigma f(x - M)^r$$

In particular $$\mu_0' = \frac{1}{N} \Sigma f(x - A)^0$$

$$= \frac{1}{N} \Sigma f = \frac{1}{N} . N = 1$$

Similarly $$\mu_0 = 1$$

$$\mu_1 = \frac{1}{N} \Sigma f(x - M) = 0$$

being the algebraic sum of the deviations from the mean

$$\mu_2 = \frac{1}{N} \Sigma f(x - M)^2 = \sigma^2 \text{ by def.}$$

The results $\mu_0 = 1$, $\mu_1 = 0$, $\mu_2 = \sigma^2$ are of fundamental importance and should be committed to memory.

3.12. RELATION BETWEEN MOMENTS ABOUT MEAN IN TERMS OF MOMENTS ABOUT ANY POINT AND VICE VERSA

By definition $\mu_r' = \frac{1}{N} \Sigma f(x - A)^r$ where A is any point

$$= \frac{1}{N} \Sigma f d^r \quad \text{where } d = x - A \qquad ...(i)$$

Putting $r = 1$ $\mu_1' = \frac{1}{N} \Sigma f d$

$$\therefore \quad M = A + \frac{1}{N} \Sigma f d = A + \mu_1'$$

or

$$\mu_1' = M - A \qquad ...(ii)$$

Now

$$\mu_r = \frac{1}{N} \Sigma f(x - M)^r$$

$$= \frac{1}{N} \Sigma f(x - A + A - M)^r$$

$$= \frac{1}{N} \Sigma f(d - \mu_1')^r \qquad | \text{ using } (ii)$$

$$= \frac{1}{N} \Sigma f [d^r - {}^rC_1 d^{r-1} \mu_1' + {}^rC_2 d^{r-2} \mu_1'^2 - {}^rC_3 d^{r-3} \mu_1'^3 + ... + (-1)^r . \mu_1'^r]$$

$$= \frac{1}{N} \Sigma f d^r - {}^rC_1 \mu_1' . \frac{1}{N} \Sigma f d^{r-1} + {}^rC_2 \mu_1'^2 \frac{1}{N} \Sigma f d^{r-2} - {}^rC_3 \mu'^3 \frac{1}{N} \Sigma f d^{r-3} + ... + (-1)^r \mu_1'^r . \frac{1}{N} \Sigma f$$

$$= \mu_r' - {}^rC_1 \mu'_{r-1} \mu_1' + {}^rC_2 \mu'_{r-2} \mu_1'^2 - {}^rC_3 \mu'_{r-3} \mu_1'^3 + ... + (-1)^r \mu_1'^r \qquad | \text{ using } (i)$$

$$\Rightarrow \quad \mu_r = \sum_{j=0}^{r} {}^rC_j \, \mu'_{r-j} (-\mu_1')^j \qquad (D.U.\ 1985)$$

In particular, putting $r = 2, 3, 4$, we get

$$\mu_2 = \mu_2' - 2\mu_1'^2 + \mu_0' \mu_1'^2 = \mu_2' - \mu_1'^2 \qquad | \because \quad \mu_0' = 1$$

$$\mu_3 = \mu_3' - 3\mu_2'\mu_1' + 3\mu_1'\mu_1'^2 - \mu_0'\mu_1'^3$$

$$= \mu'_3 - 3\mu_2'\mu_1' + 2\mu_1'^3$$

$$\mu_4 = \mu_4' - 4\mu_3'\mu_1' + 6\mu_2'\mu_1'^2 - 4\mu_1'\mu_1'^3 + \mu_0'\mu_1'^4$$

$$= \mu_4' - 4\mu_3'\mu_1' + 6\mu_2'\mu_1'^2 - 3\mu_1'^4$$

Hence $\mu_1 = 0$

$$\left.\begin{aligned}\mu_2 &= \mu_2' - \mu_1'^2\\ \mu_3 &= \mu_3' - 3\mu_2'\mu_1' + 2\mu_1'^3\\ \mu_4 &= \mu_4' - 4\mu_3'\mu_1' + 6\mu_2'\mu_1'^2 - 3\mu_1'^4\end{aligned}\right| \quad (\mu_1' = \text{M} - \text{A})$$

Conversely, $\mu_r = \frac{1}{\text{N}}\Sigma f(x - \text{M})^r$

$= \frac{1}{\text{N}}\Sigma fd^r$, where $d = x - \text{M}$ ...(*iii*)

Now $\mu_r' = \frac{1}{\text{N}}\Sigma f(x - \text{A})^r = \frac{1}{\text{N}}\Sigma f(x - \text{M} + \text{M} - \text{A})^r$

$= \frac{1}{\text{N}}\Sigma f(d + \mu_1')^r$ | using (*ii*)

$$= \frac{1}{\text{N}}\Sigma f(d^r + {}^r\text{C}_1 d^{r-1}\mu_1' + {}^r\text{C}_2 d^{r-2}\mu_1'^2 + {}^r\text{C}_3 d^{r-3}\mu_1'^3 + \ldots\ldots + \mu_1'^r)$$

$$= \frac{1}{\text{N}}\Sigma fd^r + {}^r\text{C}_1\mu_1' \cdot \frac{1}{\text{N}}\Sigma fd^{r-1} + {}^r\text{C}_2\mu_1'^2 \cdot \frac{1}{\text{N}}\Sigma fd^{r-2} + \ldots\ldots + \mu_1'^r \cdot \frac{1}{\text{N}}\Sigma f$$

$= \mu_r + {}^r\text{C}_1\mu_{r-1}\mu_1' + {}^r\text{C}_2\mu_{r-2}\mu_1'^2 + \ldots\ldots + \mu_1'^r$ | using (*iii*)

$$\Rightarrow \quad \mu_r' = \sum_{j=0}^{r} {}^r\text{C}_j\,\mu_{r-j}\,\mu_1'^j \qquad \text{(D.U. 1989)}$$

In particular, putting $r = 2, 3, 4$ and noting that $\mu_1 = 0$, $\mu_0 = 1$, we get

$$\mu_2' = \mu_2 + 2\mu_1\mu_1' + \mu_0\mu_1'^2 = \mu_2 + \mu_1'^2$$

$$\mu_3' = \mu_3 + 3\mu_2\mu_1' + 3\mu_1\mu_1'^2 + \mu_0\mu_1'^3 = \mu_3 + 3\mu_2\mu_1' + \mu_1'^3$$

$$\mu_4' = \mu_4 + 4\mu_3\mu_1' + 6\mu_2\mu_1'^2 + 4\mu_1\mu_1'^3 + \mu_0\mu_1'^4$$

$$= \mu_4 + 4\mu_3\mu_1' + 6\mu_2\mu_1'^2 + \mu_1'^4.$$

3.13. EFFECT OF CHANGE OF ORIGIN AND SCALE ON MOMENTS

Let $u = \frac{x - \text{A}}{h}$ *i.e.,* $x = \text{A} + hu$

$\therefore$ $\bar{x} = \text{A} + h\,\bar{u}$, where bar denotes the mean of the respective variable

$\therefore$ $x - \bar{x} = h(u - \bar{u})$

$$\mu_r' = \frac{1}{\text{N}}\Sigma f(x - \text{A})^r = \frac{1}{\text{N}}\Sigma f h^r u^r = h^r \cdot \frac{1}{\text{N}}\Sigma fu^r$$

Also $\mu_r = \frac{1}{\text{N}}\Sigma f(x - \bar{x})^r = \frac{1}{\text{N}}\Sigma f h^r(u - \bar{u})^r$

$$= h^r \cdot \frac{1}{\text{N}}\Sigma f(u - \bar{u})^r$$

Hence *the rth moment of the variable x is h^r times the corresponding moment of the variable u.*

3.14. SHEPPARD'S CORRECTIONS FOR MOMENTS

In the case of class intervals we assume that the frequencies are concentrated at mid-points of class intervals. Since this assumption is not true in general, some error is likely to creep into the calculation of moments. W.F. Sheppard gave the following formulae by which these errors may be corrected.

$$\mu_2 \text{ (corrected)} = \mu_2 - \frac{1}{12} h^2$$

$$\mu_3 \text{ (corrected)} = \mu_3$$

$$\mu_4 \text{ (corrected)} = \mu_4 - \frac{1}{2} h^2 \mu_2 + \frac{7}{240} h^4$$

where h is the width of class intervals.

3.15. CHARLIER'S CHECK

To check the accuracy in the calculation of first four moments, we often use the following identities known as *Charlier checks* :

$$\Sigma f(x+1) = \Sigma f x + \Sigma f = \Sigma fx + N$$

$$\Sigma f(x+1)^2 = \Sigma f x^2 + 2\Sigma fx + N$$

$$\Sigma f(x+1)^3 = \Sigma fx^3 + 3\Sigma fx^2 + 3\Sigma fx + N$$

$$\Sigma f(x+1)^4 = \Sigma fx^4 + 4\Sigma fx^3 + 6\Sigma fx^2 + 4\Sigma fx + N.$$

3.16. PEARSON'S β AND γ CO-EFFICIENTS

Karl Pearson defined the following four co-efficients based upon the first four moments about mean.

$$\beta_1 = \frac{\mu_3^2}{\mu_2^3}, \quad \gamma_1 = +\sqrt{\beta_1}$$

$$\beta_2 = \frac{\mu_4}{\mu_2^2}, \quad \gamma_2 = \beta_2 - 3$$

These co-efficients are independent of units of measurement and therefore, are pure numbers.

3.17. BASED UPON MOMENTS, CO-EFFICIENT OF SKEWNESS IS

$$S_k = \frac{\sqrt{\beta_1}\,(\beta_2 + 3)}{2(5\beta_2 - 6\beta_1 - 9)}.$$

3.18. KURTOSIS

"Given two frequency distributions which have the same variability as measured by the standard deviation, they may be *relatively* more or less flat topped than the 'normal curve'. A frequency curve may be symmetrical but it may not be equally flat topped with the normal curve. The relative flatness of the top is called *kurtosis* and is measured by β_2.

Curves which are neither flat nor sharply peaked are called *normal curves* or *mesokurtic curves* (see curve A in figure below). For such a curve $\beta_2 = 3$ and hence $\gamma_2 = 0$.

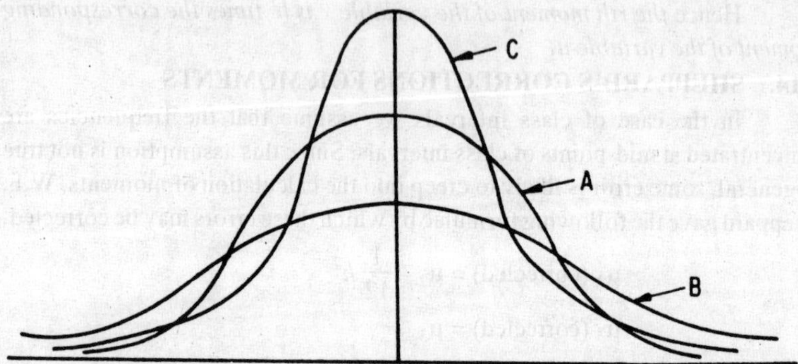

Curves which are flatter than the normal curve (see curve B in figure above) are called *platykurtic*. For such a curve $\beta_2 < 3$ and hence $\gamma < 0$.

Curves which are more sharply peaked than the normal curve (see curve C in figure above) are called *leptokurtic*. For such a curve $\beta_2 > 3$ and hence $\gamma_2 > 0$.

3.19. β_1 AS MEASURE OF SKEWNESS

For a symmetrical distribution, all the moments of odd order about the mean vanish.

Let $\bar{x}$ denote the mean of the variate x, then

$$\mu_{2r+1} = \frac{1}{N} \sum_{i=1}^{n} f_i(x_i - \bar{x})^{2r+1}, \qquad N = \Sigma f_i$$

In a symmetrical distribution, the values of the variate equidistant from the mean have equal frequencies.

$$\therefore \qquad f_1(x_1 - \bar{x})^{2r+1} + f_n(x_n - \bar{x})^{2r+1} = 0$$

[$\because$ $x_1 - \bar{x}$ and $x_n - \bar{x}$ are equal in magnitude but opposite in sign. Also $f_1 = f_n$]

Similarly $f_2(x_2 - \bar{x})^{2r+1} + f_{n-1}(x_{n-1} - \bar{x})^{2r+1} = 0$ and so on.

$\therefore$ If n is even, all the terms in $\frac{1}{N} \sum_{i=1}^{n} f_i(x_i - \bar{x})^{2r+1}$ cancel in pairs.

If n is odd, again the terms cancel in pairs and the middle term vanishes, since middle term $= \bar{x}$.

Hence $\qquad \mu_{2r+1} = 0$

In particular $\qquad \mu_3 = 0$

and hence $\qquad \beta_1 = \frac{\mu_3^2}{\mu_2^3} = 0.$

Thus, β_1 gives a measure of departure from symmetry, *i.e.*, of skewness.

Example 1. *Calculate the first four moments of the following distribution about the mean and hence find* β_1 *and* β_2 :

x :	0	1	2	3	4	5	6	7	8
f :	1	8	28	56	70	56	28	8	1

Sol. Let us first calculate moments about $x = 4$

$$\mu_r' = \frac{1}{N} \Sigma f(x-4)^r = \frac{1}{N} \Sigma f d^r \text{ where } d = x - 4.$$

x	f	$d = x - 4$	fd	fd^2	fd^3	fd^4
0	1	−4	−4	16	−64	256
1	8	−3	−24	72	−216	648
2	28	−2	−56	112	−224	448
3	56	−1	−56	56	−56	56
4	70	0	0	0	0	0
5	56	1	56	56	56	56
6	28	2	56	112	224	448
7	8	3	24	72	216	648
8	1	4	4	16	64	256
	N = 256		0	512	0	2816

$$\mu_1' = \frac{1}{N} \Sigma fd = 0$$

$$\mu_2' = \frac{1}{N} \Sigma fd^2 = \frac{512}{256} = 2$$

$$\mu_3' = \frac{1}{N} \Sigma fd^3 = 0$$

$$\mu_4' = \frac{1}{N} \Sigma fd^4 = \frac{2816}{256} = 11$$

Moments about mean are

$$\mu_1 = 0 \ (always)$$

$$\mu_2 = \mu_2' - \mu_1'^2 = 2$$

$$\mu_3 = \mu_3' - 3\mu_2' \mu_1' + 2\mu_1'^3 = 0$$

$$\mu_4 = \mu_4' - 4\mu_3'\mu_1' + 6\mu_2'\mu_1'^2 - 3\mu_1'^4 = 11$$

$$\beta_1 = \frac{\mu_3^2}{\mu_2^3} = 0.$$

$$\beta_2 = \frac{\mu_4}{\mu_2^2} = \frac{11}{4} = 2.75.$$

Example 2. *Compute the first four moments about the arithmetic mean from the following data :*

Variate	*Frequency*	*Variate*	*Frequency*
5	*8*	*25*	*23*
10	*15*	*30*	*17*
15	*20*	*35*	*5*
20	*32*		

Sol. Let us first calculate moments about $x = 20$

$$\mu_r' = \frac{1}{N} \Sigma f(x - 20)^r = \frac{1}{N} \Sigma fd^r \quad \text{where} \quad d = x - 20.$$

x	*f*	*d = x − 20*	*fd*	fd^2	fd^3	fd^4
5	8	– 15	– 120	1800	– 27000	405000
10	15	– 10	– 150	1500	– 15000	150000
15	20	– 5	– 100	500	– 2500	12500
20	32	0	0	0	0	0
25	23	5	115	575	2875	9375
30	17	10	170	1700	17000	170000
35	5	15	75	1125	16875	253125
N = 120			– 10	7200	– 7750	1000000

$$\mu_1' = \frac{1}{N} \Sigma fd = -\frac{10}{120} = -0.0833$$

$$\mu_2' = \frac{1}{N} \Sigma fd^2 = \frac{7200}{120} = 60$$

$$\mu_3' = \frac{1}{N} \Sigma fd^3 = -\frac{7750}{120} = -64.5833$$

$$\mu_4' = \frac{1}{N} \Sigma fd^4 = \frac{1000000}{120} = 8333.3333.$$

Moments about mean are

$$\mu_1 = 0 \text{ (always)}$$

$$\mu_2 = \mu_2' - \mu_1'^2 = 59.9931$$

$$\mu_3 = \mu_3' - 3\mu_2' \mu_1' + 2\mu_1'^3 = -49.5905$$

$$\mu_4 = \mu_4' - 4\mu_3' \mu_1' + 6\mu_2' \mu_1'^2 - 3\mu_1'^4 = 8314.3088.$$

Example 3. *Calculate the first four moments about the mean for the following data :*

Variate :	*1*	*2*	*3*	*4*	*5*	*6*	*7*	*8*	*9*
Frequency :	*1*	*6*	*13*	*25*	*30*	*22*	*9*	*5*	*2*

Sol. Please try yourself.

[**Ans.** $\mu_1 = 0$, $\mu_2 = 2.49$, $\mu_3 = .68$, $\mu_4 = 18.26$]

Example 4. *Calculate the first four moments about the arithmetic mean from the following data :*

Weekly wage	*No. of persons*	*Weekly wage*	*No. of persons*
0—10	*15*	*40—50*	*32*
10—20	*23*	*50—60*	*28*
20—30	*35*	*60—70*	*12*
30—40	*49*	*70—80*	*6*

Sol. Let us take the origin at 45 and the unit be taken as 10.

i.e., let $u = \frac{x - A}{h} = \frac{x - 45}{10}$.

Mid-values (x)	f	u	fu	fu^2	fu^3	fu^4
5	15	– 4	– 60	240	– 960	3840
15	23	– 3	– 69	207	– 621	1863
25	35	– 2	– 70	140	– 280	560
35	49	– 1	– 49	49	– 49	49
45	32	0	0	0	0	0
55	28	1	28	28	28	28
65	12	2	24	48	96	192
75	6	3	18	54	162	486
	N = 200		– 178	766	– 1624	7018

With respect to the variable u

$$\mu_1' = \frac{-178}{200} = -0.89$$

$$\mu_2' = \frac{766}{200} = 3.83$$

$$\mu_3' = \frac{-1624}{200} = -8.12$$

$$\mu_4' = \frac{7018}{200} = 35.09$$

$$\mu_1 = 0$$

$$\mu_2 = \mu_2' - \mu_1'^2 = 2.0379$$

$$\mu_3 = \mu_3' - 3\mu_2'\mu_1' + 2\mu_1'^3 = 0.6961$$

$$\mu_4 = \mu_4' - 4\mu_3'\mu_1' + 6\mu_2'\mu_1'^2 - 3\mu_1'^4 = 22.503$$

With respect to the variable x

$$\mu_1 = 0$$

$$\mu_2 = h^2\,(2.0379) = 100 \times 2.0379 = 203.79$$

$$\mu_3 = h^3\,(0.6961) = 1000 \times 0.6961 = 696.1$$

$$\mu_4 = h^4\,(22.503) = 10000 \times 22.503 = 225030$$

Example 5. *Calculate the first four moments about the mean of the distribution ; x-values in centimetres are the mid-points of intervals :*

x:	2.0	2.5	3.0	3.5	4.0	4.5	5.0
f:	5	38	65	92	70	40	10

Sol. Let $u = \dfrac{x - 3.5}{.5}$.

x	f	u	fu	fu^2	fu^3	fu^4
2.0	5	– 3	– 15	45	– 135	405
2.5	38	– 2	– 76	152	– 304	608
3.0	65	– 1	– 65	65	– 65	65
3.5	92	0	0	0	0	0
4.0	70	1	70	70	70	70
4.5	40	2	80	160	320	640
5.0	10	3	30	90	270	810
	320		24	582	156	2598

With respect to the variable u

$$\mu_1' = \frac{24}{320} = 0.075$$

$$\mu_2' = \frac{582}{320} = 1.81875$$

$$\mu_3' = \frac{156}{320} = 0.4875$$

$$\mu_4' = \frac{2598}{320} = 8.11875$$

$$\mu_1 = 0$$

$$\mu_2 = \mu_2' - \mu_1'^2 = 1.813125$$

$$\mu_3 = \mu_3' - 3\mu_2'\mu_1' + 2\mu_1'^3 = 0.0791$$

$$\mu_4 = \mu_4' - 4\mu_3'\mu_1' + 6\mu_2'\mu_1'^2 - 3\mu_1'^4 = 8.033.$$

With respect to the variable x

$$\mu_1 = 0$$

$$\mu_2 = h^2\,(1.813125) = (.5)^2\,(1.813125) = 0.4533$$

$$\mu_3 = h^3\,(0.0791) = (.5)^3\,(0.0791) = 0.0099$$

$$\mu_4 = h^4\,(8.033) = (.5)^4\,(8.033) = 0.502.$$

Example 6. *The first three moments of a distribution about the value 2 of the variable are 1, 16 and – 40. Show that the mean is 3, variance is 15 and $\mu_3 = -86$. Also show that the first three moments about $x = 0$ are 3, 24, and 76.* (D.U. 1984, 85, 92 ; Agra 1987)

Sol. With usual notation,

$$A = 2$$

$$\mu_1' = 1,\ \mu_2' = 16,\ \mu_3' = -40.$$

$\therefore$ Mean $\quad M = A + \mu_1' = 2 + 1 = 3$

Variance $\sigma^2 = \mu_2 = \mu_2' - \mu_1'^2 = 16 - 1 = 15$

$$\mu_3 = \mu_3' - 3\mu_2'\,\mu_1' + 2\mu_1'^3$$
$$= -40 - 48 + 2 = -86.$$

Moments about x = 0

Let μ_r' denote rth moment about $x = 0$.

Here $A = 0$.

$\because$ $M = 3$

$\therefore$ Using $M = A + \mu_1'$

$$\mu_1' = M - A = 3$$
$$\mu_2' = \mu_2 + \mu_1'^2 = 15 + 9 = 24$$
$$\mu_3' = \mu_3 + 3\mu_2\,\mu_1' + \mu_1'^3$$
$$= -86 + 135 + 27 = 76.$$

Example 7. *For a distribution, the mean is 10, variance is 16, γ_1 is 1 and β_2 is 4. Find the first four moments about the origin.* (D.U. 1990)

Sol. Mean $M = 10$

Variance $\sigma^2 = \mu_2 = 16$

$$\gamma_1 = \sqrt{\beta_1} = 1$$
$$\beta_2 = 4.$$

Since $\beta_1 = 1 \quad \therefore \quad \dfrac{\mu_3^2}{\mu_2^3} = 1 \quad \text{or} \quad \mu_3^2 = \mu_2^3 = (16)^3 = (4^3)^2$

$\therefore$ $\mu_3 = 64$

$\beta_2 = 4 \quad \therefore \quad \dfrac{\mu_4}{\mu_2^2} = 4 \quad \text{or} \quad \mu_4 = 4\,\mu_2^2 = 4 \times 256 = 1024.$

Moments about origin. Let μ_r' denote rth moment about $x = 0$

Here $A = 0 \quad \because \quad M = 10$

$\therefore$ Using $M = A + \mu_1' \qquad \mu_1' = M - A = 10$

$$\mu_2' = \mu_2 + \mu_1'^2 = 16 + 100 = 116$$
$$\mu_3' = \mu_3 + 3\,\mu_2\,\mu_1' + \mu_1'^3 = 64 + 480 + 1000 = 1544$$
$$\mu_4' = \mu_4 + 4\,\mu_3\,\mu_1' + 6\,\mu_2\,\mu_1'^2 + \mu_1'^4$$
$$= 1024 + 2560 + 9600 + 10000 = 23184.$$

Example 8. *The first four moments of a distribution about the value 4 of the variable are – 1.5, 17, – 30 and 108. Find the moments about mean, β_1 and β_2.* (Meerut 1991)

Find also the moments about (i) the origin, and (ii) the point x = 2.

Sol. Let μ_r' denote the rth moment about $x = 4$.

Here $A = 4, \mu_1' = -1.5, \mu_2' = 17, \mu_3' = -30, \mu_4' = 108.$

Moments about mean

$\mu_1 = 0$ *(always)*

$\mu_2 = \mu_2' - \mu_1'^2 = 17 - (-1.5)^2 = 14.75$

$\mu_3 = \mu_3' - 3\mu_2'\mu_1' + 2\mu_1'^3 = 39.75$

$\mu_4 = \mu_4' - 4\mu_3'\mu_1' + 6\mu_2'\mu_1'^2 - 3\mu_1'^4 = 142.3125$

Also $\beta_1 = \dfrac{\mu_3^2}{\mu_2^3} = \dfrac{(39.75)^2}{(14.75)^3} = 0.4926$

$\beta_2 = \dfrac{\mu_4}{\mu_2^2} = \dfrac{142.3125}{(14.75)^2} = 0.6543$

[Mean $M = A + \mu_1' = 4 - 1.5 = 2.5$]

Moments about the origin. Let μ_r' denote the rth moment about $x = 0$.

Using $M = A + \mu_1'$, we have $2.5 = 0 + \mu_1'$ $\therefore$ $\mu_1' = 2.5$

$\mu_2' = \mu_2 + \mu_1'^2 = 14.75 + (2.5)^2 = 21$

$\mu_3' = \mu_3 + 3\mu_2\mu_1' + \mu_1'^3 = 39.75 + 110.625 + 15.625 = 166$

$\mu_4' = \mu_4 + 4\mu_3\mu_1' + 6\mu_2\mu_1'^2 + \mu_1'^4$

$= 142.3125 + 397.5 + 553.125 + 39.0625 = 1132.$

Moments about the point x = 2. Here A = 2. Let μ_r' denote the rth moment about $x = 2$.

Using $M = A + \mu_1'$, we have $2.5 = 2 + \mu_1'$ $\therefore$ $\mu_1' = 0.5$

Proceeding as above $\mu_2' = 15$, $\mu_3' = 62$, $\mu_4' = 244$.

Example 9. *The first four moments of a distribution about the value 5 of the variable are 2, 20, 40 and 50. Find moments about the mean.* (D.U. 1983)

Sol. Please try yourself.

[**Ans.** $\mu_1 = 0$, $\mu_2 = 16$, $\mu_3 = -64$, $\mu_4 = 162$]

Example 10. *The first four moments of a distribution about x = 4 are 1, 4, 10, 45. Show that the mean is 5, the variance is 3 and μ_3 and μ_4 are 0 and 26 respectively.*

Sol. Please try yourself.

Example 11. *Show that for discrete distributions, $\beta_2 > 1$*

(D.U. 1983, 86, 91, 92 ; Kanpur 1987 ; Agra 1983, 89)

Sol. $\beta_2 = \dfrac{\mu_4}{\mu_2^2}$

$\therefore$ $\beta_2 > 1$ if $\dfrac{\mu_4}{\mu_2^2} > 1$

or if $\mu_4 > \mu_2^2$

or if $\dfrac{1}{N}\Sigma f(x - M)^4 > \left[\dfrac{1}{N}\Sigma f(x - M)^2\right]^2$

of if $\frac{1}{N}\Sigma fy^4 > \frac{1}{N^2}(\Sigma fy^2)^2$ where $y = x - M$

or if $N\Sigma fy^4 > (\Sigma fy^2)^2$

or if $(f_1 + f_2 + ... + f_n)(f_1y_1^4 + f_2y_2^4 + ... + f_ny_n^4)$

$> (f_1y_1^2 + f_2y_2^2 + ... + f_ny_n^2)^2$

or if $f_1f_2(y_1^2 - y_2^2)^2 + f_1f_3(y_1^2 - y_3^2)^2 + ... > 0$

Since f_i $(i = 1, 2, ..., n)$ is always + ve, therefore, the inequality is always true.

Hence $\beta_2 > 1$.

Example 12. *Show that for a discrete distribution, $\beta_2 > \beta_1$.*

(D.U. 1981, 90)

Sol. $\beta_2 = \frac{\mu_4}{\mu_2^2}$ $\beta_1 = \frac{\mu_3^2}{\mu_2^3}$

$\therefore$ $\beta_2 > \beta_1$

if $\frac{\mu_4}{\mu_2^2} > \frac{\mu_3^2}{\mu_2^3}$

or if $\mu_2\mu_4 > \mu_3^2$

or if $\frac{1}{N}\Sigma f_i(x_i - \bar{x})^2 . \frac{1}{N}\Sigma f_i(x_i - \bar{x})^4 > \left[\frac{1}{N}\Sigma f_i(x_i - \bar{x})^3\right]^2$

or if $\Sigma f_i y_i^2 \Sigma f_i y_i^4 > (\Sigma f_i y_i^3)^2$ where $y_i = x_i - \bar{x}$

or if $(f_1y_1^2 + f_2y_2^2 + ...)(f_1y_1^4 + f_2y_2^4 + ...) > (f_1y_1^3 + f_2y_2^3 ......)^2$

or if $f_1^2y_1^6 + f_2^2y_2^6 + f_1f_2y_1^2y_2^2(y_1^2 + y_2^2) + ... > f_1^2y_1^6 + f_2^2y_2^6$

$+ 2f_1f_2y_1^3y_2^3 + ...$

or if $f_1f_2y_1^2y_2^2(y_1^2 + y_2^2) - 2f_1f_2y_1^3y_2^3 + ... > 0$

or if $f_1f_2y_1^2y_2^2(y_1^2 + y_2^2 - 2y_1y_2) + ... > 0$

or if $f_1f_2y_1^2y_2^2(y_1 - y_2)^2 + ... > 0$

Since f_i $(i = 1, 2, ... n)$ is always +ve, the inequality is always true.

Hence $\beta_2 > \beta_1$.

4

Curve Fitting and Method of Least Squares

4.1. CURVE FITTING

Suppose we are given a data in terms of two variables x and y. The problem of finding an analytic expression of the form $y = f(x)$ which fits the given data is called *curve fitting.*

4.2. BEST FITTING CURVE

Let the given data be represented by a set of n points (x_i, y_i), $i = 1, 2, 3, \ldots\ldots, n$. Let $y = f(x)$ be an approximate curve which fits the given set of data.

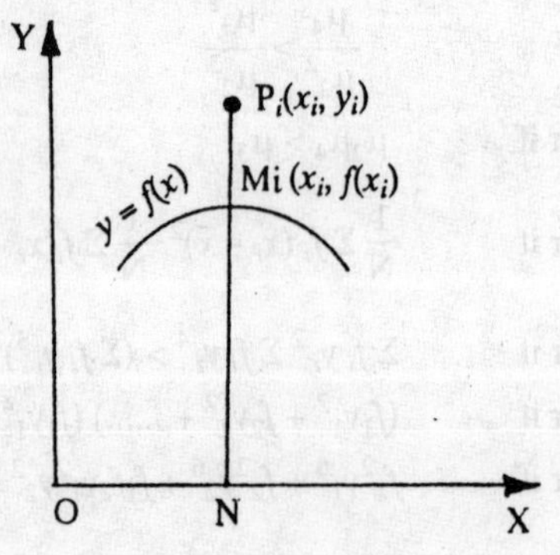

Let $Y_i = f(x_i)$ then Y_i is called the **expected value** of y corresponding to $x = x_i$. The value y_i is called the **observed value** of y corresponding to $x = x_i$.

In general $Y_i \neq y_i$, as the point $P_i(x_i, y_i)$ does not necessarily lie on the curve $y = f(x)$. $E_i = y_i - Y_i$ is called the *error of estimate* or the *residual* for y_i.

Of all curves approximating a given set of points, the curve for which

$$E = E_1^2 + E_2^2 + \ldots\ldots + E_n^2 = \sum_{i=1}^{n} E_i^2$$

is a minimum is called the best fitting curve or the least square curve.

4.3. METHOD OF LEAST SQUARES

Let (x_i, y_i), $i = 1, 2, 3, \ldots\ldots, n$ be a set of n points and let

$$y = a_0 + a_1x + a_2x^2 + \ldots.. a_kx^k \qquad \ldots(1)$$

be the polynomial of best fit to the set of n given points.

Observed value of y corresponding to $x = x_i$ is y_i.

Expected value of y corresponding to $x = x_i$ is given by

$$Y_i = a_0 + a_1x_i + a_2x_i^2 + \ldots\ldots + a_kx_i^k$$

Error of estimate for y_i is given by

$$E_i = y_i - Y_i$$
$$= y_i - a_0 - a_1x_i - a_2x_i^2 \ldots\ldots a_kx_i^k$$

Since (1) is the polynomial of best fit

$$\therefore \quad E = E_1^2 + E_2^2 + \ldots\ldots + E_n^2 = \sum_{i=1}^{n} E_i^2$$
$$= \sum_{i=1}^{n} (y_i - a_0 - a_1x_i - a_2x_i^2 \ldots\ldots a_kx_i^k)^2 \text{ is minimum.}$$

From the principle of maxima and minima, the partial derivatives of E w.r.t. $a_0, a_1, a_2, \ldots\ldots, a_k$ should vanish separately.

i.e.,
$$\frac{\partial E}{\partial a_0} = \frac{\partial E}{\partial a_1} = \frac{\partial E}{\partial a_2} = \ldots\ldots = \frac{\partial E}{\partial a_k} = 0$$

$$\left.\begin{aligned}
\frac{\partial E}{\partial a_0} &= -2\Sigma (y_i - a_0 - a_1x_i - a_2x_i^2 - \ldots\ldots - a_kx_i^k) = 0 \\
\frac{\partial E}{\partial a_1} &= -2\Sigma x_i (y_i - a_0 - a_1x_i - a_2x_i^2 - \ldots\ldots - a_kx_i^k) = 0 \\
&\vdots \\
\frac{\partial E}{\partial a_k} &= -2\Sigma x_i^k (y_i - a_0 - a_1x_i - a_2x_i^2 - \ldots\ldots - a_kx_i^k) = 0
\end{aligned}\right\}$$

which on simplification give

$$\left.\begin{aligned}
\Sigma y_i &= na_0 + a_1\Sigma x_i + a_2\Sigma x_i^2 + \ldots\ldots + a_k\Sigma x_i^k \\
\Sigma x_iy_i &= a_0\Sigma x_i + a_1\Sigma x_i^2 + a_2\Sigma x_i^3 + \ldots\ldots + a_k\Sigma x_i^{k+1} \\
&\vdots \\
\Sigma x_i^ky_i &= a_0\Sigma x_i^k + a_1\Sigma x_i^{k+1} + a_2\Sigma x_i^{k+2} + \ldots\ldots + a_k\Sigma x_i^{2k}
\end{aligned}\right\} \quad \text{(A)}$$

summation extending over i from 1 to n.

Thus we have $(k + 1)$ equations in $(k + 1)$ unknowns $a_0, a_1, a_2, \ldots\ldots, a_k$. Solving the equations of set (A) simultaneously, we get the values of the constants $a_0, a_1, a_2, \ldots\ldots, a_k$.

These equations are called *normal equations.*

4.4. PARTICULAR CASES

(*a*) **Fitting of a straight line.** When $k = 1$, the curve to be fitted is a straight line $y = a_0 + a_1 x$.

The normal equations are [from (A)]

$$\left.\begin{aligned}
\Sigma y_i &= na_0 + a_1\Sigma x_i \\
\Sigma x_iy_i &= a_0\Sigma x_i + a_1\Sigma x_i^2
\end{aligned}\right\}$$

(*b*) **Fitting of a second degree parabola.** When $k = 2$, the curve to be fitted is a second degree parabola

$$y = a_0 + a_1x + a_2x^2$$

The normal equations are

$$\left.\begin{aligned} \Sigma y_i &= na_0 + a_1\Sigma x_i + a_2\Sigma x_i^2 \\ \Sigma x_i y_i &= a_0\Sigma x_i + a_1\Sigma x_i^2 + a_2\Sigma x_i^3 \\ \Sigma x_i^2 y_i &= a_0\Sigma x_i^2 + a_1\Sigma x_i^3 + a_2\Sigma x_i^4 \end{aligned}\right\}$$

4.5. CHANGE OF ORIGIN

If the values of x and y are large, the calculation of Σx_i, Σx_i^2, Σx_i^3 etc. becomes tedious. Calculations are, to a great extent, simplified by a suitable change of origin.

Suppose that the values of x are equidistant at an interval h,

i.e., $x : a, a + h, a + 2h, \ldots\ldots, a + (n - 1)h.$

Case I. *If n is odd.* Let $n = 2m + 1$. There is only one middle term *i.e.,* $(m + 1)$th $= a + mh$.

Take $\quad X = \dfrac{x - \text{mid-value}}{h} = \dfrac{x - (a + mh)}{h}$

then the new variable X takes the values

$$X : -m, -(m - 1), \ldots\ldots, -1, 0, 1, \ldots\ldots, (m - 1), m$$

so that $\quad \Sigma X = \Sigma X^3 = \Sigma X^5 = \ldots\ldots = 0.$

If the curve of best fit w.r.t. the variable X is

$$Y = a_0 + a_1 X + a_2 X^2 + \ldots\ldots a_k X^k$$

then w.r.t. the variable x, it is obtained by substituting back the value of X in terms of x and putting $Y = y$.

Case II. *If n is even.* Let $n = 2m$. There are two middle terms, mth and $(m + 1)$th which are $a + (m - 1)h$ and $a + mh$.

Their mean $= \dfrac{a + (m - 1)h + a + mh}{2} = a + \frac{1}{2}(2m - 1)h$

Take $\quad X = \dfrac{x - (\text{mean of two middle terms})}{\frac{1}{2}h}$

$$= \frac{x - [a + \frac{1}{2}(2m - 1)h]}{\frac{1}{2}h} = \frac{2x - 2a - (2m - 1)h}{h}$$

then the new variable X takes the values

$$X : -(2m - 1), -(2m - 3) \ldots\ldots -3, -1, 1, 3, \ldots\ldots, (2m - 3), (2m - 1)$$

so that $\Sigma X = \Sigma X^3 = \Sigma X^5 = \ldots\ldots = 0.$

If the curve of best fit w.r.t. the variable X is

$$Y = a_0 + a_1 X + a_2 X^2 + \ldots\ldots + a_k X^k$$

then w.r.t. the variable x, it is obtained by substituting back the value of X in terms of x and putting $Y = y$.

Example 1. *Fit a straight line to the following data :*

x :	0	1	2	3	4
y :	1	1.8	3.3	4.5	6.3

(D.U. 1992 ; Agra 1985 ; Kanpur 1993)

Sol. Let the straight line to be fitted to the data be

$$y = a + bx,$$

then the normal equations are

$$\Sigma y = na + b\Sigma x,$$
$$\Sigma xy = a\Sigma x + b\Sigma x^2.$$

x	y	xy	x^2
0	1	0	0
1	1.8	1.8	1
2	3.3	6.6	4
3	4.5	13.5	9
4	6.3	25.2	16
Total 10	16.9	47.1	30

Here $n = 5$, $\Sigma x = 10$

$\Sigma y = 16.9$, $\Sigma xy = 47.1$

$\Sigma x^2 = 30$

On substituting these values, the normal equations become

$$16.9 = 5a + 10b \qquad ...(i)$$
$$47.1 = 10a + 30b \qquad ...(ii)$$

Solving (*i*) and (*ii*), we get

$$a = 0.72,$$
$$b = 1.33.$$

Hence the equation of the line of best fit is

$$y = 0.72 + 1.33x.$$

Example 2. *Show that the line of best fit to the following data is given by*

$$y = -.5x + 8$$

x :	6	7	7	8	8	8	9	9	10
y :	5	5	4	5	4	3	4	3	3

Sol. Let the straight line to be fitted to the data be $y = a + bx$, then the normal equations are

$$\Sigma y = na + b\Sigma x, \qquad \Sigma xy = a\Sigma x + b\Sigma x^2.$$

x	y	xy	x^2
6	5	30	36
7	5	35	49
7	4	28	49
8	5	40	64
8	4	32	64
8	3	24	64
9	4	36	81
9	3	27	81
10	3	30	100
Total 72	36	282	588

Here $n = 9$, $\Sigma x = 72$,

$\Sigma y = 36$, $\Sigma xy = 282$

$\Sigma x^2 = 588$.

On substituting these values, the normal equations become

$$36 = 9a + 72b$$

or $$4 = a + 8b \quad ...(i)$$

and $$282 = 72a + 588b$$

or $$47 = 12a + 98b \quad ...(ii)$$

Multiplying (*i*) by 12 and subtracting from (*ii*), we have

$$-1 = 2b$$

$\therefore$ $$b = -\frac{1}{2} = -.5$$

From (*i*), $$a = 4 - 8b = 4 + 4 = 8.$$

Hence the equation of required line is

$$y = 8 - .5x.$$

Example 3. *Fit a straight line to the following data :*

x :	*1*	*2*	*3*	*4*	*6*	*8*
y :	*2.4*	*3*	*3.6*	*4*	*5*	*6*

(D.U. 1987)

Sol. Please try yourself. **[Ans.** $y = 1.976 + 0.506\,x$]

Example 4. *Fit a straight line to the following data taking y as the dependent variable :*

x :	*1*	*2*	*3*	*4*	*5*	*6*	*7*	*8*	*9*
y :	*9*	*8*	*10*	*12*	*11*	*13*	*14*	*16*	*15*

Sol. Mean of *x*-series = 5, Mean of *y*-series = 12

Let $X = x - 5$ and $Y = y - 12$

Let the line of best fit be

$$Y = a + bX$$

then normal equations are

$$\Sigma Y = na + b\Sigma X$$

$$\Sigma XY = a\Sigma X + b\Sigma X^2$$

x	y	$X = x - 5$	$Y = y - 12$	XY	X^2
1	9	−4	−3	12	16
2	8	−3	−4	12	9
3	10	−2	−2	4	4
4	12	−1	0	0	1
5	11	0	−1	0	0
6	13	1	1	1	1
7	14	2	2	4	4
8	16	3	4	12	9
9	15	4	3	12	16
		0	0	57	60

Here $n = 9$, $\Sigma X = 0 = \Sigma Y$

$\Sigma XY = 57$, $\Sigma X^2 = 60$

$\therefore$ The normal equations become

$$0 = 9a + 0 \quad i.e., \quad a = 0$$

and $$57 = 0 + 60b \quad i.e., \quad b = .95$$

$\therefore$ The equation of the line of best fit is

$$Y = 0 + .95\,X$$

or $$y - 12 = 95\,(x - 5)$$

or $$y = 7.25 + .95x.$$

Example 5. *Fit a straight line to the following data taking x as the dependent variable :*

x :	1	3	4	6	8	9	11	14
y :	1	2	4	4	5	7	8	9

Sol. Mean of x-series $= \dfrac{56}{8} = 7$,

Mean of y-series $= \dfrac{40}{8} = 5$

Let $X = x - 7$ and $Y = y - 5$

Let the line of best fit be

$$X = a + bY \quad \text{(Here } x \text{ is the dependent variable)}$$

The normal equations are

$$\Sigma X = na + b\Sigma Y$$

and $$\Sigma XY = a\Sigma Y + b\Sigma Y^2$$

x	y	$X = x - 7$	$Y = y - 5$	XY	Y^2
1	1	−6	−4	24	16
3	2	−4	−3	12	9
4	4	−3	−1	3	1
6	4	−1	−1	1	1
8	5	1	0	0	0
9	7	2	2	4	4
11	8	4	3	12	9
14	9	7	4	28	16
		0	0	84	56

Here $n = 8$, $\Sigma X = 0 = \Sigma Y$, $\Sigma XY = 84$, $\Sigma Y^2 = 56$

$\therefore$ The normal equations become

$$0 = 8a + 0 \quad i.e., \quad a = 0$$

and $$84 = 0 + 56b \quad i.e., \quad b = 1.75$$

$\therefore$ The equation of the line of best fit is

$$X = 0 + 1.75Y$$

or $$x - 7 = 1.75\,(y - 5)$$

or $$x = -1.75 + 1.75y.$$

Example 6. *Find the line of best fit to the following data :*

x :	0	5	10	15	20	25
y :	12	15	17	22	24	30

(M.D.U. 1982 ; D.U. 1990)

Sol. Please try yourself. **[Ans.** $y = 0.7x + 11.28$]

Example 7. *Show that the best-fitting linear function for the points $(x_1, y_1), (x_2, y_2), \ldots\ldots, (x_n, y_n)$ may be expressed in the form*

$$\begin{vmatrix} x & y & 1 \\ \Sigma x_i & \Sigma y_i & n \\ \Sigma x_i^2 & \Sigma x_i y_i & \Sigma x_i \end{vmatrix} = 0, \ (i = 1, 2, \ldots\ldots, n).$$

Show that the line passes through the mean point $(\overline{x}, \overline{y})$.

Sol. Let the best-fitting linear function be

$$y = a + bx \qquad \ldots(1)$$

Then the normal equations are

$$\Sigma y_i = na + b\Sigma x_i \qquad \ldots(2)$$

$$\Sigma x_i y_i = a\Sigma x_i + b\Sigma x_i^2 \qquad \ldots(3)$$

Equations (1), (2) and (3) may be written as

$$bx - y + a = 0$$

$$b\Sigma x_i - \Sigma y_i + na = 0$$

$$b\Sigma x_i^2 - \Sigma x_i y_i + a\Sigma x_i = 0$$

Eliminating a and b between these equations

$$\begin{vmatrix} x & y & 1 \\ \Sigma x_i & \Sigma y_i & n \\ \Sigma x_i^2 & \Sigma x_i y_i & \Sigma x_i \end{vmatrix} = 0 \qquad \ldots(4)$$

which is the required best-fitting linear function.

For the mean point $(\overline{x}, \overline{y})$,

$$\overline{x} = \frac{1}{n} \sum x_i,$$

$$\overline{y} = \frac{1}{n} \sum y_i.$$

The line (4) passes through the point

$$(\overline{x}, \overline{y}) = \left(\frac{1}{n} \sum x_i, \frac{1}{n} \sum y_i \right)$$

if
$$\begin{vmatrix} \frac{1}{n}\sum x_i & \frac{1}{n}\sum y_i & 1 \\ \Sigma x_i & \Sigma y_i & n \\ \Sigma x_i^2 & \Sigma x_i y_i & \Sigma x_i \end{vmatrix} = 0$$

or, if
$$\begin{vmatrix} \Sigma x_i & \Sigma y_i & n \\ \Sigma x_i & \Sigma y_i & n \\ \Sigma x_i^2 & \Sigma x_i y_i & \Sigma x_i \end{vmatrix} = 0$$

which is true because the first and the second rows are identical.

Hence the line (4) passes through the mean point $(\overline{x}, \overline{y})$.

Example 8. *Fit a second degree parabola to the following data :*

x :	*0*	*1*	*2*	*3*	*4*
y :	*1*	*5*	*10*	*22*	*38*

(D.U. 1991, 93)

Sol. Let the curve of best fit be

$$y = a + bx + cx^2,$$

then the normal equations are

$$\Sigma y = na + b\Sigma x + c\Sigma x^2$$
$$\Sigma xy = a\Sigma x + b\Sigma x^2 + c\Sigma x^3$$
$$\Sigma x^2 y = a\Sigma x^2 + b\Sigma x^3 + c\Sigma x^4$$

x	y	xy	x^2	x^2y	x^3	x^4
0	1	0	0	0	0	0
1	5	5	1	5	1	1
2	10	20	4	40	8	16
3	22	66	9	198	27	81
4	38	152	16	608	64	256
10	76	243	30	851	100	354

Here $n = 5$

The normal equations become

$$76 = 5a + 10b + 30c \quad \text{...}(i)$$
$$243 = 10a + 30b + 100c \quad \text{...}(ii)$$
$$851 = 30a + 100b + 354c \quad \text{...}(iii)$$

Operating (*ii*) – 2 (*i*)

$$91 = 10b + 40c \quad \text{...}(iv)$$

Operating (*iii*) – 3 (*ii*)

$$122 = 10b + 54c \quad \text{...}(v)$$

Operating (*v*) – (*iv*)

$$31 = 14c$$

$\therefore$ $c = 2.21$

From (*iv*)

$$10b = 91 - 40\,(2.21) = 91 - 88.4 = 2.6$$

$\therefore$ $b = 0.26$

From (*i*)

$$5a = 76 - 10b - 30c$$

$$= 76 - 2.6 - 66.3 = 7.1$$

$\therefore$ $a = 1.42$

Hence the equation of the curve is

$$y = 1.42 + 0.26x + 2.21x^2.$$

(Second Method by Shift of Origin)

Since $n = 5$ is odd and the values of x are equi-distant, take the origin for the x-series at the middle value 2 and for the y-series at 20 (arbitrarily).

Let $X = x - 2, \quad Y = y - 20$

so that the curve of best fit is

$$Y = a + bX + cX^2.$$

x	y	X	Y	XY	X^2	X^2Y	X^3	X^4
0	1	– 2	– 19	38	4	– 76	– 8	16
1	5	– 1	– 15	15	1	– 15	– 1	1
2	10	0	– 10	0	0	0	0	0
3	22	1	2	2	1	2	1	1
4	38	2	18	36	4	72	8	16
		0	– 24	91	10	– 17	0	34

The normal equations become

$$\Sigma Y = na + b\Sigma X + c\Sigma X^2$$

or $$-24 = 5a + 10c \qquad ...(i)$$

$$\Sigma XY = a\Sigma X + b\Sigma X^2 + c\Sigma X^3$$

or $$91 = 10b \qquad ...(ii)$$

$$\Sigma X^2Y = a\Sigma X^2 + b\Sigma X^3 + c\Sigma X^4$$

or $$-17 = 10a + 34c \qquad ...(iii)$$

From (*ii*) $b = 9.1$

Operating (*iii*) – 2 (*i*) $31 = 14c$

$\therefore$ $c = 2.21$

$\therefore$ From (*i*) $5a = -24 - 10c = -24 - 22.1$

$$= -46.1$$

$\therefore \quad a = -9.22$

$\therefore$ Equation of the parabola is

$$Y = -9.22 + 9.1\,X + 2.21\,X^2$$

or $$y - 20 = -9.22 + 9.1\,(x-2) + 2.21\,(x-2)^2$$

or $$y = 10.78 + 9.1x - 18.2 + 2.21x^2 - 8.84x + 8.84$$

or $$y = 1.42 + 0.26x + 2.21x^2.$$

Example 9. *Fit a parabolic curve to the following data :*

x :	*1.0*	*1.5*	*2.0*	*2.5*	*3.0*	*3.5*	*4.0*
y :	*1.1*	*1.3*	*1.6*	*2.0*	*2.7*	*3.4*	*4.1*

(K.U. 1981 S ; D.U. 1984)

Sol. Let $$X = \frac{x - 2.5}{.5} = 2x - 5$$

and $$Y = y.$$

Let the curve of best fit be

$$Y = a + bX + cX^2,$$

then the normal equations are

$$\Sigma Y = na + b\Sigma X + c\Sigma X^2$$
$$\Sigma XY = a\Sigma X + b\Sigma X^2 + c\Sigma X^3$$
$$\Sigma X^2Y = a\Sigma X^2 + b\Sigma X^3 + c\Sigma X^4$$

x	y	$X = 2x - 5$	$Y = y$	X^2	XY	X^2Y	X^3	X^4
1.0	1.1	– 3	1.1	9	– 3.3	9.9	– 27	81
1.5	1.3	– 2	1.3	4	– 2.6	5.2	– 8	16
2.0	1.6	– 1	1.6	1	– 1.6	1.6	– 1	1
2.5	2.0	0	2.0	0	0	0	0	0
3.0	2.7	1	2.7	1	2.7	2.7	1	1
3.5	3.4	2	3.4	4	6.8	13.6	8	16
4.0	4.1	3	4.1	9	12.3	36.9	27	81
Total		0	16.2	28	14.3	69.9	0	196

$$16.2 = 7a \qquad ...(i)$$
$$14.3 = 28b \qquad ...(ii)$$
$$69.9 = 28a + 196c \qquad ...(iii)$$

Solving (*i*), (*ii*) and (*iii*)

$$a = 2.07, \quad b = .511, \quad c = .061.$$

Hence the curve of best fit is

$$Y = 2.07 + .511\,X + .061\,X^2$$

or $$y = 2.07 + .511\,(2x - 5) + .061\,(2x - 5)^2$$

or $$y = 1.04 - .198x + .244x^2.$$

Example 10. *Fit a second degree parabola to the following data :*

x :	0	1	2	3	4
y :	1	1.8	1.3	2.5	6.3

(M.D.U. 1981, 82 S ; K.U. 1981 ; G.N.D.U. 1981 ; D.U. 1982, 88, 93)

Sol. Please try yourself. [**Ans.** $y = 1.42 - 1.07x + 0.55x^2$]

[**Hint.** $X = x - 2, Y = y$]

Example 11. *Fit a second degree parabola to the following data taking x as the independent variable.*

x :	1	2	3	4	5	6	7	8	9
y :	2	6	7	8	10	11	11	10	9

(M.D.U. 1981 S)

Sol. Let $X = x - 5,$

$Y = y - 7$

and let the curve of best fit be

$$Y = a + bX + cX^2$$

The normal equations are

$$\Sigma Y = na + b\Sigma X + c\Sigma X^2$$

$$\Sigma XY = a\Sigma X + b\Sigma X^2 + c\Sigma X^3$$

$$\Sigma X^2Y = a\Sigma X^2 + b\Sigma X^3 + c\Sigma X^4$$

x	y	X	Y	XY	X^2	X^3	X^4
1	2	−4	−5	20	16	−64	256
2	6	−3	−1	3	9	−27	81
3	7	−2	0	0	4	−8	16
4	8	−1	1	−1	1	−1	1
5	10	0	3	0	0	0	0
6	11	1	4	4	1	1	1
7	11	2	4	8	4	8	16
8	10	3	3	9	9	27	81
9	9	4	2	8	16	64	256
Total		0	11	51	60	0	708

∴ The normal equations become

$$11 = 9a + 60c$$

$$51 = 60b$$

$$-9 = 60a + 708c.$$

Solving these equations

$$a = 3, \quad b = 0.85, \quad c = -0.27.$$

Hence the curve of best fit is

$$Y = 3 + 0.85\,X - 0.27\,X^2$$

or
$$y - 7 = 3 + 0.85\,(x - 5) - 0.27\,(x - 5)^2$$
$$= 3 + 0.85x - 4.25 - 0.27x^2 + 2.7x - 6.75$$

or
$$y = -1 + 3.55x - 0.27x^2.$$

Example 12. *Fit a parabola of the type $y = a + bx + cx^2$ to the following data :*

x :	10	15	20	25	30	35	40
y :	11	13	16	20	27	34	41

(K.U. 1982)

Sol. Let
$$X = \frac{x - 25}{5}$$

and
$$Y = y - 27$$

and let the curve of best fit be

$$Y = a + bX + cX^2.$$

The normal equations are

$$\Sigma Y = na + b\Sigma X + c\Sigma X^2$$
$$\Sigma XY = a\Sigma X + b\Sigma X^2 + c\Sigma X^2$$
$$\Sigma X^2Y = a\Sigma X^2 + b\Sigma X^3 + c\Sigma X^4.$$

x	y	X	Y	XY	X^2	X^2Y	X^3	X^4
10	11	−3	−16	48	9	−144	−27	81
15	13	−2	−14	28	4	−56	−8	16
20	16	−1	−11	11	1	−11	−1	1
25	20	0	−7	0	0	0	0	0
30	27	1	0	0	1	0	1	1
35	34	2	7	14	4	28	8	16
40	41	3	14	42	9	126	27	81
Total		0	−27	143	28	−57	0	196

The normal equations become

$$-27 = 7a + 28c \qquad ...(i)$$
$$143 = 28b \qquad ...(ii)$$
$$-57 = 28a + 196c \qquad ...(iii)$$

From (*ii*)

$$b = 5.107$$

(*iii*) − 4 (*i*) gives $51 = 84c$

$$c = \frac{17}{28} = 0.607$$

From (*i*) $7a = -27 - 28(0.607) = -27 - 16.996$

$= -10.004$

$a = -1.429.$

Hence the curve of best fit is

$$Y = -1.429 + 5.107\,X + 0.607\,X^2$$

or $$y - 27 = -1.429 + 5.107\left(\frac{x-25}{5}\right) + 0.607\left(\frac{x-25}{5}\right)^2$$

$$= -1.429 + 1.0214x - 25.535 + 0.0243\,x^2 - 1.214x + 15.175$$

or $$y = 15.211 - 0.1926x + 0.0243x^2.$$

Example 13. *Fit a second degree parabola to the following data :*

x :	*1*	*2*	*3*	*4*	*5*
y :	*1090*	*1220*	*1390*	*1625*	*1915*

(K.U. 1982 S ; G.N.D.U. 1981 S)

Sol. Let $X = x - 3$ and $Y = \dfrac{y - 1450}{5}$ and let the parabola of best fit be

$$Y = a + bX + cX^2$$

The normal equations are

$$\Sigma Y = 5a + b\Sigma X + c\Sigma X^2$$
$$\Sigma XY = a\Sigma X + b\Sigma X^2 + c\Sigma X^3$$
$$\Sigma X^2Y = a\Sigma X^2 + b\Sigma X^3 + c\Sigma X^4.$$

x	*y*	*X*	*Y*	*XY*	X^2	X^2Y	X^3	X^4
1	1090	− 2	− 72	144	4	− 288	− 8	16
2	1220	− 1	− 46	46	1	− 46	− 1	1
3	1390	0	− 12	0	0	0	0	0
4	1625	1	35	35	1	35	1	1
5	1915	2	93	186	4	372	8	16
Total		0	− 2	411	10	73	0	34

∴ The normal equations become

$$-2 = 5a + 10c$$
$$411 = 10b$$
$$73 = 10a + 34c$$

Solving these equations $a = -11.4$, $b = 41.1$, $c = 5.5$. Hence the curve of best fit is

$$Y = -11.4 + 41.1X + 5.5X^2$$

or $$\frac{y-1450}{5} = -11.4 + 41.1(x-3) + 5.5(x-3)^2$$

or $$y = 1024 + 40.5x + 27.5x^2.$$

Example 14. *Fit a parabola $y = a + bx + cx^2$ for the following data :*

x :	1	2	3	4	5	6	7
y :	2.3	5.2	9.7	16.5	29.4	35.5	54.4

Sol. Let $X = x - 4$, $Y = y$

Let the parabola of best fit be $Y = a + bX + cX^2$, then normal equations are

$$\Sigma Y = na + b\Sigma X + c\Sigma X^2 \quad ...(1)$$

$$\Sigma XY = a\Sigma X + b\Sigma X^2 + c\Sigma X^3 \quad ...(2)$$

$$\Sigma X^2Y = a\Sigma X^2 + b\Sigma X^3 + c\Sigma X^4 \quad ...(3)$$

x	$y = Y$	$X = x - 4$	XY	X^2	X^2Y	X^3	X^4
1	2.3	−3	−6.9	9	20.7	−27	81
2	5.2	−2	−10.4	4	20.8	−8	16
3	9.7	−1	−9.7	1	9.7	−1	1
4	16.5	0	0	0	0	0	0
5	29.4	1	29.4	1	29.4	1	1
6	35.5	2	71.0	4	142.0	8	16
7	54.4	3	163.2	9	489.6	27	81
	153.0	0	236.6	28	712.2	0	196

From (2) $236.6 = 28b \Rightarrow b = 8.45$

From (1) $153 = 7a + 28c$ ...(4)

From (3) $712.2 = 28a + 196c$ ...(5)

Multiplying (4) by 7 $1071 = 49a + 196c$

Subtracting (5) from it $358.8 = 21a \Rightarrow a = 17.09$

Multiplying (4) by 4 and subtracting from (5)

$$.2 = 84c \Rightarrow c = .002$$

∴ Equation of parabola of best fit is

$$Y = 17.09 + 8.45X + .002X^2$$

or $$y = 17.09 + 8.45(x-4) + .002(x-4)^2$$
$$= 17.09 - 33.80 + .032 + 8.45x - .016x + .002x^2$$

or $$y = -16.678 + 8.434x + .002x^2.$$

Example 15. *Fit a parabola of second degree to the following data taking x as the independent variable*

x :	1	2	3	4	5
y :	1.8	5.1	9.0	14	19

(D.U. 1981)

Sol. Please try yourself. [**Ans.** $y = 32.13 + 5.18x + 3.81\,x^2$]

4.6. (*a*) **Fitting of a Power Curve $y = ax^b$** (*Raj. 1985*)

Let (x_i, y_i) $i = 1, 2, \ldots, n$ be the set of n points.

The curve to be fitted is $y = ax^b$ ...(1)

Taking log of both sides $\log y = \log a + b \log x$

or $$Y = A + bX$$

where $Y = \log y$, $A = \log a$ and $X = \log x$

This is a linear equation in X and Y.

Normal equations for estimating A and b are

$$\Sigma Y = nA + b\Sigma X$$

and $$\Sigma XY = A\Sigma X + b\Sigma X^2$$

Solving these equations for A and b, we can find a which is = antilog A

($\because$ $A = \log a$)

Substituting the values of a and b so obtained in (1), we get the curve of best fit to the given data in the desired form

(*b*) **Fitting of Exponential Curves** (*i*) $y = ab^x$ (*ii*) $y = ae^{bx}$

(*i*) The curve to be fitted is $y = ab^x$ ...(1)

Taking log of both sides $\log y = \log a + x \log b$

or $$Y = A + Bx$$

where $Y = \log y$, $A = \log a$ and $B = \log b$.

This is a linear equation in x and Y.

Normal equations for estimating A and B are

$$\Sigma Y = nA + B\Sigma x$$

and $$\Sigma xY = A\Sigma x + B\Sigma x^2$$

Solving these equations for A and B, we can find a = antilog A and b = antilog B.

Substituting the values of a and b so obtained in (1), we get the curve of best fit to the given data in the desired form.

(*ii*) The curve to be fitted is $y = ae^{bx}$ ...(1)

Taking log of both sides

$$\log y = \log a + bx \log e$$

$$\Rightarrow \quad \log y = \log a + (b \log e)\, x$$

$$\Rightarrow \quad Y = A + Bx$$

where $Y = \log y$, $A = \log a$ and $B = b \log e$

This is a linear equation in x and Y.

Normal equations for estimating A and B are

$$\Sigma Y = nA + B\Sigma x$$

and
$$\Sigma xY = A\Sigma x + B\Sigma x^2$$

Solving these equations for A and B, we can find a = antilog A and $b = \dfrac{B}{\log e}$.

Substituting the values of a and b so obtained in (1), we get the curve of best fit to the given data in the desired form.

Note. In numerical problems in Statistics, we deal with common logarithms *i.e.,* logarithms to the base 10. $\therefore$ $A = \log a \Rightarrow A = \log_{10} a$

☞**Caution.** In natural logarithms *i.e.,* logarithms to the base e, $\log e = \log_e e = 1$ but in common logarithms, $\log e = \log_{10} e \neq 1$ because $e \neq 10$ rather $e = 2.71828$.

Example 1. *Fit the curve* $y = ax^b$ *to the following data* :

x :	1	2	3	4	5	6
y :	2.98	4.26	5.21	6.10	6.80	7.50

Sol. The curve to be fitted is $y = ax^b$ ...(*i*)

Taking logarithms, $\log y = \log a + b \log x$

or $Y = A + bX$

where $Y = \log y$, $A = \log a$ and $X = \log x$

The normal equations for estimating A and b are

$$\Sigma Y = nA + b\Sigma X \quad \text{...(ii)}$$

and
$$\Sigma XY = A\Sigma X + b\Sigma X^2 \quad \text{...(iii)}$$

Here $n = 6$

x	$X (= \log x)$	y	$Y (= \log y)$	XY	X^2
1	0	2.98	0.4742	0	0
2	0.3010	4.26	0.6294	0.1894	0.0906
3	0.4771	5.21	0.7168	0.3420	0.2276
4	0.6021	6.10	0.7853	0.4728	0.3625
5	0.6990	6.80	0.8325	0.5819	0.4886
6	0.7782	7.50	0.8751	0.6810	0.6056
	2.8574		4.3133	2.2671	1.7749

$\therefore$ (*ii*) $\Rightarrow$ $4.3133 = 6A + 2.8574\,b$ ...(*iv*)

(*iii*) $\Rightarrow$ $2.2671 = 2.8574A + 1.7749b$ ...(*v*)

Solving (*iv*) and (*v*) $b = .5143$

$\therefore$ From (*iv*) $4.3133 = 6A + 2.8574\,(.5143)$

$\Rightarrow$ $6A = 4.3133 - 1.4696 = 2.8437$

$\Rightarrow$ $A = 0.4739$

$\Rightarrow$ $\log a = 0.4739$

$\therefore \quad a = \text{antilog } 0.4739 = 2.978$

Hence from (*i*), the equation of the curve is

$$y = 2.978\, x^{.5143}.$$

Example 2. *Fit an equation of the form $y = ab^x$ to the following data :*

x :	2	3	4	5	6
y :	144	172.8	207.4	248.8	298.6

Sol. The curve to be fitted is $\quad y = ab^x \qquad \ldots(i)$

Taking logarithms, $\quad \log y = \log a + x \log b$

or $\quad Y = A + Bx$

where $Y = \log y, \quad A = \log a \quad$ and $\quad B = \log b$

The normal equations for estimating A and B

$$\Sigma Y = nA + B\Sigma x \qquad \ldots(ii)$$

and
$$\Sigma xY = A\Sigma x + B\Sigma x^2 \qquad \ldots(iii)$$

Here $\quad n = 5$

x	y	$Y (= \log y)$	xY	x^2
2	144	2.1584	4.3168	4
3	172.8	2.2375	6.7125	9
4	207.4	2.3168	9.2672	16
5	248.8	2.3959	11.9795	25
6	298.6	2.4751	14.8506	36
20		11.5837	47.1266	90

(*ii*) $\Rightarrow \quad 11.5837 = 5A + 20B \qquad \ldots(iv)$

(*iii*) $\Rightarrow \quad 47.1266 = 20A + 90B \qquad \ldots(v)$

Solving (*iv*) and (*v*) $\quad A = 2.0001, B = 0.0792$

$\therefore \quad a = \text{antilog } A = 100, \quad b = \text{antilog } B = 1.2$

Hence from (*i*), the equation of the curve is

$$y = 100(1.2)^x.$$

Example 3. *Fit an exponential curve of the form $y = ab^x$ to the following data :*

x :	1	2	3	4	5	6	7	8
y :	1.0	1.2	1.8	2.5	3.6	4.7	6.6	9.1

Sol. Please try yourself. **[Ans.** $y = 0.68\,(1.38)^x$]

Example 4. *For the data given below, find the equation to the best fitting exponential curve of the form $y = ae^{bx}$.*

x :	1	2	3	4	5	6
y :	1.6	4.5	13.8	40.2	125	300

Sol. The curve to be fitted is $y = ae^{bx}$ ...(i)

Taking logarithms, $\log y = \log a + bx \log e$

or $Y = A + Bx$

where $Y = \log y$, $A = \log a$ and $B = b \log e$

The normal equations for estimating A and B are

$$\Sigma Y = nA + B\Sigma x \quad ...(ii)$$

and $$\Sigma xY = A\Sigma x + B\Sigma x^2 \quad ...(iii)$$

Here $n = 6$

x	y	$Y (= \log y)$	xY	x^2
1	1.6	0.2041	0.2041	1
2	4.5	0.6532	1.3064	4
3	13.8	1.1399	3.4197	9
4	40.2	1.6042	6.4168	16
5	125	2.0969	10.4845	25
6	300	2.4771	14.8626	36
21		8.1754	36.6941	91

(ii) $\Rightarrow$ $8.1754 = 6A + 21B$ ...(iv)

(iii) $\Rightarrow$ $36.6941 = 21A + 91B$ ...(v)

$2(v) - 7(iv)$ $\Rightarrow$ $16.1604 = 35B$ $\Rightarrow$ $B = 0.4617$

$3(v) - 13(iv)$ $\Rightarrow$ $3.8021 = -15A$ $\Rightarrow$ $A = -0.2535$

$\therefore$ $A = -1 + 1 - 0.2535 = \bar{1}.7465$

$$a = \text{antilog } \bar{1}.7465 = .5578$$

$$b = \frac{B}{\log e} = \frac{0.4617}{\log 2.718} = \frac{0.4617}{0.4343} = 1.06$$

Hence from (i), the equation of the curve is

$$y = (.5578)e^{1.06x}.$$

4.7. MOST PLAUSIBLE SOLUTION OF A SYSTEM OF INDEPENDENT LINEAR EQUATIONS

Suppose we have m independent linear equations in n variables x, y, z......

$$a_1x + b_1y + c_1z + ... = A_1$$
$$a_2x + b_2y + c_2z + ... = A_2$$
$$..................................$$
$$..................................$$
$$a_mx + b_my + c_mz + ... = A_m$$

where a's, b's, c's, ... and A's are constants.

If $m = n$ *i.e.*, if the number of equations is equal to the number of variables, we can find a set of values of x, y, z,....., to satisfy them.

However, if $m > n$, *i.e.,* if the number of equations is greater than the number of variables, there may exist no such solution. In this case, we try to find out those values of $x, y, z,\ldots$ which satisfy the given system of equations as nearly as possible.

Let E_i be the residual or the error of *i*th equation, then

$$E_i = a_i x + b_i y + c_i z + \ldots\ldots - A_i$$

The principle of least squares asserts that the values of $x, y, z, \ldots\ldots$ which satisfy the given system of equations as nearly as possible are those which make

$$E = \sum_{i=1}^{m} E_i^2 = \sum_{i=1}^{m} (a_i x + b_i y + c_i z + \ldots\ldots - A_i)^2 \text{ minimum.}$$

$$\Rightarrow \quad \frac{\partial E}{\partial x} = \frac{\partial E}{\partial y} = \frac{\partial E}{\partial z} = \ldots\ldots = 0$$

Thus

$$\left.\begin{aligned} \frac{\partial E}{\partial x} &= \sum_{i=1}^{m} a_i (a_i x + b_i y + c_i z + \ldots\ldots - A_i) = 0 \\ \frac{\partial E}{\partial y} &= \sum_{i=1}^{m} b_i (a_i x + b_i y + c_i z + \ldots\ldots - A_i) = 0 \\ \frac{\partial E}{\partial z} &= \sum_{i=1}^{m} c_i (a_i x + b_i y + c_i z + \ldots\ldots - A_i) = 0 \\ &\ldots\ldots\ldots\ldots\ldots\ldots\ldots\ldots\ldots\ldots\ldots\ldots \\ &\ldots\ldots\ldots\ldots\ldots\ldots\ldots\ldots\ldots\ldots\ldots\ldots \end{aligned}\right\}$$

These are known as normal equations for $x, y, z,\ldots$ respectively. Thus we have n equations in n variables and their unique solution gives *the best or the most plausible* solution of the system.

Note. The normal equation for any variable is obtained by multiplying L.H.S. of each equation (R.H.S. being zero) by the co-efficient of that variable in that equation and then adding up all the resulting equations.

Example 1. *Find the most plausible values of x and y from the four equations :*

$$x + y = 3, \qquad x - y = 2, \qquad x + 2y - 4 = 0, \qquad x = 2y + 1.$$

Sol. Let $E = (x + y - 3)^2 + (x - y - 2)^2 + (x + 2y - 4)^2 + (x - 2y - 1)^2$

The most plausible values are given by

$$\frac{\partial E}{\partial x} = \frac{\partial E}{\partial y} = 0$$

$$\Rightarrow \quad (x + y - 3) + (x - y - 2) + (x + 2y - 4) + (x - 2y - 1) = 0$$

and $\quad (x + y - 3) - (x - y - 2) + 2(x + 2y - 4) - 2\,(x - 2y - 1) = 0$

or $\quad 4x - 10 = 0$

and $\quad 10y - 7 = 0$

$$\therefore \qquad x = 2.5, \qquad y = 0.7.$$

Example 2. *Find the most plausible values of x and y from the equations :*

$$x + y = 301,\ 2x - y = 3,\ x + 3y = 702,\ 3x + y = 497.$$

Sol. Let

$$E = (x + y - 301)^2 + (2x - y - 3)^2 + (x + 3y - 702)^2 + (3x + y - 497)^2$$

The most plausible values are given by

$$\frac{\partial E}{\partial x} = \frac{\partial E}{\partial y} = 0$$

$$\Rightarrow (x + y - 301) + 2(2x - y - 3) + (x + 3y - 702) + 3(3x + y - 497) = 0$$

and $$(x + y - 301) - (2x - y - 3) + 3(x + 3y - 702) + (3x + y - 497) = 0$$

i.e., $$15x + 5y - 2500 = 0 \qquad ...(i)$$

and $$5x + 12y - 2901 = 0 \qquad ...(ii)$$

Operating $(i) - 3(ii)$ $\quad -31y + 6203 = 0 \quad \therefore \quad y = 200$

From (i) $\quad 15x = 2500 - 1000 = 1500 \quad \therefore \quad x = 100$

Hence $\quad x = 100, \quad y = 200.$

Example 3. *Find the most plausible values of x and y from the following equations :*

$$x - 5y + 4 = 0, \qquad 2x - 3y + 5 = 0$$
$$x + 2y - 3 = 0, \qquad 4x + 3y + 1 = 0.$$ (G.N.D.U. 1981 S)

Sol. Please try yourself. [**Ans.** $x = -0.8, y = 0.86$]

Example 4. *Find the most plausible values of x, y and z from the following equations :*

$$x + 2y + z = 1, \qquad 2x + y + z = 4$$
$$-x + y + 2z = 3, \qquad 4x + 2y - 5z = -7.$$

Sol. Let $E = (x + 2y + z - 1)^2 + (2x + y + z - 4)^2 + (-x + y + 2z - 3)^2 + (4x + 2y - 5z + 7)^2$

The most plausible values are given by

$$\frac{\partial E}{\partial x} = \frac{\partial E}{\partial y} = \frac{\partial E}{\partial z} = 0$$

$$\Rightarrow \begin{cases} (x + 2y + z - 1) + 2(2x + y + z - 4) - (-x + y + 2z - 3) + 4(4x + 2y - 5z + 7) = 0 \\ 2(x + 2y + z - 1) + (2x + y + z - 4) + (-x + y + 2z - 3) + 2(4x + 2y - 5z + 7) = 0 \\ (x + 2y + z - 1) + (2x + y + z - 4) + 2(-x + y + 2z - 3) - 5(4x + 2y - 5z + 7) = 0 \end{cases}$$

or $$\begin{cases} 22x + 11y - 19z + 22 = 0 & ...(i) \\ 11x + 10y - 5z + 5 = 0 & ...(ii) \\ -19x - 5y + 31z - 46 = 0 & ...(iii) \end{cases}$$

Operating (*i*) – 2 (*ii*) $\quad -9y - 9z + 12 = 0$

or $\quad 3y + 3z = 4 \quad$ ...(*iv*)

Operating 19 (*ii*) + 11 (*iii*) $\quad 135y + 246z = 411$

or $\quad 45y + 82z = 137 \quad$ ...(*v*)

Operating (*v*) – 15 (*iv*) $\quad 37z = 77 \Rightarrow z = 2.08$

From (*iv*) $\quad 3y = 4 - 3(2.08) = 4 - 6.24 = -2.24$

$\therefore \quad y = -0.75$

From (*ii*) $\quad 11x = -10y + 5z - 5 = 7.5 + 10.40 - 5 = 12.90$

$x = 1.17$

Hence $x = 1.17, \quad y = -0.75, \quad z = 2.08.$

Example 5. *Find the most plausible values of x, y and z from the following equations :*

$$x - y + 2z = 3, \qquad 3x + 2y - 5z = 5.$$
$$4x + y + 4z = 21, \qquad -x + 3y + 3z = 14.$$

Sol. Please try yourself. [**Ans.** $x = 2.47, y = 3.55, z = 1.92$]

5
Theory of Probability

5.1. DEFINITIONS OF VARIOUS TERMS

Here we define and explain certain terms which are used frequently.

(*a*) **Die.** It is a small cube used in gambling. On its faces are marked.

(dots)

Plural of die is dice. The outcome of throwing a die is the number of dots on its upper face.

(*b*) **Cards.**(ताश)A pack of cards consists of four suits (रंग) called Spades (हुक्म), Hearts (पान), Diamonds (ईंट) and Clubs (चिड़ी). Each suit consists of 13 cards, nine cards numbered 2 to 10, an Ace (इक्का) a King (बादशाह), a Queen (बेगम) and a Jack or Knave (गुलाम). Spades and Clubs are black faced cards while Hearts and Diamonds are red faced cards. The Aces, Kings, Queens and Jacks are called face cards.

(*c*) **Trial and Event.** Let an experiment (real or conceptual) be repeated under essentially the same conditions and let it result in any one of the several possible outcomes. Then, the experiment is called a *trial* and the possible outcomes are known as *events* or *cases*.

For example. (*i*) Tossing of a coin is a trial and the turning up of head or tail is an event.

(*ii*) Throwing a die is a trial and getting 1 or 2 or 3 or 4 or 5 or 6 is an event.

(*d*) **Exhaustive Events** (*K.U. 1983 S*)

The total number of all possible outcomes in any trial is known as *exhaustive events* or *exhaustive cases*.

For example. (*i*) In tossing a coin, there are two exhaustive cases, head and tail.

(*ii*) In throwing a die, there are 6 exhaustive cases, for any one of the six faces may turn up.

(*iii*) In throwing two dice, the exhaustive cases are $6 \times 6 = 6^2$ for any of the six numbers from 1 to 6 on one die can be associated with any of the six numbers on the other die.

In general, in throwing *n* dice, the exhaustive cases are 6^n.

(*e*) **Favourable Events or Cases.** The cases which entail the happening of an event are said to be *favourable* to the event. It is the total number of possible outcomes in which the specified event happens.

For example. (*i*) In throwing a die, the number of cases favourable to the appearance of a multiple of 3 are two, *viz.*, 3 and 6 while the number of cases favourable to the appearance of an even number are three, *viz.*, 2, 4 and 6.

(*ii*) In a throw of two dice, the number of cases favourable to getting a sum 6 is 5, *viz.*, (1, 5) ; (5, 1) ; (2, 4) ; (4, 2) ; (3, 3).

(*f*) **Mutually Exclusive Events** (*M.D.U. 1983*)

Events are said to be *mutually exclusive* or *incompatible* if the happening of any one of them precludes (*i.e.*, rules out) the happening of all others, *i.e.*, if no two or more than two of them can happen simultaneously in the same trial.

For example (*i*) In tossing a coin, the events head and tail are mutually exclusive, since if the outcome is head, the possibility of getting tail in the same trial is ruled out.

(*ii*) In throwing a die, all the six faces numbered 1, 2, 3, 4, 5, 6 are mutually exclusive since any outcome rules out the possibility of getting any other.

(*g*) **Equally Likely Events** (*K.U. 1983 S*)

Events are said to be *equally likely* if there is no reason to expect any one in preference to any other.

For example (*i*) When a card is drawn from a well shuffled pack, any card may appear in the draw so that the 52 different cases are equally likely.

(*ii*) In throwing a die, all the six faces are equally likely to come.

(*h*) **Independent and Dependent Events**

(*M.D.U. 1981 S ; K.U. 1983 S*)

Two or more events are said to be *independent* if the happening or non-happening of any one does not depend (or is not affected) by the happening or non-happening of any other. Otherwise they are said to be *dependent*.

For example. If a card is drawn from a pack of well shuffled cards and replaced before drawing the second card, the result of the second draw is independent of the first draw. However, if the first card drawn is not replaced, then, the second draw is dependent on the first draw.

5.2. MATHEMATICAL (OR CLASSICAL OR A PRIORI) DEFINITION OF PROBABILITY

If a trial results in *n exhaustive, mutually exclusive* and *equally likely cases* and *m* of them are favourable to the happening of an event E, then the probability of happening of E is given by

$$p \text{ or } P(E) = \frac{\text{Favourable number of cases}}{\text{Exhaustive number of cases}} = \frac{m}{n}$$

Note 1. Since the number of cases favourable to happening of E is m and the exhaustive number of cases is n, therefore, the number of cases unfavourable to happening of E are $n - m$.

Note 2. The probability that the event E will not happen is given by

$$q \text{ or } P(\bar{E}) = \frac{\text{Unfavourable number of cases}}{\text{Exhaustive number of cases}} = \frac{n-m}{n}$$

$$= 1 - \frac{m}{n} = 1 - p$$

$$p + q = 1 \text{ i.e., } P(E) + P(\bar{E}) = 1$$

Obviously, p and q are non-negative and cannot exceed unity,

i.e., $0 \le p \le 1, \qquad 0 \le q \le 1.$

Note 3. If $P(E) = 1$, E is called a *certain event i.e.*, the chance of its happening is cent per cent.

If $P(E) = 0$, then $P(\bar{E}) = 1$.

⇒ The chance of non-happening of E is cent per cent.

⇒ E is an *impossible event*.

Note 4. If n cases are favourable to E and m cases are favourable to $\bar{E}$ (*i.e.*, unfavourable to E), then exhaustive number of cases $= n + m$

$$P(E) = \frac{n}{n+m} \quad \text{and} \quad P(\bar{E}) = \frac{m}{n+m}$$

We say that "odds in *favour of E*" are $n : m$ and "odds *against E*" are $m : n$.

5.3. STATISTICAL (OR EMPIRICAL) DEFINITION OF PROBABILITY

If in n trials, an event E happens m times, then the probability of happening of E is given by

$$p = P(E) = \operatorname*{Lt}_{n \to \infty} \frac{m}{n}.$$

5.4. FACTORIAL NOTATION, PERMUTATIONS AND COMBINATIONS (RECAPITULATION)

(*i*) The continued product of the first n natural numbers is denoted by the symbol $\underline{|n}$ or $\underline{|n}$ and is read as "*factorial n*".

Thus $\underline{|n}$ or $n! = 1 . 2 . 3 \ldots\ldots (n-1)\, n$

$$= n\,(n-1)\,(n-2) \ldots\ldots 3 . 2 . 1$$

$$6! = 6.5.4.3.2.1 = 720.$$

Also $\quad 6! = 6(5!) = 6.5.\ (4!)$ etc.

$$n! = n\,.\,(n-1)! = n\,(n-1)\,.\,(n-2)! \text{ etc.}$$

We define $\quad 0! = 1.$

(*ii*) **Principle of Association of Operations.** If one operation can be performed in m ways and if corresponding to each of these m ways of performing this operation, there are n ways of performing a second (independent) operation, then the number of ways of performing the two operations together is $m \times n$.

(*iii*) **Permutations.** Each of the arrangements that can be made by taking some or all of a number of things at a time is called a *permutation.*

(*a*) The number of permutations of n different things taken r at a time is denoted by ^nP_r or $P(n, r)$ and

$$^nP_r = \begin{cases} n(n-1)(n-2) \ldots (n-r+1) & \text{[Product Form]} \\ \dfrac{n!}{(n-r)!} & \text{[Factorial Form]} \end{cases}$$

Thus $^7P_3 = 7.6.5$ (product of three factors)

$^6P_6 = 6.5.4.3.2.1$ (product of six factors)

$= 6!$

In general $^nP_n = n!$

(*b*) The number of permutations of n different things taken r at a time in which 3 particular things are always

Included is $^rP_3 \times {}^{n-3}P_{r-3}$

Excluded is $^{n-3}P_r$.

(*c*) The number of permutations of n things taken all at a time when p of them are alike of one kind and q of them are alike of another kind, all other things being different, is

$$= \frac{n!}{p!\,q!}$$

For example, the word 'CHANDIGARH' contains 10 letters. H occurs twice, A occurs twice and the rest are all different.

∴ The number of arrangements of the letters of the word 'CHANDIGARH'

$$= \frac{10!}{2!\,2!} = 907200.$$

(*d*) The number of permutations of n different things taken r at a time when each thing may be repeated any number of times in any arrangement is n^r.

(*e*) The number of ways in which n different things can be arranged round a circle is $(n-1)!$

(*iv*) **Combinations.** Each of the groups or selections that can be made by taking some or all of a number of things at a time is called a *combination.*

(*a*) The number of combinations of n different things taken r at a time is denoted by ^nC_r or $C(n, r)$ or $\dbinom{n}{r}$ and

$$^nC_r = \frac{^nP_r}{r!} = \frac{n!}{(n-r)!\,r!}.$$

Thus $\quad {}^8C_3 = \frac{{}^8P_3}{3\,!} = \frac{8.7.6}{3.2.1} = 56$

(*b*) **Complementary Combination.** $\quad {}^nC_r = {}^nC_{n-r}$

This result is especially useful when $\quad r > \frac{n}{2}$

For example, $\quad {}^{16}C_{13} = {}^{16}C_{16-13} = {}^{16}C_3 = \frac{16 \times 15 \times 14}{3 \times 2 \times 1} = 560.$

ILLUSTRATIVE EXAMPLES

Example 1. *In a class of 10 students, 4 are boys and the rest are girls. Find the probability that a students selected will be a girl.*

Sol. Number of students = 10, Number of boys = 4

⇒ Number of girls = 6

The probability that a girl is selected

$$= \frac{\text{Favourable number of cases}}{\text{Exhaustive number of cases}} = \frac{6}{10} = \frac{3}{5}.$$

Example 2. *What is the chance of throwing 4 with an ordinary die ?*

Sol. Number of favourable cases = 1

Exhaustive number of cases = 6

$$\text{Probability} = \tfrac{1}{6}.$$

Example 3. *Find the chance that if a card is drawn at random from an ordinary pack, it is one of the court cards.* (G.N.D.U. 1982)

Sol. There are 4 × 3 = 12 court cards (kings, queens, jacks) in a pack (of 52 cards).

Number of favourable cases = 12

Exhaustive number of cases = 52

$$\therefore \quad \text{Probability} = \frac{12}{52} = \frac{3}{13}.$$

Example 4. *A bag contains 7 white and 9 red balls. Find the probability of drawing a white ball.*

Sol. One white ball can be drawn out of 7 in ${}^7C_1 = 7$ ways.

∴ Number of favourable cases = 7

∴ Number of exhaustive cases = 7 + 9 = 16

$$\text{Probability} = \frac{7}{16}.$$

Example 5. *What is the probability of throwing a number greater than 2 with an ordinary die ?*

Sol. Number of exhaustive cases = 6

Number of favourable cases = 4

[$\because$ a die can fall in 6 ways and of these (3, 4, 5, 6) are favourable to the event.]

$$\text{Probability} = \frac{4}{6} = \frac{2}{3}.$$

Example 6. *What is the chance that a non-leap year should have fifty three Sundays.* (D.U. 1984)

Sol. A non-leap year consists of 365 days. Therefore, in a non-leap year, there are 52 complete weeks and *1 day over* which can be any one of the seven days of a week but there is only one favourable case that it is Sunday.

Number of favourable cases = 1

Exhaustive number of cases = 7

$\therefore$ Probability = $\frac{1}{7}$.

Example 7. *What is the chance that a leap year, selected at random, will contain 53 Sundays ?* (D.U. 1984 ; Kanpur 1985)

Sol. A leap year consists of 366 days. Therefore, in a leap year, there are 52 complete weeks and *2 days over*. The following are the likely cases for these two over (successive) days :

(*i*) Monday and Tuesday (*ii*) Tuesday and Wednesday

(*iii*) Wednesday and Thursday (*iv*) Thursday and Friday

(*v*) Friday and Saturday (*vi*) Saturday and Sunday

(*vii*) Sunday and Monday.

Of these seven likely cases, only the last two are favourable.

$\therefore$ Probability = $\frac{2}{7}$.

Example 8. *A card is drawn from an ordinary pack and a gambler bets that it is a spade or an ace. What are the odds against his winning this bet ?*

Sol. From a pack of 52 cards, one card can be drawn in 52 ways.

$\therefore$ Exhaustive number of cases = 52.

Numbers of ways in which a card can be a spade = 13

Number of ways in which a card can be an ace = 3

($\because$ One ace is in spades)

$\therefore$ Number of favourable cases = Number of ways in which a card can be spade or an ace = 13 + 3 = 16.

$$\text{The probability of winning the bet} = \frac{16}{52} = \frac{4}{13}$$

$$\therefore \text{ The probability of losing the bet} = 1 - \frac{4}{13} = \frac{9}{13}$$

$$\text{Hence the odds against winning} = \frac{9}{13} : \frac{4}{13} = 9 : 4.$$

Example 9. *Prove that the probability of obtaining a total of 9 in a single throw with two dice is one by nine.*

Sol. Exhaustive number of cases = $6^2 = 36$

Favourable cases are (3, 6) ; (6, 3) ; (4, 5) ; (5, 4).

Their number = 4

$\therefore$ Probability $= \frac{4}{36} = \frac{1}{9}$.

Example 10. *Prove that in a single throw with a pair of dice, the probability of getting the sum of 7 is equal to $\frac{1}{6}$ and the probability of getting the sum of 10 is equal to $\frac{1}{12}$.*

Sol. Please try yourself.

Example 11. *A bag contains 7 white, 6 red and 5 black balls. Two balls are drawn at random. Find the probability that they will both be white.*

Sol. Total number of balls = 7 + 6 + 5 = 18.

Out of 18 balls, 2 can be drawn in $^{18}C_2$ ways.

$\therefore$ Exhaustive number of cases $= {}^{18}C_2 = \frac{18 \times 17}{2 \times 1} = 153$

Out of 7 white balls, 2 can be drawn in $^7C_2$

$= \frac{7 \times 6}{2 \times 1} = 21$ ways.

$\therefore$ Favourable number of cases = 21

Probability $= \frac{21}{153} = \frac{7}{51}$.

Example 12. *A bag contains 6 white, 7 red and 5 black balls. Find the chance that three balls drawn at random are all white.*

Sol. Proceeding as in example 11.

Exhaustive number of cases $= {}^{18}C_3 = \frac{18 \times 17 \times 16}{3 \times 2 \times 1} = 816$

Favourable number of cases $= {}^6C_3 = \frac{6 \times 5 \times 4}{3 \times 2 \times 1} = 20$

$\therefore$ Reqd. probability $= \frac{20}{816} = \frac{5}{204}$.

Example 13. *A bag contains 3 red, 6 white and 7 black balls. What is the probability that two balls drawn are white and black ?*

Sol. Total number of balls = 3 + 6 + 7 = 16.

Out of 16 balls, 2 can be drawn in $^{16}C_2$ ways.

Exhaustive number of cases $= {}^{16}C_2 = \frac{16 \times 15}{2 \times 1} = 120$.

Out of 6 white balls, 1 ball can be drawn in $^6C_1 = 6$ ways and out of 7 black balls, 1 ball can be drawn in $^7C_1 = 7$ ways. Since each of the former cases can be associated with each of the latter cases, the total number of favourable cases $= 6 \times 7 = 42$.

$\therefore$ Required probability $= \dfrac{42}{120} = \dfrac{7}{20}$.

Example 14. *A bag contains 10 white, 6 red, 4 black and 7 blue balls. 5 balls are drawn at random. What is the probability that 2 of them are red and one black ?*

Sol. Total number of balls = 10 + 6 + 4 + 7 = 27.

Out of 27 balls, 5 can be drawn in $^{27}C_5$ ways.

$\therefore$ Exhaustive number of cases

$$= {}^{27}C_5 = \frac{27 \times 26 \times 25 \times 24 \times 23}{5 \times 4 \times 3 \times 2 \times 1} = 80730.$$

Out of 6 red balls, 2 can be drawn in $^6C_2$ ways and out of 4 black balls, 1 can be drawn in $^4C_1$ ways. Since each of the former cases can be associated with each of the latter cases, the total number of favourable cases $= {}^6C_2 \times {}^4C_1 = 15 \times 4 = 60$.

$\therefore$ Required probability $= \dfrac{60}{80730} = \dfrac{2}{2691}$.

Example 15. *A bag contains 4 white, 5 red and 6 green balls. Three balls are drawn at random. What is the probability that a white, a red and a green balls are drawn ?*

Sol. Please try yourself.

Exhaustive number of cases = $^{15}C_3 = 455$

Favourable cases = $^4C_1 \times {}^5C_1 \times {}^6C_1 = 120$

Reqd. probability $= \dfrac{120}{455} = \dfrac{24}{91}$.

Example 16. *From a set of 17 cards, numbered 1, 2, 3, 4,....., 16, 17, one is drawn at random. Show that the chance that its number is divisible by 3 or 7 is* $\frac{7}{17}$.

Sol. Exhaustive number of cases = 17

Numbers divisible by 3 are 3, 6, 9, 12, 15 and those divisible by 7 are 7, 14.

$\therefore$ Favourable number of cases = 5 + 2 = 7

Required probability $= \dfrac{7}{17}$.

Example 17. *An integer is chosen at random from the first two hundred digits. What is the probability that the integer chosen is divisible by 6 or 8 ?*

Sol. One integer can be chosen out of 200 in 200 ways.

$\therefore$ Exhaustive number of cases = 200

Integers from 1 to 200 divisible by 6 are 6, 12,....., 198.

Their number is $\frac{198}{6} = 33$

Integers from 1 to 200 divisible by 8 are 8, 16,...., 200.

Their number is $\frac{200}{8} = 25$

Integers from 1 to 200 divisible by both 6 and 8 (*i.e.*, divisible by their L.C.M. 24) are 24, 48, ..., 192.

Their number $= \frac{192}{24} = 8.$

These 8 numbers have been counted twice, in multiples of 6 and also in multiples of 8.

$\therefore$ Favourable number of cases = 33 + 25 – 8 = 50.

Required probability $= \frac{50}{200} = \frac{1}{4}.$

Example 18. *The chance of one event happening is the square of the chance of a second event, but the odds against the first are the cube of the odds against the second. Find the chance of each.*

Sol. Let the two events be E_1 and E_2. Let the chances of their happening be p_1 and p_2 respectively. Then

$$p_1 = p_2^2 \qquad ...(1) \qquad |\,given$$

The chances of not happening of the events are $1 - p_1$ and $1 - p_2$ respectively.

Odds against the first event are $1 - p_1 : p_1 = \frac{1 - p_1}{p_1}$

" " " second " " $1 - p_2 : p_2 = \frac{1 - p_2}{p_2}$

By the given condition $\frac{1 - p_1}{p_1} = \left(\frac{1 - p_2}{p_2}\right)^3$

or $$\frac{1 - p_2^2}{p_2^2} = \frac{(1 - p_2)^3}{p_2^3} \qquad |\,using\ (1)$$

or $$1 + p_2 = \frac{(1 - p_2)^2}{p_2} \qquad |\because\ p_2 \neq 1$$

or $$p_2 + p_2^2 = 1 - 2p_2 + p_2^2$$

or $$3p_2 = 1 \qquad \therefore\ p_2 = \tfrac{1}{3} \text{ and } p_1 = \tfrac{1}{9}$$

Hence the chances of happening of E_1 and E_2 are $\frac{1}{9}$ and $\frac{1}{3}$ respectively.

Example 19. *Two cards are drawn at random from a well-shuffled pack of 52 cards. Show that the chance of drawing two aces is 1/221.*

Sol. From a pack of 52 cards, 2 cards can be drawn in $^{52}C_2$ ways.

$\therefore$ Exhaustive number of cases = $^{52}C_2$

2 aces can be drawn out of 4 in $^4C_2$ ways.

$\therefore$ Required probability $= \frac{^4C_2}{^{52}C_2} = \frac{4 \times 3}{2 \times 1} \times \frac{2 \times 1}{52 \times 51} = \frac{1}{221}$.

Example 20. *From a pack of 52 cards, three are drawn at random. Find the chance that they are a king, a queen and a knave.*

Sol. From a pack of 52 cards, 3 cards can be drawn in

$$^{52}C_3 = \frac{52 \times 51 \times 50}{3 \times 2 \times 1} = 22100 \text{ ways.}$$

A king can be drawn out of 4 kings in $^4C_1$ ways.

A queen can be drawn out of 4 queens in $^4C_1$ ways.

A knave can be drawn out of 4 knaves in $^4C_1$ ways.

Since each way of drawing a king can be associated with each of the ways of drawing a queen and a knave.

$\therefore$ Favourable number of cases = $^4C_1 \times {}^4C_1 \times {}^4C_1 = 64$.

Required probability $= \frac{64}{22100} = \frac{16}{5525}$

Example 21. *Four cards are drawn from a pack of cards. Find the probability that (i) all are diamonds, (ii) there is one card of each suit, and (iii) there are two spades and two hearts.*

Sol. 4 cards can be drawn from a pack of 52 cards in $^{52}C_4$ ways.

$\therefore$ Exhaustive number of cases

$$= {}^{52}C_4 = \frac{52 \times 51 \times 50 \times 49}{4 \times 3 \times 2 \times 1} = 270725.$$

(*i*) There are 13 diamonds in the pack and 4 can be drawn out of them in $^{13}C_4$ ways.

$\therefore$ Favourable number of cases

$$= {}^{13}C_4 = \frac{13 \times 12 \times 11 \times 10}{4 \times 3 \times 2 \times 1} = 715$$

Required probability $= \frac{715}{270725} = \frac{143}{54145} = \frac{11}{4165}$.

(*ii*) There are 4 suits, each containing 13 cards.

$\therefore$ Favourable number of cases

$$= {}^{13}C_1 \times {}^{13}C_1 \times {}^{13}C_1 \times {}^{13}C_1 = 13 \times 13 \times 13 \times 13.$$

Required probability $= \frac{13 \times 13 \times 13 \times 13}{270725} = \frac{2197}{20825}$.

(*iii*) 2 spades out of 13 can be drawn in $^{13}C_2$ ways.

2 hearts out of 13 can be drawn in $^{13}C_2$ ways.

$\therefore$ Favourable number of cases $= {}^{13}C_2 \times {}^{13}C_2 = 78 \times 78$

Required probability $= \dfrac{78 \times 78}{270725} = \dfrac{468}{20825}$.

Example 22. *From a pack of 52 cards, two are drawn at random. Find the chance that one is a king and the other is a queen.*

Sol. Please try yourself. $\left[\textbf{Ans}.\ \dfrac{8}{663}\right]$

Example 23. *If three cards are drawn from a pack of 52, what is the probability that all the three will be kings ?*

Sol. Exhaustive number of cases $= {}^{52}C_3$

Favourable number of cases $= {}^{4}C_3 = {}^{4}C_1 = 4$

$\therefore$ Required probability $= \dfrac{4}{{}^{52}C_3} = \dfrac{4 \times 3 \times 2 \times 1}{52 \times 51 \times 50} = \dfrac{1}{5525}$.

Example 24. *Six cards are drawn at random from a pack of 52 cards. What is the probability that 3 will be red and 3 black ?*

Sol. 6 cards can be drawn from a pack of 52 cards in ${}^{52}C_6$ ways.

Diamonds and Hearts are red. Total number of red cards = 26 and 3 cards can be drawn out of them in ${}^{26}C_3$ ways.

Spades and Clubs are black. $\therefore$ Total number of black cards = 26 and 3 cards can be drawn out of them in ${}^{26}C_3$ ways.

Favourable number of cases $= {}^{26}C_3 \times {}^{26}C_3$

Required probability $= \dfrac{{}^{26}C_3 \times {}^{26}C_3}{{}^{52}C_6} = \dfrac{13000}{39151}$.

Example 25. *In a hand at whist, what is the chance that the four kings are held by a specified player ?*

Sol. [*Game of whist* (ताश का खेल). It is a game of cards played by two pairs of players with a pack of 52 cards. The cards are dealt out equally].

In a hand at whist, the specified player has 13 cards including 4 kings. Four kings can be drawn out of 4 in ${}^{4}C_4$ ways. The remaining 9 cards held by him can be drawn out of the remaining 48 cards in the pack in ${}^{48}C_9$ ways. The total number of ways in which 4 kings and 9 other cards can be drawn is ${}^{4}C_4 \times {}^{48}C_9$.

Favourable number of cases $= {}^{4}C_4 \times {}^{48}C_9$.

Also, exhaustive number of cases $= {}^{52}C_{13}$.

Required probability $= \dfrac{{}^{4}C_4 \times {}^{48}C_9}{{}^{52}C_{13}} = \dfrac{11}{4165}$.

Example 26. *A number is chosen from each of two sets :*

{1, 2, 3 ,4, 5, 6, 7, 8, 9} ; {1, 2, 3, 4, 5, 6, 7, 8, 9}.

If p_1 denotes the probability that the sum of the two numbers be 10 and p_2 the probability that their sum be 8 ; find $p_1 + p_2$.

Sol. In each set, there are 9 numbers. The total number of ways of chosing one from each = ${}^9C_1 \times {}^9C_1 = 81$.

∴ Exhaustive number of cases = 81.

A sum of 10 can be obtained in the following ways :

(1, 9) ; (9, 1) ; (2, 8) ; (8, 2) ; (3, 7) ; (7, 3) ; (4, 6) ; (6, 4) ; (5, 5) which are 9 in all.

∴ Favourable number of cases = 9

$$p_1 = \frac{9}{81}$$

A sum of 8 can be obtained in the following ways :

(1, 7) ; (7, 1) ; (2, 6) ; (6, 2) ; (3, 5) ; (5, 3) ; (4, 4) which are 7 in all.

∴ Favourable number of cases = 7.

$$p_2 = \frac{7}{81}$$

Hence $$p_1 + p_2 = \frac{9}{81} + \frac{7}{81} = \frac{16}{81}.$$

Example 27. *From a set of raffle tickets numbered 1 to 100, three are drawn at random. What is the probability that all are odd numbered ?*

Sol. Three tickets can be drawn out of 100 in ${}^{100}C_3$ ways.

∴ Exhaustive number of cases = ${}^{100}C_3$.

Out of the 100 tickets, 50 are odd numbered (1, 3, 5,....., 99). Three tickets out of these 50 can be drawn in ${}^{50}C_3$ ways.

∴ Favourable number of cases = ${}^{50}C_3$

Required probability

$$= \frac{{}^{50}C_3}{{}^{100}C_3} = \frac{50 \times 49 \times 48}{3 \times 2 \times 1} \times \frac{3 \times 2 \times 1}{100 \times 99 \times 98} = \frac{4}{33}.$$

Example 28. *Two different digits are chosen at random from the set 1, 2, 3,...., 8. Show that the probability that the sum of the digits will be equal to 5 is the same as the probability that their sum will exceed 13, each being $\frac{1}{14}$. Also show that the chance of both digits exceeding 5 is $\frac{3}{28}$.*

Sol. Two different digits can be chosen out of 8 in ${}^8C_2 = 28$ ways.

(*i*) The sum of the two chosen digits can be equal to 5 in the following ways :

(1, 4) ; (2, 3) which are 2 in all.

Required probability $= \frac{2}{28} = \frac{1}{14}$.

(*ii*) The sum of the two chosen digits can exceed 13 in the following ways :

(6, 8) ; (7, 8) which are 2 in all.

Required probability $= \frac{2}{28} = \frac{1}{14}$.

(*iii*) Both the chosen digits can exceed 5 in the following ways : (6, 7) ; (6, 8) ; (7, 8) which are 3 in all.

Required probability $= \frac{3}{28}$.

Example 29. *If from a lottery of 30 tickets, marked 1, 2, 3,... 30, four tickets be drawn, what is the chance that those marked 1 and 2 are among them ?*

Sol. Four tickets can be drawn out of 30 in ${}^{30}C_4$ ways.

If the tickets marked 1 and 2 are to be included always, the remaining two can be drawn out of remaining 28 in ${}^{28}C_2$ ways.

Hence required probability $= \frac{{}^{28}C_2}{{}^{30}C_4} = \frac{2}{145}$.

Example 30. *A bag contains 50 tickets numbered 1, 2, 3,..., 50, of which five are drawn at random and arranged in ascending order of magnitude ($x_1 < x_2 < x_3 < x_4 < x_5$). What is the probability that $x_3 = 30$?*

Sol. Exhaustive number of cases $= {}^{50}C_5$.

If $x_3 = 30$, then the two tickets with numbers x_1 and x_2 must come out of 29 tickets numbered 1 to 29 and this can be done in ${}^{29}C_2$ ways. The other two tickets with numbers x_4 and x_5 must come out of the 20 tickets numbered 31 to 50 and this can be done in ${}^{20}C_2$ ways.

$\therefore$ Favourable number of cases $= {}^{29}C_2 \times {}^{20}C_2$.

Required probability $= \frac{{}^{29}C_2 \times {}^{20}C_2}{{}^{50}C_5} = \frac{551}{15134}$.

Example 31. *What is the probability of getting 9 cards of the same suit in one hand at a game of bridge ?*

Sol. One hand in a game of bridge consists of 13 cards.

$\therefore$ Exhaustive number of cases $= {}^{52}C_{13}$.

Number of ways in which, in one hand, a particular player gets 9 cards of one suit are ${}^{13}C_9$ and the number of ways in which the remaining 4 cards are of some other suit are ${}^{39}C_4$. Since there are 4 suits in a pack of cards, total number of favourable cases.

$$= 4 \times {}^{13}C_9 \times {}^{39}C_4$$

$\therefore$ Required probability $= \frac{4 \times {}^{13}C_9 \times {}^{39}C_4}{{}^{52}C_{13}}$

(There is no need to simplify).

Example 32. *A party of n persons sit at a round table. Find the odds against two specified individuals sitting next to each other.*

Sol. n persons can sit at a *round table in* $(n-1)!$ ways.

$\therefore$ Exhaustive number of cases $= (n-1)!$

If two persons always sit together, then $(n-2)+1 = n-1$ persons can sit in $(n-2)!$ ways. But these two persons can interchange their positions in $2!$ ways.

$\therefore$ Favourable number of cases of 2 persons sitting next to each other $= 2!\,(n-2)!$

$\therefore$ The probability of happening of the event

$$= p = \frac{2!\,(n-2)!}{(n-1)!} = \frac{2\,.\,(n-2)!}{(n-1)\,.\,(n-2)!} = \frac{2}{n-1}$$

The probability of non-happening of the event

$$= q = 1 - \frac{2}{n-1} = \frac{n-3}{n-1}$$

$\therefore$ The odds against its happening are $q : p = n-3 : 2 = \dfrac{n-3}{2}$.

Example 33. *A party of 23 persons take their seats at a round table. Show that the odds are 10 to 1 against two specified persons sitting together.*

Sol. Please try yourself.

Example 34. *There are n letters and n addressed envelopes. If the letters are placed in the envelopes at random, what is the probability that all the letters are not placed in the right envelopes ?*

Sol. The total number of ways of placing n letters in n envelopes $= n!$.

All the letters can be placed correctly in only one way.

$\therefore$ The probability that the letters are placed in right envelopes

$$= \frac{1}{n!}.$$

Hence the probability that the letters are not placed in right envelopes

$$= 1 - \frac{1}{n!}.$$

Example 35. *Eight letters, to each of which corresponds an envelope, are placed in the envelopes at random. What is the probability that all letters are not placed in the right envelopes ?*

Sol. Please try yourself. $\left[\textbf{Ans}.\ 1 - \dfrac{1}{8!}\right]$

Example 36. *Five cards are drawn from a pack of 52. What is the chance that these five will contain (a) just one ace, (b) at least one ace ?*

Sol. Exhaustive number of cases $= {}^{52}C_5$.

(*a*) One ace can be drawn out of 4 aces in $^4C_1$ ways.

The remaining 4 cards can be drawn out of the remaining 48 cards in $^{48}C_4$ ways.

∴ Favourable number of cases when the five cards drawn contain just one ace $^4C_1 \times {}^{48}C_4$.

Required probability $= \dfrac{^4C_1 \times {}^{48}C_4}{^{52}C_5}$.

(*b*) If the five cards contain none of the four aces, then favourable number of cases $= {}^{48}C_5$.

∴ The probability that the five cards contain no ace $= \dfrac{^{48}C_5}{^{52}C_5}$.

Hence the probability that they contain at least one ace

$$= 1 - \frac{^{48}C_5}{^{52}C_5}.$$

Example 37. *From 25 tickets, marked with the first 25 numerals, one is drawn at random. Find the chance that*

(*a*) *it is a multiple of 5 or 7.*

(*b*) *it is a multiple of 3 or 7.*

Sol. Exhaustive number of cases $= {}^{25}C_1 = 25$.

(*a*) Numbers from 1 to 25 which are multiples of 5 are 5, 10,...., 25.

i.e., $\dfrac{25}{5} = 5$ in all.

Numbers from 1 to 25 which are multiples of 7 are 7, 14, 21 *i.e.,* 3 in all.

[Numbers which are multiples of the L.C.M. of 5 and 7 *i.e.,* 35 is none].

∴ Favourable number of cases = 5 + 3 = 8

Required probability $= \dfrac{8}{25}$.

(*b*) Numbers from 1 to 25 which are multiples of 3 are 3, 6, 9,....., 24, *i.e.,* $\dfrac{24}{3} = 8$ in all.

Numbers from 1 to 25 which are multiples of 7 are 7, 14, 21, *i.e.,* 3 in all.

Numbers from 1 to 25 which are multiples of L.C.M. of 3 and 7, *i.e.,* 21 is only one (21 itself).

∴ Favourable number of cases = 8 + 3 − 1 = 10.

∴ Required probability $= \dfrac{10}{25} = \dfrac{2}{5}$.

Example 38. *A and B stand in a ring with 10 other persons. If the arrangement of the twelve persons is at random ; find the chance that there are exactly three persons between A and B.*

Sol. Total number of persons = 10 + 2 = 12.

12 persons can stand in a ring in (12 – 1) ! = 11 ! ways.

∴ Exhaustive number of cases = 11 !

When the positions of A and B are so fixed that there are exactly three persons between them

(*i*) A and B can interchange places in 2 ! ways.

(*ii*) Three persons out of 10 can be arranged between them in ${}^{10}P_3$ ways.

(*iii*) The remaining 7 persons can be arranged among themselves in 7 ! ways.

$$\therefore \quad \text{Favourable number of cases} = 2\,! \times {}^{10}P_3 \times 7\,!$$
$$= 2 \times 10 \times 9 \times 8 \times 7\,!$$
$$= 2 \times (10\,!)$$

$$\therefore \quad \text{Required probability} = \frac{2 \times (10\,!)}{11\,!} = \frac{2 \times (10\,!)}{11 \times (10\,!)} = \frac{2}{11}.$$

Example 39. *The first twelve letters of the alphabet are written down at random. What is the probability that there are four letters between the letters A and B ?*

Sol. Exhaustive number of cases = 12 !

When the positions of A and B are so fixed that there are four letters between them

(*i*) A and B can interchange places in 2 ! ways.

(*ii*) Four letters out of the remaining 10 can be arranged between them in ${}^{10}P_4$ ways.

(*iii*) The remaining 6 letters can be arranged among themselves in 6 ! ways.

$$\therefore \quad \text{Favourable number of cases} = 2\,! \times {}^{10}P_4 \times 6\,!$$
$$= 2 \times 10 \times 9 \times 8 \times 7 \times (6\,!) = 2\,(10\,!)$$

$$\therefore \quad \text{Required probability} = \frac{2\,(10\,!)}{12\,!} = \frac{2\,(10\,!)}{12 \times 11 \times (10\,!)} = \frac{1}{66}.$$

Example 40. (*a*) *A five-figured number is formed by the digits 0, 1, 2, 3, 4 (without repetition). Find the probability that the number formed is divisible by 4.*

Sol. The digits 0, 1, 2, 3, 4 can be arranged among themselves in 5 ! ways out of which 4 ! numbers will start with 0.

Exhaustive number of cases

= total number of five-figured numbers

= 5 ! – 4 ! = 120 – 24 = 96.

Now, only those numbers are divisible by 4 in which the number formed by their last two digits is divisible by 4.

Thus, the numbers ending with 04, 12, 20, 24, 32, 40 will be divisible by 4.

(*i*) If the numbers end in 04, the remaining three digits 1, 2, 3, can be arranged in 3 ! = 6 ways.

(*ii*) If the numbers end in 20, the remaining three digits 1, 3, 4 can be arranged in 3 ! = 6 ways.

(*iii*) If the numbers end in 40, the remaining three digits 1, 2, 3 can be arranged in 3 ! = 6 ways.

(*iv*) If the numbers end in 12, the remaining three digits 0, 3, 4 can be arranged in 3 ! ways but we are to discard those cases when 0 is the extreme left digit and such numbers are 2 !.

$\therefore$ The numbers ending in 12 are 3 ! – 2 ! = 6 – 2 = 4

Similarly the numbers ending in 24 or 32 are 3 ! – 2 ! = 4

$\therefore$ Total number of favourable cases = 6 + 6 + 6 + 4 + 4 + 4 = 30

$\therefore$ Required probability $= \dfrac{30}{96} = \dfrac{5}{16}$.

Example 40. *(b) A five digit number is formed by the digits 1, 2, 3, 4, 5 without repetition. Find the probability that the number formed is divisible by 4.*

Sol. Please try yourself. $\left[\textbf{Ans. } \dfrac{1}{5}\right]$

Example 41. *Out of (2n + 1) tickets consecutively numbered, three are drawn at random. Find the chance that the numbers on them are in A.P.*

(D.U. 1984, 87, 90, 93)

Sol. Out of $(2n + 1)$ tickets, 3 tickets can be drawn in ${}^{2n+1}C_3$ ways.

$$\therefore \quad \text{Exhaustive number of cases} = {}^{2n+1}C_3 = \frac{(2n+1)(2n)(2n-1)}{3.2.1}$$

$$= \frac{n(4n^2-1)}{3}.$$

To find the favourable number of cases, we have to take into account all the cases in which the numbers on the three drawn tickets are in A.P.

If common difference **d = 1,** possible cases are as follows :

$$\left.\begin{matrix} 1, & 2, & 3 \\ 2, & 3, & 4 \\ & \vdots & \\ 2n-1, & 2n, & 2n+1 \end{matrix}\right\} (2n-1) \text{ in all}$$

If **d = 2**, possible cases are as follows :

$$\left.\begin{matrix} 1, & 3, & 5 \\ 2, & 4, & 6 \\ & \vdots & \\ 2n-3, & 2n-1, & 2n+1 \end{matrix}\right\} (2n-3) \text{ in all}$$

If **d** = 3, possible cases are as follows :

$$\left.\begin{matrix} 1, & 4, & 7 \\ 2, & 5, & 8 \\ & \vdots & \\ 2n-5, & 2n-2, & 2n+1 \end{matrix}\right\} (2n-5) \text{ in all}$$

and so on.

If **d = n – 1**, possible cases are as follows :

$$\left.\begin{matrix} 1, & n, & 2n-1 \\ 2, & n+1, & 2n \\ 3, & n+2, & 2n+1 \end{matrix}\right\} 3 \text{ in all}$$

If **d = n**, there is only one favourable case *viz.*, 1, $n + 1$, $2n + 1$

Hence total number of favourable cases

$$= (2n-1) + (2n-3) + (2n-5) + \ldots\ldots + 3 + 1$$

which is a series in A.P. with n terms

$$= \frac{n}{2}(2n - 1 + 1) = n^2$$

$$\text{Required probability} = \frac{n^2}{\dfrac{n(4n^2-1)}{3}} = \frac{3n}{4n^2-1}.$$

Example 42. *The sum of two non-negative quantities is equal to 2n. Find the chance that their product is not less than $\frac{3}{4}$ times their greatest product.*

Sol. Let $x > 0$ and $y > 0$ be the given quantities so that

$$x + y = 2n \qquad \ldots(1)$$

Let $\quad P = xy = x(2n - x) = 2nx - x^2 \qquad$ | *using* (1)

$$\frac{dP}{dx} = 2n - 2x, \quad \frac{d^2P}{dx^2} = -2 < 0$$

$\therefore$ The value of x obtained from $\frac{dP}{dx} = 0$ makes P greatest

$$\frac{dP}{dx} = 0 \quad \Rightarrow \quad x = n$$

$\therefore$ Greatest value of product $P = n(2n - n) = n^2$

Now xy is not less than $\frac{3}{4}n^2$

$$\Rightarrow \quad xy \geq \frac{3}{4}n^2$$

$$\Rightarrow \quad 4x(2n - x) \geq 3n^2$$

$$\Rightarrow \quad -4x^2 + 8nx - 3n^2 \geq 0$$

$$\Rightarrow \quad 4x^2 - 8nx + 3n^2 \leq 0$$

$$\Rightarrow \quad (2x - n)(2x - 3n) \leq 0$$

$$\Rightarrow \quad 2\left(x - \frac{n}{2}\right) . 2\left(x - \frac{3n}{2}\right) \leq 0$$

$$\Rightarrow \quad \left(x - \frac{n}{2}\right)\left(x - \frac{3n}{2}\right) \le 0$$

$$\Rightarrow \quad \frac{n}{2} \le x \le \frac{3n}{2}$$

$$[\because \quad (x - \alpha)(x - \beta) \le 0 \quad \Rightarrow \quad \alpha \le x \le \beta]$$

Also $x > 0, \quad y > 0$ and $x + y = 2n$

$\Rightarrow \quad 0 < x + y = 2n$

$\therefore$ Exhaustive range $= 2n - 0 = 2n$

Favourable range $= \dfrac{3n}{2} - \dfrac{n}{2} = n.$

$\therefore$ Required probability $= \dfrac{n}{2n} = \dfrac{1}{2}.$

Example 43. *If 8 biscuits be distributed among 11 beggars, find the chance that a particular beggar receives 5 biscuits.*

Sol. Every biscuit can be given to any one of the 11 beggars. Thus first biscuit can be distributed in 11 ways, second also in 11 ways and so on.

$\therefore$ The total number of ways in which 8 biscuits can be distributed at random among 11 beggars $= 11 \times 11 \times 11 \times \ldots\ldots \times 11$ (8 times) $= 11^8$. 5 biscuits can be given to any particular beggar in ${}^{11}C_5$ ways. Now we are left with $8 - 5 = 3$ biscuits which are to be distributed among the remaining $11 - 1 = 10$ beggars and this can be done in $10 \times 10 \times 10 = 10^3$ ways.

$\therefore$ Favourable number of cases $= {}^{11}C_5 \times 10^3$

Hence required probability $= \dfrac{{}^{11}C_5 \times 10^3}{11^8} = \dfrac{462000}{11^8}.$

Example 44. *Each co-efficient in the equation $ax^2 + bx + c = 0$ is determined by throwing an ordinary die. Find the possibility that the equation will have real roots.*

Sol. The roots of the equation $ax^2 + bx + c = 0$ will be real if its discriminant $b^2 - 4ac \ge 0$ *i.e.,* $b^2 \ge 4ac$.

Since each co-efficient is determined by throwing an ordinary die, each of the co-efficients a, b, c can take the values from 1 to 6.

$\therefore$ Exhaustive number of cases $= 6 \times 6 \times 6 = 216.$

The number of favourable cases can be enumerated as follows :

ac		*a*	*c*	*4ac*	*b* (*s.t.* $b^2 \ge 4ac$)	*No. of cases*
1		1	1	4	2, 3, 4, 5, 6	$1 \times 5 = 5$
2	(*i*)	1	2	8	3, 4, 5, 6	$2 \times 4 = 8$
	(*ii*)	2	1			
3	(*i*)	1	3	12	4, 5, 6	$2 \times 3 = 6$
	(*ii*)	3	1			

4	(*i*)	1	4	16	4, 5, 6	$3 \times 3 = 9$
	(*ii*)	4	1			
	(*iii*)	2	2			
5	(*i*)	1	5	20	5, 6	$2 \times 2 = 4$
	(*ii*)	5	1			
6	(*i*)	1	6	24	5, 6	$4 \times 2 = 8$
	(*ii*)	6	1			
	(*iii*)	2	3			
	(*iv*)	3	2			
7	$ac = 7$ is not possible					0
8	(*i*)	2	4	32	6	$2 \times 1 = 2$
	(*ii*)	4	2			
9		3	3	36	6	$1 \times 1 = 1$
						Total = 43

Since maximum value of b is 6, $\therefore$ Max. value of b^2 is 36.

$\because$ $ac \neq 10, 11, \ldots\ldots$, etc.

$\therefore$ Total number of favourable cases = 43

Hence the required probability $= \dfrac{43}{216}$.

5.5. RANDOM EXPERIMENT

Occurrences which can be repeated a number of times, essentially under the same conditions, and whose result cannot be predicted before hand are known as **random experiments.**

For example, rolling of a die, tossing a coin, taking out balls from an urn.

Sample Space. Out of the several possible outcomes of a random experiment, one and only one can take place in a trial. The set of all these possible outcomes is called the **sample space** for the particular experiment and is denoted by S.

For example, if a coin is tossed, the possible outcomes are H (Head) and T (Tail).

Thus $\quad S = \{H, T\}$.

Sample Point. The elements of S, the sample space, are called **sample points.**

For example, if a coin is tossed and H and T denote 'Head' and 'Tail' respectively, then $S = \{H, T\}$.

The two sample points are H and T.

Finite Sample Space. If the number of sample points in a sample space is finite, we call it a **finite sample space.** (In this chapter, we shall deal with finite sample spaces only).

Event. Every subset of S, the sample space, is called an event.

Since $S \subset S$, S itself is an event ; called a **certain event.**

Also $\phi \subset S$, the null set is also an event, called an **impossible event.**

If $e \in S$, then e is called an **elementary event.** Every elementary event contains only one sample point.

5.6. AXIOMS

(*i*) With each event E (*i.e.,* a sample point) is associated a real number between 0 and 1, called the probability of that event and is denoted by P(E). Thus $0 \leq P(E) \leq 1$.

(*ii*) The sum of the probabilities of all simple (elementary) events constituting the sample space is 1. Thus $P(S) = 1$.

(*iii*) The probability of a compound event (*i.e.,* an event made up of two or more simple events) is the sum of the probabilities of the simple events comprising the compound event.

Thus, if there are *n equally likely* possible outcomes of a random experiment, then the sample space S contains *n* sample points and the probability associated with each sample point is $\frac{1}{n}$. [By Axiom (*ii*)]

Now, if an event E consists of *m* sample points, then the probability of E is

$$P(E) = \frac{1}{n} + \frac{1}{n} + \ldots\ldots m \text{ times} = \frac{m}{n}$$

$$= \frac{\text{Number of sample points in E}}{\text{Number of sample points in S}}$$

This closely agrees with the classical definition of probability.

5.7. PROBABILITY OF THE IMPOSSIBLE EVENT IS ZERO, i.e., $P(\phi) = 0$.

Proof. Impossible event contains no sample point. As such, the sample space S and the impossible event ϕ are *mutually exclusive.*

$$\Rightarrow \quad S \cup \phi = S$$

$$\Rightarrow \quad P(S \cup \phi) = P(S)$$

$$\Rightarrow \quad P(S) + P(\phi) = P(S)$$

$$\Rightarrow \quad P(\phi) = 0.$$

5.8. PROBABILITY OF THE COMPLEMENTARY EVENT $\bar{A}$ OF A IS GIVEN BY $P(\bar{A}) = 1 - P(A)$.

Proof. A and $\bar{A}$ are disjoint events. Also $A \cup \bar{A} = S$

$\therefore \quad P(A \cup \bar{A}) = P(S)$

$\Rightarrow \quad P(A) + P(\bar{A}) = 1$. Hence $P(\bar{A}) = 1 - P(A)$

5.9. FOR ANY TWO EVENTS A AND B, $P(\bar{A} \cap B) = P(B) - P(A \cap B)$

Proof. $\bar{A} \cap B = \{p : p \in B \text{ and } p \notin A\}$

Now $\bar{A} \cap B$ and $A \cap B$ are disjoint sets

and $(\overline{A} \cap B) \cup (A \cap B) = B$

$\Rightarrow$ $P[(\overline{A} \cap B) \cup (A \cap B)] = P(B)$

$\Rightarrow$ $P(\overline{A} \cap B) + P(A \cap B) = P(B)$

$\Rightarrow$ $P(\overline{A} \cap B) = P(B) - P(A \cap B).$

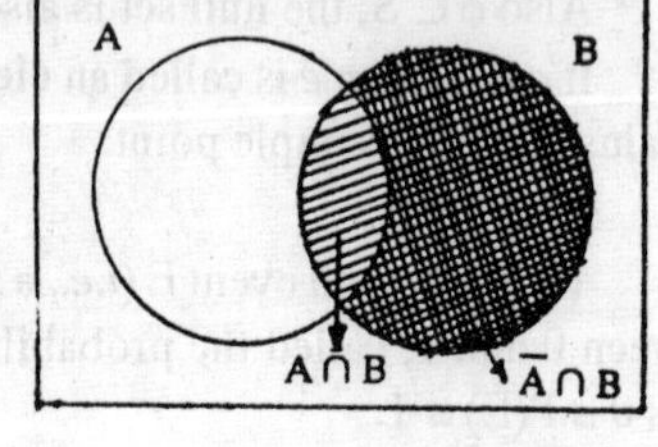

Note. Similarly, it can be proved that $P(A \cap \overline{B}) = P(A) - P(A \cap B)$.

5.10. If $B \subset A$, then

(i) $P(A \cap \overline{B}) = P(A) - P(B)$

(ii) $P(B) \leq P(A)$

Proof. When $B \subset A$, B and $A \cap \overline{B}$ are disjoint and their union is A.

$\Rightarrow$ $B \cup (A \cap \overline{B}) = A$

$\Rightarrow$ $P[B \cup (A \cap \overline{B})] = P(A)$

$\Rightarrow$ $P(B) + P(A \cap \overline{B}) = P(A)$

$\Rightarrow$ $P(A \cap \overline{B}) = P(A) - P(B)$ ...(I)

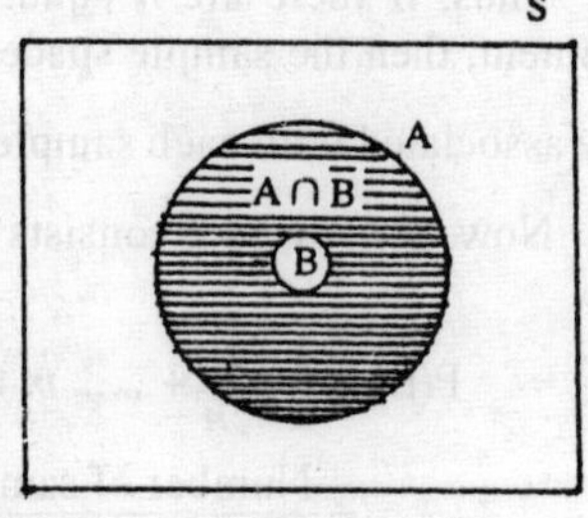

Now, if E is any event,

then $0 \leq P(E) \leq 1$, *i.e.* $P(E) \geq 0$

$\therefore$ $P(A \cap \overline{B}) \geq 0$

$\Rightarrow$ $P(A) - P(B) \geq 0$ [using I]

$\Rightarrow$ $P(B) \leq P(A)$.

5.11. $P(A \cap B) \leq P(A)$ AND $P(A \cap B) \leq P(B)$.

Proof. By article 5.10, $B \subset A \Rightarrow P(B) \leq P(A)$

Since $(A \cap B) \subset A$ and $(A \cap B) \subset B$

$\therefore$ $P(A \cap B) \leq P(A)$ and $P(A \cap B) \leq P(B)$.

☞5.12. ADDITION THEOREM OF PROBABILITY (OR THEOREM OF TOTAL PROBABILITY)

(M.D.U. 1981 S, 83 ; K.U. 1981, 82 ; D.U. 1982, 84, 85, 87, 88, 89, 92, 93 ; M.U. 1990 ; G.N.D.U. 1982)

Statement. *If A and B are any two events, then*

$$P(A \cup B) = P(A) + P(B) - P(A \cap B)$$

i.e., $P(A \text{ or } B) = P(A) + P(B) - P(A \text{ and } B)$.

Proof. A and $\overline{A} \cap B$ are *disjoint sets* and their union is $A \cup B$.

$\Rightarrow$ $A \cup B = A \cup (\overline{A} \cap B)$

$\Rightarrow$ $P(A \cup B) = P[A \cup (\overline{A} \cap B)] = P(A) + P(\overline{A} \cap B)$

$= P(A) + [P(\overline{A} \cap B) + P(A \cap B)] - P(A \cap B)$

$= P(A) + P[(\overline{A} \cap B) \cup (A \cap B)] - P(A \cap B)$

[$\because$ $\overline{A} \cap B$ and $A \cap B$ are disjoint]

$= P(A) + P(B) - P(A \cap B) \quad [\because \quad (\bar{A} \cap B) \cup (A \cap B) = B]$

Hence $P(A \cup B) = P(A) + P(B) - P(A \cap B)$.

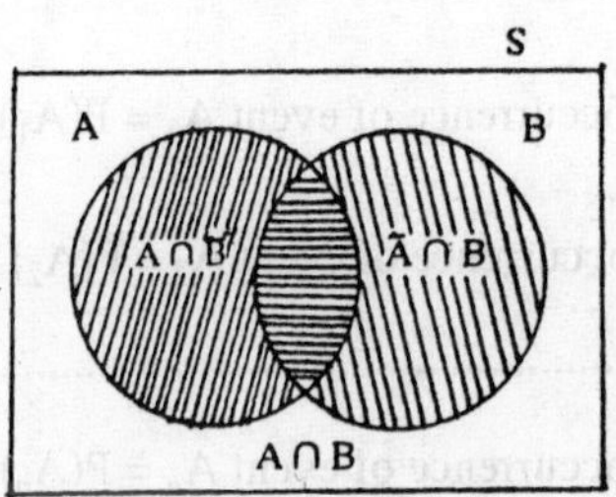

Note 1. If A and B are two mutually disjoint events, then $A \cap B = \phi$, so that $P(A \cap B) = P(\phi) = 0$

$\therefore \quad P(A \cup B) = P(A) + P(B)$.

Note 2. $P(A \cup B)$ is also written as $P(A + B)$. Thus, for mutually disjoint events A and B, $P(A + B) = P(A) + P(B)$

$P(A \cap B)$ is also written as $P(AB)$.

5.13. IF A, B AND C ARE ANY THREE EVENTS, THEN

$$\mathbf{P(A \cup B \cup C) = P(A) + P(B) + P(C) - P(A \cap B) - P(B \cap C) - P(C \cap A) + P(A \cap B \cap C)}$$

Or

$$\mathbf{P(A + B + C) = P(A) + P(B) + P(C) - P(AB) - P(BC) - P(CA) + P(ABC).}$$

Proof. Using the above theorem 5.12 for two events, we have

$$\begin{aligned}
P(A \cup B \cup C) &= P[(A \cup B) \cup C] \\
&= P(A \cup B) + P(C) - P[(A \cup B) \cap C] \\
&= [P(A) + P(B) - P(A \cap B)] + P(C) - P[(A \cap C) \cup (B \cap C)] \quad \text{[By distributive law]} \\
&= P(A) + P(B) + P(C) - P(A \cap B) - [P(A \cap C) \\
&\quad + P(B \cap C) - P\{(A \cap C) \cap (B \cap C)\} \quad \text{[By theorem 5.12]} \\
&= P(A) + P(B) + P(C) - P(A \cap B) - P(A \cap C) \\
&\quad - P(B \cap C) + P(A \cap B \cap C) \\
&\qquad [\because \quad (A \cap C) \cap (B \cap C) = A \cap B \cap C] \\
&= P(A) + P(B) + P(C) - P(A \cap B) - P(B \cap C) \\
&\quad - P(C \cap A) + P(A \cap B \cap C) \quad [\because \quad A \cap C = C \cap A]
\end{aligned}$$

or $\quad P(A + B + C) = P(A) + P(B) + P(C) - P(AB) - P(BC) - P(CA) + P(ABC)$.

5.14. IF $A_1, A_2, \ldots\ldots, A_n$ ARE n MUTUALLY EXCLUSIVE EVENTS, THEN THE PROBABILITY OF THE HAPPENING OF ONE OF THEM IS

$$\mathbf{P(A_1 \cup A_2 \cup \ldots\ldots \cup A_n) = P(A_1 + A_2 + \ldots\ldots + A_n) = P(A_1) + P(A_2) + \ldots\ldots + P(A_n).}$$

Proof. Let N be the total number of mutually exclusive, exhaustive and equally likely cases, of which m_1 are favourable to A_1, m_2 are favourable to A_2 and so on.

$$\left.\begin{aligned} &\text{Probability of occurrence of event } A_1 = P(A_1) = \frac{m_1}{N} \\ &\text{Probability of occurrence of event } A_2 = P(A_2) = \frac{m_2}{N} \\ &\cdots\cdots\cdots\cdots\cdots\cdots\cdots\cdots\cdots\cdots \\ &\text{Probability of occurrence of event } A_n = P(A_n) = \frac{m_n}{N} \end{aligned}\right\} \quad ...(1)$$

The events being mutually exclusive and equally likely, the number of cases favourable to the event

A_1 or A_2 or ...... or A_n is $m_1 + m_2 + ... + m_n$.

$\therefore$ Probability of occurrence of one of the events $A_1, A_2, ..., A_n$ is $P(A_1 + A_2 + ... + A_n)$

$$= \frac{m_1 + m_2 + ... + m_n}{N}$$

$$= \frac{m_1}{N} + \frac{m_2}{N} + ... + \frac{m_n}{N}$$

$$= P(A_1) + P(A_2) + ... + P(A_n) \quad \text{[Using (1)]}$$

Important Note. The student should not get confused with Theorems 5.12, 5.13 and 5.14. Theorems 5.12 and 5.13 are for ANY events (mutually exclusive or not) whereas Theorem 5.14 is for mutually exclusive events. (Read the statements carefully).

Example 1. *In a given race, the odds in favour of four horses A, B, C, D are 1 : 3, 1 : 4, 1 : 5, 1 : 6 respectively. Assuming that a dead heat is impossible ; find the chance that one of them wins the race.*

Sol. Let p_1, p_2, p_3, p_4 be the probabilities of winning of the horses A, B, C, D respectively.

Since a *dead heat* (in which all the four horses cover *same distance in same time*) is not possible, the events are mutually exclusive.

Odds in favour of A are 1 : 3 $\quad \therefore \quad p_1 = \frac{1}{1+3} = \frac{1}{4}$

Similarly $p_2 = \frac{1}{5}, \quad p_3 = \frac{1}{6}, \quad p_4 = \frac{1}{7}$.

If p is the chance that one of them wins, then

$$p = p_1 + p_2 + p_3 + p_4$$

$$= \frac{1}{4} + \frac{1}{5} + \frac{1}{6} + \frac{1}{7} = \frac{319}{420}.$$

Example 2. *What is the chance of throwing a total of 3 or 5 or 11 with two dice ?*

Sol. The three events are mutually exclusive.

Exhaustive number of cases = $6^2 = 36$.

Let p_1, p_2, p_3 be the probabilities of throwing a total of 3, 5, 11 respectively.

Favourable cases of throwing a total of 3 are (1, 2) ; (2, 1).

$$\therefore \qquad p_1 = \frac{2}{36}.$$

Favourable cases of throwing a total of 5 are (1, 4) ; (4, 1) ; (2, 3) ; (3, 2).

$$\therefore \qquad p_2 = \frac{4}{36}.$$

Favourable cases of throwing a total of 11 are (5, 6) ; (6, 5).

$$\therefore \qquad p_3 = \frac{2}{36}.$$

If p is the probability of throwing a total of 3 or 5 or 11, then

$$p = p_1 + p_2 + p_3 = \frac{2}{36} + \frac{4}{36} + \frac{2}{36} = \frac{8}{36} = \frac{2}{9}.$$

Example 3. *If a card is drawn from a deck of playing cards, what is the probability that it will be either the King of Diamonds or the Queen of Hearts ?*

Sol. Let A = the event of drawing the King of Diamonds

and B = the event of drawing the Queen of Hearts.

The two events are mutually exclusive.

$$P(A) = \frac{1}{52}, \qquad P(B) = \frac{1}{52}$$

$$\therefore \qquad P(A + B) = P(A) + P(B) = \frac{1}{52} + \frac{1}{52} = \frac{1}{26}.$$

Example 4. *A bag contains 6 white, 5 black and 4 yellow balls. Find the chance of getting either a white or a black ball in a single draw.*

Sol. Let A = the event of getting a white ball

and B = the event of getting a black ball

Total number of balls = 6 + 5 + 4 = 15

$$P(A) = \frac{6}{15}, \qquad P(B) = \frac{5}{15}$$

Since the two events are mutually exclusive

$$P(A + B) = P(A) + P(B) = \frac{6}{15} + \frac{5}{15} = \frac{11}{15}.$$

Example 5. *A card is drawn from a well-shuffled pack of playing cards. What is the probability that it is either a spade or an ace ?*

(M.D.U. 1981 S)

Sol. Let A = the event of drawing a spade

and B = the event of drawing an ace

A and B are *not* mutually exclusive.

AB = the event of drawing the ace of spades

$$P(A) = \frac{13}{52}, \qquad P(B) = \frac{4}{52}, \qquad P(AB) = \frac{1}{52}$$

$\therefore\ P(A+B) = P(A) + P(B) - P(AB)$

$$= \frac{13}{52} + \frac{4}{52} - \frac{1}{52} = \frac{16}{52} = \frac{4}{13}.$$

Example 6. *From a pack of cards, one card is selected at random. What is the probability that the card is a spade, an honour card or a jack ?*

Sol. [Ace, King and Queen are called honour cards]

Let A = the event of getting a spade
B = the event of getting an honour card
C = the event of getting a jack

There are 13 spades in a pack of cards

$\therefore\ P(A) = \frac{13}{52}$

There are 3 honour cards in each suit. Therefore, the number of honour cards in 3 × 4 = 12

$\therefore\ P(B) = \frac{12}{52}$

There are 4 jacks in a pack of cards.

$\therefore\ P(C) = \frac{4}{52}$

The events A, B, C are *not* mutually exclusive.

AB = the event of getting an honour card of spades (which are 3 in number)

$\therefore\ P(AB) = \frac{3}{52}$

BC = the event of getting an honour card which is a jack = ϕ

$\therefore\ P(BC) = 0$

CA = the event of getting a jack of spades

$\therefore\ P(CA) = \frac{1}{52}$

ABC = the event of getting a jack of spades which is an honour card = ϕ

$\therefore\ P(ABC) = 0$

$\therefore\ P(A+B+C) = P(A) + P(B) + P(C) - P(AB) - P(BC) - P(CA) + P(ABC)$

$$= \frac{13}{52} + \frac{12}{52} + \frac{4}{52} - \frac{3}{52} - 0 - \frac{1}{52} + 0 = \frac{25}{52}.$$

Example 7. *Three newspapers A, B, C are published in a city and a survey of readers indicates the following :*

20% read A, 16% read B, 14% read C.
8% read both A and B, 5% read both A and C.
4% read both B and C, 2% read all the three.

For a person chosen at random, find the probability that he reads none of the papers.

Sol. Here $P(A) = 20\% = \frac{20}{100}, P(B) = \frac{16}{100}, P(C) = \frac{14}{100}$

$$P(AB) = \frac{8}{100}, \quad P(AC) = \frac{5}{100}, \quad P(BC) = \frac{4}{100},$$

$$P(ABC) = \frac{2}{100}$$

$\therefore$ $P(A + B + C)$ = the probability that the person reads A or B or C

$$= P(A) + P(B) + P(C) - P(AB) - P(BC) - P(CA) + P(ABC)$$

$$= \frac{20}{100} + \frac{16}{100} + \frac{14}{100} - \frac{8}{100} - \frac{5}{100} - \frac{4}{100} + \frac{2}{100}$$

$$= \frac{35}{100} = \frac{7}{20}$$

Hence the probability that he reads none of the papers

$$= P(\overline{A + B + C}) = 1 - P(A + B + C)$$

$$= 1 - \frac{7}{20} = \frac{13}{20} = \frac{13}{20} \times 100\ \% = 65\ \%.$$

Example 8. *There are three events A, B, C, one of which must, and only one can happen ; the odds are 7 to 3 against A and 6 to 4 against B. Find the odds against C.*

Sol. Odds against A are 7 : 3 $\Rightarrow$ $P(A) = \frac{3}{10}$

Odds against B are 6 : 4 $\Rightarrow$ $P(B) = \frac{4}{10}$

Since only one of the three events can happen, the events are mutually exclusive.

$\therefore$ $P(A) + P(B) + P(C) = 1$

$\Rightarrow$ $\frac{3}{10} + \frac{4}{10} + P(C) = 1$ $\Rightarrow$ $P(C) = 1 - \frac{7}{10} = \frac{3}{10}$

Also $P(\overline{C}) = 1 - P(C) = 1 - \frac{3}{10} = \frac{7}{10}$

$\therefore$ Odds against C = $P(\overline{C}) : P(C) = 7 : 3.$

Example 9. *Only one of the three events A, B, C can happen. Given that the chance of A is one-third that of B and the odds are 2 : 1 against C, find the odds in favour of A.*

Sol. Since only one of the events can happen, the events are mutually exclusive.

$\therefore$ $P(A) + P(B) + P(C) = 1$ ...(1)

Also $P(A) = \frac{1}{3} P(B)$ $\Rightarrow$ $P(B) = 3\ P(A)$ ...(2)

Odds against C are 2 : 1 $\Rightarrow$ $P(C) = \frac{1}{3}$ ...(3)

Putting the values of P(B) and P(C) from (2) and (3) in (1),

we have $P(A) + 3\,P(A) + \frac{1}{3} = 1 \quad \Rightarrow \quad 4\,P(A) = \frac{2}{3}$

$\Rightarrow \quad P(A) = \frac{1}{6} \quad \therefore \quad P(\overline{A}) = 1 - P(A) = \frac{5}{6}$

$\therefore$ Odds in favour of A = $P(A) : P(\overline{A}) = 1 : 5$.

Example 10. *Discuss and criticise the following :*

$P(A) = \frac{2}{3}$, $P(B) = \frac{1}{4}$, $P(C) = \frac{1}{3}$ are the probabilities of three mutually exclusive events A, B and C.

Sol. Since A, B, C are mutually exclusive events,

$$P(A + B + C) = P(A) + P(B) = P(C)$$
$$= \frac{2}{3} + \frac{1}{4} + \frac{1}{3} = \frac{5}{4} > 1$$

But $P(A + B + C)$ must be ≤ 1

$\therefore$ The given statement is false.

5.15. SIMPLE AND COMPOUND EVENTS (*M.D.U. 1981, 82 S, 83 S*)

A simple event is a single event. When two or more simple events occur in connection with each other, their simultaneous occurrence is called a *Compound Event.*

For example, if a bag contains 6 red and 3 white balls and we are required to find the chance in which 4 balls can be drawn, all red, then it is a simple event. However, if we are required to find the chance of drawing 4 red balls and then 2 white balls, then it is a compound event because it is made up of two simple events.

Notation. If E_1 and E_2 are two simple events, then $E_1\,E_2$ denotes the simultaneous occurrence of E_1 and E_2. Thus $E_1\,E_2$ is called a compound event.

5.16. CONDITIONAL PROBABILITY (*M.D.U. 1983 S*)

The probability of the happening of an event E_1 when another event E_2 is known to have already happened is called *Conditional Probability* and is denoted by $P(E_1/E_2)$.

5.17. MUTUALLY INDEPENDENT EVENTS

An event E_1 is said to be independent of an event E_2 if

$$P(E_1/E_2) = P(E_1).$$

i.e., if the probability of happening of E_1 is independent of the happening of E_2.

5.18. MULTIPLICATIVE LAW OF PROBABILITY (OR THEOREM OF COMPOUND PROBABILITY)

(*D.U. 1983, 86, 91 ; M.U. 1987, 89, 90*)

The probability of simultaneous occurrence of two events is equal to the probability of one of the events multiplied by the conditional probability of the other, *i.e.*, **for two events A and B,**

$$\mathbf{P(A \cap B) = P(A) \times P(B/A)}$$

where P(B / A) represents the conditional probability of occurrence of B when the event A has already happened.

Proof. Suppose a trial results in n exhaustive, mutually exclusive and equally likely outcomes, m of them being favourable to the happening of the event A.

$\therefore$ Probability of happening of the event A

$$= P(A) = \frac{m}{n} \qquad ...(1)$$

Out of m outcomes favourable to the happening of A, let m_1 be favourable to the happening of the event B.

$\therefore$ Conditional probability of B, given that A has happened

$$= P(B/A) = \frac{m_1}{m} \qquad ...(2)$$

Now out of n exhaustive, mutually exclusive and equally likely outcomes, m_1 are favourable to the happening of *'A and B'*.

$\therefore$ Probability of simultaneous occurrence of A and B

$$= P(A \cap B) = \frac{m_1}{n}$$

$$= \frac{m_1}{m} \times \frac{m}{n} = \frac{m}{n} \times \frac{m_1}{m}$$

$$= P(A) \times P(B/A) \qquad \text{[Using (1) and (2)]}$$

Hence $P(A \cap B) = P(A) \times P(B/A)$

Note. $P(A \cap B)$ is also written as $P(AB)$.

Thus $P(AB) = P(A) \times P(B/A)$.

Cor. 1. Interchanging A and B

$$P(BA) = P(B) \times P(A/B)$$

or $$P(AB) = P(B) \times P(A/B). \qquad [\because B \cap A = A \cap B]$$

☞**Cor. 2.** If A and B are independent events, then

$$P(B/A) = P(B)$$

$\therefore$ $P(AB) = P(A) \times P(B)$.

Generalisation. If $A_1, A_2, ..., A_n$ are n independent events, then $P(A_1 A_2 ... A_n) = P(A_1) \times P(A_2) \times ..... \times P(A_n)$.

Cor. 3. If p is the chance that an event will happen in one trial then the chance that it will happen in a succession of r trials is

$$p.p \, ... \, p \ (r \text{ times}) = p^r.$$

☞**Cor. 4.** If $p_1, p_2, ..., p_n$ are the probabilities that certain events happen, then the probabilities of their non-happening are $1 - p_1, 1 - p_2, ..., 1 - p_n$ and, therefore, the probability of all of these failing is

$$(1 - p_1)(1 - p_2) \, ... \, (1 - p_n).$$

Hence the chance in which at least one of these events must happen is $1 - (1 - p_1)(1 - p_2) \, ... \, (1 - p_n)$. *(D.U. 1984, 90 ; M.U. 1986, 88, 89)*

Example 1. *Find the chance of drawing a king, a queen and a knave in that order from a pack of cards in three consecutive draws, the cards drawn not being replaced.*

Sol. There are 4 kings in a pack of 52 cards.

The probability p_1 of drawing a king $= \frac{4}{52}$.

Since the cards are not replaced, there remain 51 cards containing 4 queens.

The probability p_2 of drawing a queen in the second draw $= \frac{4}{51}$.

Now there remain 50 cards containing 4 knaves.

The probability p_3 of drawing a knave in the third draw $= \frac{4}{50}$.

Therefore, the required probability

$$= p_1p_2p_3 = \frac{4}{52} \times \frac{4}{51} \times \frac{4}{50} = \frac{8}{16775}.$$

Example 2. *Four cards are drawn without replacement. What is the probability that (a) they are all aces (b) they are all of different suits ?*

Sol. (*a*) There are 4 aces in a pack of 52 cards.

The probability p_1 of drawing in ace in the first draw $= \frac{4}{52}$.

Since the cards are not replaced, there remain 51 cards containing 3 aces.

The probability p_2 of drawing an ace in the second draw $= \frac{3}{51}$.

Similarly, the probability p_3 of drawing an ace in the third draw $= \frac{2}{50}$

and the probability p_4 of drawing an ace in the fourth draw $= \frac{1}{49}$.

Therefore, the required probability

$$= p_1p_2p_3p_4 = \frac{4}{52} \times \frac{3}{51} \times \frac{2}{50} \times \frac{1}{49} = \frac{1}{270725}.$$

(*b*) The probability p_1 of drawing a card of any suit in the first draw $= \frac{52}{52}$.

Since the cards are not replaced, there remain 51 cards containing 12 cards of the suit already drawn. Thus these 51 cards contain 39 cards of other three suits.

The probability p_2 of drawing a card of other suit in the second draw $= \frac{39}{51}$.

Now there remain 50 cards containing 26 cards of the remaining two suits.

The probability p_3 of drawing a card of other suit in the third draw $= \frac{26}{50}$.

Similarly, the probability p_4 of drawing a card of fourth suit in the fourth draw $= \frac{13}{49}$.

The required probability

$$= \frac{52}{52} \times \frac{39}{51} \times \frac{26}{50} \times \frac{13}{49} = \frac{2197}{20825}.$$

Example 3. *The odds against a certain event are 5 to 2 and the odds in favour of another (independent) event are 6 to 5. Find the chance that at least one of the events to will happen.*

Sol. The probability p_1 of happening of first event

$$= \frac{2}{5+2} = \frac{2}{7}.$$

The probability p_2 of happening of second event $= \frac{6}{5+5} = \frac{6}{11}$

The probabilities of non-happening of the two events are $1 - p_1 = \frac{5}{7}$ and $1 - p_2 = \frac{5}{11}$ respectively.

The probability of non-happening of both the events

$$= \frac{5}{7} \times \frac{5}{11} = \frac{25}{77}.$$

Hence the chance in which at least one of the events must happen

$$= 1 - \frac{25}{77} = \frac{52}{77}.$$

Example 4. *The odds against A solving a certain problem are 8 to 6 and the odds in favour of B solving the same problem are 14 to 10. What is the probability that if both of them try, the problem would be solved ?*

Sol. The probability of failure of A to solve the problem

$$= \frac{8}{8+6} = \frac{4}{7}.$$

The probability of failure of B to solve the problem

$$= \frac{10}{14+10} = \frac{5}{12}.$$

∴ The probability that both A and B fail to solve the problem

$$= \frac{4}{7} \times \frac{5}{12} = \frac{5}{21}.$$

Hence the probability that the problem will be solved by at least one of them

$$= 1 - \frac{5}{21} = \frac{16}{21}.$$

Example 5. *A problem in Statistics is given to three students A, B, C whose chances of solving it are $\frac{1}{2}, \frac{1}{3}, \frac{1}{4}$ respectively. What is the probability that the problem will be solved ?* (D.U. 1983, 86)

Sol. The probabilities of A, B, C solving the problem are $\frac{1}{2}, \frac{1}{3}, \frac{1}{4}$.

The probabilities of A, B, C not solving the problem are

$$1 - \frac{1}{2}, 1 - \frac{1}{3}, 1 - \frac{1}{4} \quad \textit{i.e.,} \quad \frac{1}{2}, \frac{2}{3}, \frac{3}{4}.$$

$\therefore$ The probability that the problem is not solved by any of them

$$= \frac{1}{2} \times \frac{2}{3} \times \frac{3}{4} = \frac{1}{4}.$$

Hence the probability that the problem is solved by at least one of them

$$= 1 - \frac{1}{4} = \frac{3}{4}.$$

Example 6. *A problem in Statistics is given to five students. Their chances of solving it are $\frac{1}{2}, \frac{1}{3}, \frac{1}{4}, \frac{1}{4}$ and $\frac{1}{5}$. What is the probability that the problem will be solved ?* (M.D.U. 1981, 83 S)

Sol. Please try yourself. $\left[\textbf{Ans. } \frac{17}{20}\right]$

Example 7. *A work is given to 6 men and their probabilities for doing the work are $\frac{1}{7}, \frac{1}{4}, \frac{1}{6}, \frac{1}{8}, \frac{1}{5}$ and $\frac{1}{3}$. What is the probability that the work is done ?* (K.U. 1983)

Sol. Please try yourself. [**Ans.** $\frac{3}{4}$]

Example 8. *The odds that a book will be favourably reviewed by three independent critics are 5 to 2, 4 to 3 and 3 to 4 respectively. What is the probability that, of the three reviews, a majority will be favourable ?*

Sol. Let the three critics be A, B, C. The probabilities p_1, p_2, p_3 of the book being favourably reviewed by A, B, C are $\frac{5}{7}, \frac{4}{7}, \frac{3}{7}$ respectively.

$\therefore$ The probabilities that the book is unfavourably reviewed by A, B, C are

$$1 - \frac{5}{7} = \frac{2}{7}, \quad 1 - \frac{4}{7} = \frac{3}{7}, \quad 1 - \frac{3}{7} = \frac{4}{7}.$$

A majority will be favourable if the reviews of at least two are favourable.

(*i*) If A, B, C all review favoruably, the probability is

$$\frac{5}{7} \times \frac{4}{7} \times \frac{3}{7} = \frac{60}{343} \qquad | \; p_1 p_2 p_3$$

(*ii*) If A, B review favourably and C reviews unfavourably, the probability is

$$\frac{5}{7} \times \frac{4}{7} \times \frac{4}{7} = \frac{80}{343} \qquad \Big| \; p_1p_2(1-p_3)$$

(*iii*) If A, C review favourably and B reviews unfavourably, the probability is

$$\frac{5}{7} \times \frac{3}{7} \times \frac{3}{7} = \frac{45}{343} \qquad \Big| \; p_1(1-p_2)p_3$$

(*iv*) If B, C review favourably and A reviews unfavourably, the probability is

$$\frac{2}{7} \times \frac{4}{7} \times \frac{3}{7} = \frac{24}{343} \qquad \Big| \; (1-p_1)p_2p_3$$

Hence the probability that a majority will be favourable is

$$\frac{60}{343} + \frac{80}{343} + \frac{45}{343} + \frac{24}{343} = \frac{209}{343}.$$

☞ **Example 9.** *A can hit a target 3 times in 6 shots ; B 2 times in 6 shots and C 4 times in 4 shots. They fire a volley. What is the probability that at least 2 shots hit ?*

Sol. Probability of A's hitting the target $= \frac{3}{6}$

" " B's " " " $= \frac{2}{6}$

" " C's " " " $= \frac{4}{4}$.

For at least two hits, we may have

(*i*) A, B, C all hit the target, the probability for which is

$$\frac{3}{6} \times \frac{2}{6} \times \frac{4}{4} = \frac{24}{144}.$$

(*ii*) B, C hit the target and A misses it, the probability for which is

$$\left(1 - \frac{3}{6}\right) \times \frac{2}{6} \times \frac{4}{4} = \frac{3}{6} \times \frac{2}{6} \times \frac{4}{4} = \frac{24}{144}.$$

(*iii*) A, C hit the target and B misses it, the probability for which is

$$\frac{3}{6} \times \left(1 - \frac{2}{6}\right) \times \frac{4}{4} = \frac{3}{6} \times \frac{4}{6} \times \frac{4}{4} = \frac{48}{144}.$$

☞ **C never misses the target**

Since these are mutually exclusive events, required probability

$$= \frac{24}{144} + \frac{24}{144} + \frac{48}{144} = \frac{96}{144} = \frac{2}{3}.$$

Example 10. *A can hit a target 4 times in 5 shots ; B 3 times in 4 shots ; C twice in 3 shots. They fire a volley. What is the probability that at least two shots hit ?* (Meerut 1983)

Sol. Probability of A's hitting the target $= \frac{4}{5}$

" " B's " " " $= \frac{3}{4}$

" " C's " " " $= \frac{2}{3}$.

For at least two hits, we may have

(*i*) A, B, C all hit the target, the probability for which is

$$\frac{4}{5} \times \frac{3}{4} \times \frac{2}{3} = \frac{24}{60}.$$

(*ii*) A, B hit the target and C misses it, the probability for which is

$$\frac{4}{5} \times \frac{3}{4} \times \left(1 - \frac{2}{3}\right) = \frac{4}{5} \times \frac{3}{4} \times \frac{1}{3} = \frac{12}{60}.$$

(*iii*) A, C hit the target and B misses it, the probability for which is

$$\frac{4}{5}\left(1 - \frac{3}{4}\right) \times \frac{2}{3} = \frac{4}{5} \times \frac{1}{4} \times \frac{2}{3} = \frac{8}{60}.$$

(*iv*) B, C hit the target and A misses it, the probability for which is

$$\left(1 - \frac{4}{5}\right) \times \frac{3}{4} \times \frac{2}{3} = \frac{1}{5} \times \frac{3}{4} \times \frac{2}{3} = \frac{6}{60}$$

Since these are mutually exclusive events, required probability

$$= \frac{24}{60} + \frac{12}{60} + \frac{8}{60} + \frac{6}{60} = \frac{50}{60} = \frac{5}{6}.$$

Example 11. *A can hit a target 5 times in 6 shots, B 4 times in 5 shots and C 3 times in 4 shots. They fire a volley. What is the probability that two shots at least hit ?* (K.U. 1982)

Sol. Please try yourself. $\left[\textbf{Ans. } \frac{107}{120}\right]$

Example 12. *Three groups of children contain respectively 3 girls and 1 boy ; 2 girls and 2 boys ; 1 girl and 3 boys. One child is selected at random from each group. Show that the chance that the three selected consist of 1 girl and 2 boys is* $\frac{13}{32}$. (M.U. 1988 ; K.U. 1982 S)

Sol. The required event of getting 1 girl and 2 boys among the three selected children can materialise in the following three mutually exclusive ways :

Group	*I*	*II*	*III*
(*i*)	Girl	Boy	Boy
(*ii*)	Boy	Girl	Boy
(*iii*)	Boy	Boy	Girl

The probability of selecting a girl from the first group is $\frac{3}{4}$, of selecting a boy from the second group is $\frac{2}{4}$ and of selecting a boy from the third group is $\frac{3}{4}$. Since these three events of selecting children from three groups are independent of each other, we have

$$P_{(i)} = \frac{3}{4} \times \frac{2}{4} \times \frac{3}{4} = \frac{9}{32}.$$

Similarly $$P_{(ii)} = \frac{1}{4} \times \frac{2}{4} \times \frac{3}{4} = \frac{3}{32}.$$

$$P_{(iii)} = \frac{1}{4} \times \frac{2}{4} \times \frac{1}{4} = \frac{1}{32}.$$

Required probability $= P_{(i)} + P_{(ii)} + P_{(iii)}$

$$= \frac{9}{32} + \frac{3}{32} + \frac{1}{32} = \frac{13}{32}.$$

Example 13. *Four persons are chosen at random from a group containing 3 men, 2 women and 4 children. Show that the chance that exactly two of them will be children is* $\frac{10}{21}$. (M.D.U. 1983 ; D.U. 1988)

Sol. 4 persons can be selected out of 9 in ${}^9C_4$ ways.

2 children can be selected out of 4 in ${}^4C_2$ ways. The remaining two persons are to be selected out of 3 men and 2 women *i.e.*, 5 persons and this can be done in ${}^5C_2$ ways.

$\therefore$ 4 persons containing exactly 2 children can be selected in ${}^4C_2 \times {}^5C_2$ ways.

$$\therefore \quad \text{Required probability} = \frac{{}^4C_2 \times {}^5C_2}{{}^9C_4} = \frac{6 \times 10}{126} = \frac{10}{21}$$

Example 14. *From a group of 8 children, 5 boys and 3 girls, three children are selected at random. Calculate the probabilities that the selected group contains (i) no girl, (ii) only one girl, (iii) only one particular girl, (iv) at least one girl and (v) more girls than boys.*

Sol. 3 children can be selected out of 8 in ${}^8C_3 = 56$ ways.

(*i*) The selection contains no girl. Therefore, all the three selected children are boys. 3 boys can be selected out of 5 in ${}^5C_3 = {}^5C_2 = 10$ ways.

$$\text{Required probability} = \frac{10}{56} = \frac{5}{28}.$$

(*ii*) The selection contains only one girl and therefore the remaining two children must be boys. One girl can be selected out of 3 in ${}^3C_1$ ways and 2 boys can be selected out of 5 in ${}^5C_2$ ways.

$\therefore$ One girl and 2 boys can be selected in

$${}^3C_1 \times {}^5C_2 = 3 \times 10 = 30 \text{ ways.}$$

$\therefore$ Required probability $= \frac{30}{56} = \frac{15}{28}$.

(*iii*) The selection contains one particular girl. The remaining two children must be boys. 2 boys can be selected out of 5 in

$${}^5C_2 = 10 \text{ ways.}$$

$\therefore$ Required probability $= \frac{10}{56} = \frac{5}{28}$.

(*iv*) The selection contains at least one girl. This event can materialise in the following mutually exclusive ways :

(*a*) 1 girl and 2 boys
(*b*) 2 girls and 1 boy
(*c*) 3 girls.

$$P_{(a)} = \frac{{}^3C_1 \times {}^5C_2}{{}^8C_3} = \frac{3 \times 10}{56} = \frac{30}{56}$$

$$P_{(b)} = \frac{{}^3C_2 \times {}^5C_1}{{}^8C_3} = \frac{3 \times 5}{56} = \frac{15}{56}$$

$$P_{(c)} = \frac{{}^3C_3}{{}^8C_3} = \frac{1}{56}$$

$\therefore$ Required probability $= \frac{30}{56} + \frac{15}{56} + \frac{1}{56} = \frac{46}{56} = \frac{23}{28}$.

(*v*) The selection contains more girls than boys. This event can materialise in the following mutually exclusive ways :

(*a*) 2 girls and 1 boy
(*b*) 3 girls (and no boy).

$$P_{(a)} = \frac{{}^3C_2 \times {}^5C_1}{{}^8C_3} = \frac{3 \times 5}{56} = \frac{15}{56}.$$

$$P_{(b)} = \frac{{}^3C_3}{{}^8C_3} = \frac{1}{56}.$$

$\therefore$ Required probability $= \frac{15}{56} + \frac{1}{56} = \frac{16}{56} = \frac{2}{7}$.

Example 15. *From a group of 10 children, containing 6 boys and 4 girls, three children are selected at random. Calculate the probabilities that the selected group contains : (i) no girl, (ii) only one girl, (iii) only one particular girl, (iv) at least one girl, (v) more girls than boys.* (K.U. 1983 S)

Sol. Please try yourself. [**Ans.** (*i*) $\frac{1}{6}$ (*ii*) $\frac{1}{2}$ (*iii*) $\frac{1}{8}$ (*iv*) $\frac{5}{6}$ (*v*) $\frac{1}{3}$]

Example 16. *One shot is fired from each of the three guns. E_1, E_2, E_3 denote the events that the target is hit by the first, second and third guns respectively. If $P(E_1) = 0.5$, $P(E_2) = 0.6$ and $P(E_3) = 0.8$ and E_1, E_2, E_3 are independent events, find the probability that (a) exactly one hit is registered (b) at least two hits are registered.*

Sol. (*a*) Exactly one hit can be registered in the following mutually exclusive ways :

(*i*) E_1 happens and E_2, E_3 do not happen, the probability for which is

$$P(E_1)[1 - P(E_2)][1 - P(E_3)] = 0.5 \times 0.4 \times 0.2 = 0.040.$$

(*ii*) E_2 happens and E_1, E_3 do not happen, the probability for which is

$$[1 - P(E_1)]P(E_2)[1 - P(E_3)] = 0.5 \times 0.6 \times 0.2 = 0.060.$$

(*iii*) E_3 happens and E_1, E_2 do not happen, the probability for which is

$$[1 - P(E_1)][1 - P(E_2)]P(E_3) = 0.5 \times 0.4 \times 0.8 = 0.160$$

Required probability = 0.04 + 0.06 + 0.16 = 0.26.

(*b*) Proceeding as in Example 10, required probability = 0.70.

Example 17. *The chances of winning of two race-horses are* $\frac{1}{3}$ *and* $\frac{1}{6}$ *respectively. What is the probability that at least one will win when the horses are running (a) in different races and (b) in the same race ?*

Sol. (*a*) Here the two events are mutually exclusive.

The chances of losing of two horses are $1 - \dfrac{1}{3} = \dfrac{2}{3}$ and $1 - \dfrac{1}{6} = \dfrac{5}{6}$ respectively.

The probability that both lose $= \dfrac{2}{3} \times \dfrac{5}{6} = \dfrac{5}{9}$.

$\therefore$ The probability that at least one wins $= 1 - \dfrac{5}{9} = \dfrac{4}{9}$.

(*b*) The probability that *either the first or the second* horse wins, by Additive Law of probability, is $\dfrac{1}{3} + \dfrac{1}{6} = \dfrac{1}{2}$.

Example 18. *A bag contains 6 white and 9 black balls. Four balls are drawn at a time. Find the probability for the first draw to give 4 white and the second draw to give 4 black balls in each of the following cases :*

(*i*) *The balls are replaced before the second draw.*

(*ii*) *The balls are not replaced before the second draw.*

Sol. (*i*) In the first draw, any 4 balls can be drawn out of 6 + 9 = 15 balls in ${}^{15}C_4$ ways and 4 white balls can be drawn out of 6 white balls in ${}^{6}C_4$ ways.

$\therefore$ The probability that the 4 balls drawn in the first draw are white is

$$\frac{{}^{6}C_4}{{}^{15}C_4} = \frac{15}{1365} = \frac{1}{91}.$$

Before the second draw, the balls are replaced. Therefore, we again have 6 white and 9 black balls.

Any four balls can be drawn out of 6 + 9 = 15 balls in ${}^{15}C_4$ ways and 4 black balls can be drawn out of 9 in ${}^{9}C_4$ ways.

$\therefore$ The probability that 4 black balls are drawn in the second draw

$$= \frac{{}^{6}C_4}{{}^{15}C_4} = \frac{126}{1365} = \frac{6}{65}.$$

$$\therefore \quad \text{Required probability} = \frac{1}{91} \times \frac{6}{65} = \frac{6}{5915}.$$

(*ii*) The probability of drawing 4 white balls in the first draw is

$$\frac{{}^{9}C_4}{{}^{15}C_4} = \frac{1}{91}.$$

Since the balls drawn are not replaced, we are left with 2 white balls and 9 black balls. The probability of drawing 4 black balls out of these 11 balls is

$$\frac{{}^{9}C_4}{{}^{11}C_4} = \frac{21}{55}.$$

$$\therefore \quad \text{Required probability} = \frac{1}{91} \times \frac{21}{55} = \frac{3}{715}.$$

Example 19. *A bag contains 4 white and 10 black balls. Two draws of 3 balls are made such that (i) the balls are not replaced before the second draw and (ii) the balls are replaced before the second draw. Find the probability that the first draw will give 3 white and second 3 black balls in each case.*

Sol. Please try yourself.

$$\left[\textbf{Ans}.\ (i)\ \frac{{}^{4}C_3}{{}^{14}C_3} \times \frac{{}^{10}C_3}{{}^{11}C_3}, \quad (ii)\ \frac{{}^{4}C_3}{{}^{14}C_3} \times \frac{{}^{10}C_3}{{}^{14}C_3} \right]$$

Example 20. *A bag contains 10 balls, two of which are red, three blue and five black. Three balls are drawn at random from the bag. What is the probability that*

(*i*) *the three balls are of different colours* ;

(*ii*) *two balls are of the same colour* ;

(*iii*) *the balls are all of the same colour.*

Sol. 3 balls can be drawn at random from 10 balls in

$${}^{10}C_3 = 120 \text{ ways.}$$

(*i*) The three balls are of different colours. Therefore one is red, one is blue and one is black. One red, one blue and one black ball can be drawn out of 2 red, 3 blue and 5 black balls in

$${}^{2}C_1 \times {}^{3}C_1 \times {}^{5}C_1 = 30 \text{ ways.}$$

$$\therefore \quad \text{Required probability} = \frac{30}{120} = \frac{1}{4}.$$

(*ii*) 2 balls of the same colour can be drawn in following three ways :

(*a*) **2 red** out of two and one blue or black out of 3 blue and 5 black *i.e.,* 8 balls.

The probability for which is $\frac{{}^2C_2 \times {}^8C_1}{120} = \frac{8}{120}$.

(*b*) **2 blue** out of 3 and one red or black out of 2 red and 5 black *i.e.,* 7 balls.

The probability for which is $\frac{{}^3C_2 \times {}^7C_1}{120} = \frac{21}{120}$.

(*c*) **2 black** out of 5 and one red or blue out of 2 red and 3 blue *i.e.,* 5 balls.

The probability for which is $\frac{{}^5C_2 \times {}^5C_1}{120} = \frac{50}{120}$.

$\therefore$ Required probability $= \frac{8}{120} + \frac{21}{120} + \frac{50}{120} = \frac{79}{120}$.

(*iii*) The three balls cannot be red as there are only two red balls.

$\therefore$ Three balls of the same colour can be drawn in following ways :

(*a*) **3 blue** out of a total of 3 blue balls.

The probability for which is $\frac{{}^3C_3}{120} = \frac{1}{120}$.

(*b*) **3 black** out of a total of 5 black balls.

The probability for which is $\frac{{}^5C_3}{120} = \frac{10}{120}$.

$\therefore$ Required probability $= \frac{1}{120} + \frac{10}{120} = \frac{11}{120}$.

Example 21. *The probability that a 50-year old man will be alive at 60 is 0.83 and the probability that a 45-year old woman will be alive at 55 is 0.87. What is the probability that a man who is 50 and his wife who is 45 will both be alive 10 years hence ?*

Sol. The probability that the 50 years old man will live 10 years hence (*i.e.,* till he is 60) = 0.83.

The probability that the 45 years old wife will live 10 years hence (*i.e.,* till she is 55) = 0.87.

Hence the probability that both husband and wife will live 10 years hence = 0.83 × 0.87 = 0.7221.

Example 22. *It is 8 : 5 against a person who is 40 years old living till he is 70 and 4 : 3 against a person now 50 living till he is 80. Find the probability that one at least of these persons will be alive 30 years hence.*

Sol. The probability that the 40 years old man will not live 30 years hence (*i.e.,* till he is 70) $= \frac{8}{8+5} = \frac{8}{13}$.

The probability that the 50 years old man will not live 30 years hence (*i.e.*, till he is 80) $= \frac{4}{4+3} = \frac{4}{7}$.

The probability that both will not live 30 years hence

$$= \frac{8}{13} \times \frac{4}{7} = \frac{32}{91}.$$

$\therefore$ The probability that at least one of them lives 30 years hence

$$= 1 - \frac{32}{91} = \frac{59}{91}.$$

Example 23. *It is 9 to 7 against a person A who is now 35 years of age living till he is 65, and 3 to 2 against a person B now 45 living till he is 75. Find the chance that one at least of these persons will be alive 30 years hence.*

Sol. Please try yourself. $\left[\textbf{Ans. } \frac{53}{80}\right]$

Example 24. *It is 8 : 5 against a husband who is 55 years old living till he is 75 and 4 : 3 against his wife who is now 48, living till she is 68. Find the probability that (i) the couple will be alive 20 years hence and (ii) at least one of them will be alive 20 years hence.*

Sol. Please try yourself. $\left[\textbf{Ans. } (i)\ \frac{15}{91},\quad (ii)\ \frac{59}{91}\right]$

Example 25. *There are two bags, one of which contains 5 red and 7 white balls and the other 3 red and 12 white balls. One ball is to be drawn from one or other of the two bags. Find the chance of drawing a red ball.*

Sol. Required event can happen in the following mutually exclusive ways :

(*i*) First bag is chosen and then a red ball is drawn.

(*ii*) Second bag is chosen and then a red ball is drawn.

Since the probability of choosing any bag is $\frac{1}{2}$, the required probability 'p' is given by

$$p = \frac{1}{2} \times \frac{^5C_1}{^{12}C_1} + \frac{1}{2} \times \frac{^3C_1}{^{15}C_1}$$

$$= \frac{1}{2} \times \frac{5}{12} + \frac{1}{2} \times \frac{3}{15} = \frac{5}{24} + \frac{1}{10} = \frac{37}{120}.$$

Example 26. *Three ships A, B and C sail from England to India. Odds is favour of their arriving safely are 2 : 5, 3 : 7 and 6 : 11 respectively. Find the chance that they all arrive safely.*

Sol. Probability of A arriving safely $= \frac{2}{2+5} = \frac{2}{7}$

Probability of B arriving safely $= \frac{3}{3+7} = \frac{3}{10}$

" " C " " $= \frac{6}{6+11} = \frac{6}{17}$.

$\therefore$ Required probability $= \frac{2}{7} \times \frac{3}{10} \times \frac{6}{17} = \frac{18}{595}$.

Example 27. *One bag contains 3 white balls and 2 black balls, another contains 5 white and 3 black balls. If a bag is chosen at random and a ball is drawn from it, what is the chance that it is white ?* (K.U. 1981 S)

Sol. Proceeding as in Example 25, required probability

$$= \frac{1}{2} \times \frac{^3C_1}{^5C_1} + \frac{1}{2} \times \frac{^5C_1}{^8C_1} = \frac{1}{2} \times \frac{3}{5} + \frac{1}{2} \times \frac{5}{8}$$

$$= \frac{3}{10} + \frac{5}{16} = \frac{24+25}{80} = \frac{49}{80}.$$

Example 28. *One purse contains 5 sovereigns and 4 shillings, another contains 5 sovereigns and 3 shillings. One purse is taken at random and a coin drawn out. What is the chance that it will be a sovereign ?*

Sol. Please try yourself. $\left[\textbf{Ans. } \frac{85}{144}\right]$

Example 29. *There are three events A, B, C one of which must, and only one can happen. The odds are 8 to 3 against A, 5 to 2 against B ; find the odds against C.*

Sol. The probability of A's happening $= \frac{3}{8+3} = \frac{3}{11}$

" " " B's " " $= \frac{2}{5+2} = \frac{2}{7}$

Because one and only one of the events can and must happen

$\therefore$ The probability of C's happening $= 1 - \left(\frac{3}{11} + \frac{2}{7}\right)$

$$= 1 - \frac{43}{77} = \frac{34}{77}.$$

The probability of C's non-happening $= 1 - \frac{34}{77} = \frac{43}{77}$

Hence the odds against C are 43 : 34.

Example 30. *There are four events A, B, C and D of which only one must, and only one can happen. The odds are 8 to 3 against A, 5 to 3 against B, 4 to 1 against C. Find the odds against D.*

Sol. Please try yourself. [**Ans.** 373 : 67]

Example 31. *If n biscuits be distributed among N beggars, find the chance that a particular beggar receives r (< n) biscuits.* (D.U. 1983)

Sol. Any biscuit can be given to any one of the N beggars. Thus, there are N ways of distributing any biscuit.

$\therefore$ Total number of ways in which n biscuits can be distributed at random among N beggars = N.N.N.....N (n times) = N^n.

A particular beggar can receive r biscuits in ${}^nC_r$ ways. The remaining $(n - r)$ biscuits can be distributed among the remaining $(N - 1)$ beggars in $(N - 1)^{n-r}$ ways.

Favourable number of cases = ${}^nC_r \times (N - 1)^{n-r}$.

Exhaustive number of cases = N^n.

Required probability $= \dfrac{{}^nC_r(N-1)^{n-r}}{N^n}$.

Example 32. *Find the chance of getting at least one six in a throw of four dice.*

Sol. The probability of throwing a six with one die $= \frac{1}{6}$.

The probability of not throwing a six with one die $= 1 - \frac{1}{6} = \frac{5}{6}$

" " " " " " " " " four dice $= (\frac{5}{6})^4$

Hence the probability of throwing at least one six with four dice

$= 1 - (\frac{5}{6})^4$.

Example 33. *A person throws 3 coins. Find the probability that at least one head turns up.* (M.D.U. 1984)

Sol. The probability of throwing a head with one coin $= \frac{1}{2}$.

The probability of not throwing a head with one coin $= 1 - \frac{1}{2} = \frac{1}{2}$.

$\therefore$ The probability of not throwing a head with 3 coins

$= (\frac{1}{2})^3 = \frac{1}{8}$.

Hence the probability of throwing a head at least once

$= 1 - \dfrac{1}{8} = \dfrac{7}{8}$.

Example 34. *Find the chance of throwing a 5 or a 6 at least once in four throws of a single die.* (M.D.U. 1982)

Sol. The probability of throwing a 5 or a 6 in a single throw

$= \dfrac{2}{6} = \dfrac{1}{3}$.

The probability of not throwing a 5 or a 6 in a single throw

$= 1 - \dfrac{1}{3} = \dfrac{2}{3}$.

The probability of not throwing a 5 or a 6 in four throws

$= \left(\dfrac{2}{3}\right)^4 = \dfrac{16}{81}$.

∴ The probability of throwing a 5 or a 6 at least once in four throws

$$= 1 - \frac{16}{81} = \frac{65}{81}.$$

Example 35. *A has 2 shares in a lottery in which there are 3 prizes and 5 blanks ; B has 3 shares in a lottery in which there are 4 prizes and 6 blanks. Show that A's chance of success is to B's as 27 : 35.*

Sol. A can draw two tickets (out of 3 + 5 = 8) in ${}^8C_2 = 28$ ways.

A will get all blanks in ${}^5C_2 = 10$ ways.

∴ A can win a prize in 28 – 10 = 18 ways

Hence A's chance of success $= \frac{18}{28} = \frac{9}{14}$

B can draw 3 tickets in ${}^{10}C_3 = 120$ ways.

B will get all blanks in ${}^6C_3 = 20$ ways.

∴ B can win a prize in 120 – 20 = 100 ways.

Hence B's chance of success $= \frac{100}{120} = \frac{5}{6}$

∴ A's chance : B's chance $= \frac{9}{14} : \frac{5}{6} = 27 : 35.$

Example 36. *A has 3 shares in a lottery where there are 3 prizes and 6 blanks. B has one share in another, where there is but one prize and two blanks. Show that A has a better chance of winning a prize than B in the ratio of 16 to 7.*

Sol. A can draw 3 tickets (out of 3 + 6 = 9) in ${}^9C_3 = 84$ ways.

A will get all blanks in ${}^6C_3 = 20$ ways.

∴ A can win a prize in 84 – 20 = 64 ways.

⇒ A's chance of winning a prize $= \frac{64}{84} = \frac{16}{21}$

B can draw 1 ticket (out of 1 + 2 = 3) in ${}^3C_1 = 3$ ways.

B will get all blanks in ${}^2C_1 = 2$ ways.

∴ B can win a prize in 3 – 2 = 1 way.

⇒ B's chance of winning a prize $= \frac{1}{3}$

Ratio of the chances of A and B winning a prize

$$= \frac{16}{21} : \frac{1}{3} = 16 : 7.$$

Example 37. *A and B throw alternately with a single die, A having the first throw. The person who first throws ace is to win. What are their respective chances of winning ?*

Sol. The chance of throwing an ace with a single die $= \frac{1}{6}$

The chance of not throwing an ace with a single die

$$= 1 - \frac{1}{6} = \frac{5}{6}.$$

If A is to win, he should throw an ace in the first or third or fifth,....., throws.

If B is to win, he should throw an ace in the second or fourth or sixth,....., throws.

The chances that an ace is thrown in the first, second, third,....., throws are

$$\frac{1}{6}, \frac{5}{6}.\frac{1}{6}, \frac{5}{6}.\frac{5}{6}.\frac{1}{6}, \frac{5}{6}.\frac{5}{6}.\frac{5}{6}.\frac{1}{6} \ldots\ldots$$

or $$\frac{1}{6}, \frac{5}{6}.\frac{1}{6}, \left(\frac{5}{6}\right)^2.\frac{1}{6}, \left(\frac{5}{6}\right)^3.\frac{1}{6}, \ldots\ldots$$

$$\therefore \quad \text{A's chance} = \frac{1}{6} + \left(\frac{5}{6}\right)^2.\frac{1}{6} + \left(\frac{5}{6}\right)^4.\frac{1}{6} + \ldots\ldots$$

$$= \frac{1}{6}\left[1 + \left(\frac{5}{6}\right)^2 + \left(\frac{5}{6}\right)^4 + \ldots\ldots\right]$$

$$= \frac{1}{6} \times \frac{1}{1 - \left(\frac{5}{6}\right)^2} \qquad \left|\ \text{Sum of an infinite G.P.} = \frac{a}{1-r}\right.$$

$$= \frac{6}{11}$$

B's chance $= 1 - \frac{6}{11} = \frac{5}{11}$

Note : Briefly, B's chance $= 1 - \frac{6}{11} = \frac{5}{11}$

Example 38. *A and B throw with one die for a prize of Rs. 44 which is to be won by a player who first throws 6. If A has the first throw, what are their respective expectations ?*

Sol. Proceeding as in Example 37 above,

A's chance : B's chance = 6 : 5

Total prize = Rs. 44

$$\therefore \quad \left.\begin{aligned} \text{A's expectation} &= \text{Rs. } \frac{6}{11} \times 44 = \text{Rs. } 24 \\ \text{B's expectation} &= \text{Rs. } \frac{5}{11} \times 44 = \text{Rs. } 20 \end{aligned}\right\}$$

Example 39. *A, B and C, in order, toss a coin. The first one to throw a head wins. If A starts, find their respective chances of winning.*

(Meerut 1983)

Sol. The chance of throwing a head with a single coin $= \frac{1}{2}$

The chance of not throwing a head with a single coin

$$= 1 - \frac{1}{2} = \frac{1}{2}$$

If A is to win, he should throw a head in the 1st or 4th or 7th,..... throws.

If B is to win, he should throw a head in the 2nd or 5th or 8th,... throws.

If C is to win, he should throw a head in the 3rd or 6th or 9th,..... throws.

The chances that a head is thrown in the 1st, 2nd, 3rd, 4th, 5th, 6th,..... throws are

$$\frac{1}{2}, \left(\frac{1}{2}\right)^2, \left(\frac{1}{2}\right)^3, \left(\frac{1}{2}\right)^4, \left(\frac{1}{2}\right)^5, \left(\frac{1}{2}\right)^6, \ldots\ldots$$

$\therefore$ A's chance $= \frac{1}{2} + \left(\frac{1}{2}\right)^4 + \left(\frac{1}{2}\right)^7 + \ldots\ldots$ [*Infinite G.P.*]

$$= \frac{\frac{1}{2}}{1 - \left(\frac{1}{2}\right)^3} = \frac{1}{2} \times \frac{8}{7} = \frac{4}{7}$$

B's chance $= \left(\frac{1}{2}\right)^2 + \left(\frac{1}{2}\right)^5 + \left(\frac{1}{2}\right)^8 + \ldots\ldots$

$$= \frac{\left(\frac{1}{2}\right)^2}{1 - \left(\frac{1}{2}\right)^3} = \frac{1}{4} \times \frac{8}{7} = \frac{2}{7}$$

C's chance $= \left(\frac{1}{2}\right)^3 + \left(\frac{1}{2}\right)^6 + \left(\frac{1}{2}\right)^9 + \ldots\ldots$

$$= \frac{\left(\frac{1}{2}\right)^3}{1 - \left(\frac{1}{2}\right)^3} = \frac{1}{8} \times \frac{8}{7} = \frac{1}{7}.$$

Note : Briefly, C's chance $= 1 - \frac{4}{7} - \frac{2}{7} = \frac{1}{7}$.

Example 40. *A and B toss a coin alternately on the understanding that the first who obtains the head wins. If A starts, show that their respective chances of winning are $\frac{2}{3}$ and $\frac{1}{3}$.*

Sol. Please try yourself.

Example 41. *If the probability of success is* $\frac{1}{100}$, *how many trials are necessary in order that probability of at least one success is greater than half ?*

Sol. p, the probability of success $= \frac{1}{100}$

$\therefore$ q, the probability of failure $= 1 - p = \frac{99}{100} = .99$

Let the reqd. number of trials $= n$.

The probability of failures in all trials $= \left(\frac{99}{100}\right)^n$

$\therefore$ The probability of at least one success $= 1 - \left(\frac{99}{100}\right)^n$

By the given condition $1 - (.99)^n > \frac{1}{2}$

$\Rightarrow$ $(.99)^n < \frac{1}{2}$ $\Rightarrow$ $\log (.99)^n < \log .5$

$\Rightarrow$ $n \log .99 < \log .5$

$\Rightarrow$ $n(\bar{1}.9956) < \bar{1}.6990$

$\Rightarrow$ $n(-1 + .9956) < -1 + .6990$

$\Rightarrow$ $-.0044n < -.3010$

$\Rightarrow$ $n > \frac{.3010}{.0044}$ $\Rightarrow$ $n > 68.4$

Hence $n = 69$.

Example 42. *A speaks truth in 60% cases and B in 70% cases. In what percentage of cases are they likely to contradict each other in stating the same fact ?*

Sol. They will contradict each other only if one of them speaks the truth and the other tells a lie.

The probability of A speaking the truth and B telling a lie

$= \frac{60}{100} \times \frac{30}{100} = \frac{9}{50}$

The probability of A telling a lie and B speaking the truth

$= \frac{40}{100} \times \frac{70}{100} = \frac{14}{50}$

$\therefore$ The probability of their contradicting each other

$= \frac{9}{50} + \frac{14}{50} = \frac{23}{50}$

$= \frac{23}{50} \times 100\% = 46\%$.

Example 43. *An urn contains 3 red and 4 black balls. Two balls are taken out at random. Find the probability that the balls are of (i) different colours (ii) black colour (iii) red colour.*

Sol. Exhaustive number of cases $= {}^7C_2 = 21$

(*i*) When the balls are of different colours, one is red and the other is black.

Favourable number of cases = $^3C_1 \times {}^4C_1 = 12$

Reqd. probability $= \frac{12}{21} = \frac{4}{7}$.

(*ii*) Favourable number of cases = $^4C_2 = 6$

Reqd. probability $= \frac{6}{21} = \frac{2}{7}$.

(*iii*) Favourable number of cases = $^3C_2 = 3$

Reqd. probability $= \frac{3}{21} = \frac{1}{7}$.

Example 44. *A hand of 13 cards is dealt out randomly from a full deck of 52 cards. Find the probability that the hand contains no spade card. Also show that the probability that the hand contains 4 cards of one suit and 3 cards each of the other three suits is*

$$4\binom{13}{4}\binom{13}{3}^3 \Big/ \binom{52}{13}.$$

Sol. $\left[\textbf{Note}.\ \binom{n}{r} \Rightarrow {}^nC_r\right]$

Exhaustive no. of cases = $^{52}C_{13} = \binom{52}{13}$

(*i*) Favourable number of cases when the hand contains all the 13 spade cards

$$= {}^{13}C_{13} = 1$$

∴ The probability of the hand containing all spade cards

$$= 1 \Big/ \binom{52}{13}$$

The probability of the hand containing no spade card

$$= 1 - 1 \Big/ \binom{52}{13}$$

(*ii*) There are 4 suits and one suit can be selected out of 4 in 4 ways.

∴ The hand will contain 4 cards of some suit in $4 \times {}^{13}C_4$

$$= 4\binom{13}{4} \text{ ways.}$$

3 cards each of other suits can be had in $\binom{13}{3}^3$ ways.

∴ Favourable number of cases

$$= 4\binom{13}{4}\binom{13}{3}^3$$

$\therefore$ Reqd. probability $= 4\binom{13}{4}\binom{13}{3}^3\Big/\binom{52}{13}$.

Example 45. *Cards are dealt one by one from a well shuffled pack until an ace appears. Show that the probability that exactly n cards are dealt before the first ace appears is*

$$\frac{4(51-n)(50-n)(49-n)}{52.51.50.49}.\quad \text{(D.U. 1985, 89, 90, 92)}$$

Sol. Let A be the event of drawing n non-ace cards and B, the event of drawing an ace in the $(n + 1)$th draw.

Consider the event A

n cards can be drawn out of 52 cards in ${}^{52}C_n$ ways.

$\Rightarrow$ Exhaustive cases $= {}^{52}C_n$

n non-ace cards can be drawn out of 52 cards is ${}^{48}C_n$ ways.

$\Rightarrow$ Favourable cases $= {}^{48}C_n$

$$\therefore \quad P(A) = {}^{48}C_n / {}^{52}C_n = \frac{48\,!}{(48-n)\,!\,n\,!} \times \frac{(52-n)\,!\,n\,!}{52\,!}$$

$$= \frac{48\,!\,.\,(52-n)(51-n)(50-n)(49-n)\,(48-n)\,!}{(48-n)\,!\,.\,52.51.50.49.4\,!}$$

$$= \frac{(52-n)(51-n)(50-n)(49-n)}{52\,.\,51\,.\,50\,.\,49}$$

Consider the event B

n cards have already been drawn in the first n draws.

Exhaustive cases $= {}^{52-n}C_1 = 52 - n$

Favourable cases $= {}^{4}C_1 = 4$

$$\therefore \quad P(B/A) = \frac{4}{52-n}$$

Reqd. Probability = P(A) . P(B/A)

$$= \frac{(52-n)(51-n)(50-n)(49-n)}{52.51.50.49} \times \frac{4}{52-n}.$$

$$= \frac{4(51-n)(50-n)(49-n)}{52.51.50.49}$$

Example 46. *A and B take turns in throwing two dice, the first to throw 9 being awarded the prize. Show that if A has the first throw, their chances of winning are in the ratio 9 : 8.* (D.U. 1991)

Sol. Exhaustive no. of cases $= 6^2 = 36$

Favourable cases (6, 3) ; (3, 6) ; (4, 5) ; (5, 4) *i.e.,* 4 in all.

The probability of throwing 9 with two dice $= \frac{4}{36} = \frac{1}{9}$

$\therefore$ The probability of not throwing 9 with two dice

$$= 1 - \frac{1}{9} = \frac{8}{9}$$

If A is to win, he should throw 9 in 1st or 3rd or 5th.... throws.

If B is to win, he should throw 9 in 2nd or 4th or 6th..... throws.

The chances that 9 is thrown in the 1st, 2nd, 3rd, 4th..... throws are

$$\frac{1}{9}, \frac{8}{9} \times \frac{1}{9}, \left(\frac{8}{9}\right)^2 \times \frac{1}{9}, \left(\frac{8}{9}\right)^3 \times \frac{1}{9} \ldots\ldots$$

A's chance $= \frac{1}{9} + \left(\frac{8}{9}\right)^2\left(\frac{1}{9}\right) + \left(\frac{8}{9}\right)^4\left(\frac{1}{9}\right) + \ldots\ldots$ [*Infinite G.P.*]

$$= \frac{\frac{1}{9}}{1 - \left(\frac{8}{9}\right)^2} = \frac{1}{9} \times \frac{81}{17} = \frac{9}{17}$$

B's chance $= \left(\frac{8}{9}\right)\left(\frac{1}{9}\right) + \left(\frac{8}{9}\right)^3\left(\frac{1}{9}\right) + \left(\frac{8}{9}\right)^5\left(\frac{1}{9}\right) + \ldots$

$$= \frac{\frac{8}{9} \times \frac{1}{9}}{1 - \left(\frac{8}{9}\right)^2} = \frac{8}{81} \times \frac{81}{17} = \frac{8}{17}$$

$\therefore$ A's chance : B's chance $= \frac{9}{17} : \frac{8}{17} = 9 : 8.$

Example 47. *A and B take turns in throwing two dice, the first to throw 10 being awarded the prize. Show that if A has the first throw, their chances of winning are in the ratio 12 : 11.*

Sol. Please try yourself.

Example 48. *A and B alternately cut a pack of cards and the pack is shuffled after each cut. If A starts and the game is continued until one cuts a diamond, what are the respective chances of A and B first cutting a diamond.*

Sol. The probability of cutting a diamond

$$= \frac{13}{52} = \frac{1}{4}$$

The probability of not cutting a diamond

$$= 1 - \frac{1}{4} = \frac{3}{4}$$

Proceeding as in Example 46

A's chance $= \frac{4}{7}$, B's chance $= \frac{3}{7}$.

Example 49. *A and B throw alternately with a pair of ordinary dice. A wins if he throws 6 before B throws 7 and B wins if he throws 7 before A throws 6. If A begins, find their respective chances of winning.*

(Meerut 1984, 86)

[Huyghen's Problem]

Sol. 6 can be obtained with two dice in the following ways.

(1, 5) ; (5, 1) ; (2, 4) ; (4, 2) ; (3, 3) *i.e.,* in 5 ways.

∴ The probability of throwing 6 with two dice

$$= \frac{5}{36}$$

The probability of not throwing 6 with two dice

$$= 1 - \frac{5}{36} = \frac{31}{36}$$

7 can be obtained with two dice in the following ways :

(1, 6) ; (6, 1) ; (2, 5) ; (5, 2) ; (3, 4) ; (4, 3) *i.e.,* in 6 ways.

∴ The probability of throwing 7 with two dice

$$= \frac{6}{36} = \frac{1}{6}$$

The probability of not throwing 7 with two dice

$$= 1 - \frac{1}{6} = \frac{5}{6}$$

If A is to win, he should throw 6 in 1st or 3rd or 5th ...... throws.

If B is to win, he should throw 7 in 2nd or 4th or 6th ...... throws.

$$\text{A's chance} = \frac{5}{36} + \frac{31}{36} \cdot \frac{5}{6} \cdot \frac{5}{36} + \left(\frac{31}{36}\right)^2 \left(\frac{5}{6}\right)^2 \frac{5}{36}$$

+ ...... *[Infinite G.P.]*

$$= \frac{\frac{5}{36}}{1 - \frac{31}{36} \cdot \frac{5}{6}} = \frac{5}{36} \times \frac{216}{61} = \frac{30}{61}$$

$$\text{B's chance} = \frac{31}{36} \cdot \frac{1}{6} + \left(\frac{31}{36}\right)^2 \cdot \frac{5}{6} \cdot \frac{1}{6} + \left(\frac{31}{36}\right)^3 \left(\frac{5}{6}\right)^2 \frac{1}{6} + \ldots\ldots$$

$$= \frac{\frac{31}{36} \cdot \frac{1}{6}}{1 - \frac{31}{36} \cdot \frac{5}{6}} = \frac{31}{216} \times \frac{216}{61} = \frac{31}{61}.$$

5.19. USE OF BINOMIAL THEOREM

If the probabilities of the happening and of the non-happening (or failure) of an event in a single trial are p and q respectively, then the probability of its happening exactly r times in n trials is $^nC_r p^r q^{n-r}$, *i.e.,* the (r + 1)th term in the expansion of $(q + p)^n$.

Proof. As p is the probability of the happening of the event in a single trial, p^r is the probability of its happening in a particular set of r trials. Also, q

being the probability of its failure in a single trial, q^{n-r} is the probability of its failure in the remaining $(n-r)$ trials. Now the happening of the event in r trials and its failure in the remaining $(n-r)$ trials are themselves two independent events and, therefore, the probability of their concurrence is $p^r q^{n-r}$. Further, a set of r trials can be selected out of n trials in ${}^nC_r$ ways.

Hence, the probability that the event will happen exactly r times in n trials is ${}^nC_r p^r q^{n-r}$ which is the $(r+1)$th term in the expansion of $(q+p)^n$.

| ☞ not $(p+q)^n$

☞ **Note 1.** **It is a very useful result** and shows that the probabilities that the event will happen **exactly** once, **exactly** twice, **exactly** thrice, ... , in n trials are ${}^nC_1 pq^{n-1}$, ${}^nC_2 p^2 q^{n-2}$, ${}^nC_3 p^3 q^{n-3}$,....., respectively.

Note 2. The probability of happening of the event **at least r times** in n trials = the sum of the probabilities of its happening $r, r+1, r+2, \ldots, n$ times.

$$= {}^nC_r p^r q^{n-r} + {}^nC_{r+1} p^{r+1} q^{n-r-1} + \ldots + {}^nC_n p^n.$$

Example 1. *In tossing 10 coins, what is the probability of having exactly 5 heads ?*

Sol. In tossing one coin, the probability p of having a head is $\frac{1}{2}$ and the probability q of not having a head is $\frac{1}{2}$.

In tossing 10 coins, the probability of having **exactly** 5 heads

= 6th term in the expansion of $(q+p)^{10}$

i.e.,
$$\left(\frac{1}{2}+\frac{1}{2}\right)^{10} = {}^{10}C_5\left(\frac{1}{2}\right)^5\left(\frac{1}{2}\right)^5 = \frac{63}{256}.$$

Example 2. *In five throws with a pair of dice, what is the chance of throwing doublets :*

(*i*) *exactly three times* ;

(*ii*) *at least three times* ? (K.U. 1982 S)

Sol. (1, 1), (2, 2), (3, 3), (4, 4), (5, 5), (6, 6) are called doublets. In one throw with a pair of dice, the chance p of throwing doublets is $\frac{6}{36} = \frac{1}{6}$ and the chance q of not throwing doublets is $1 - \frac{1}{6} = \frac{5}{6}$.

(*i*) The chance of throwing doublets **exactly** three times in five throws

$$= \text{4th term in the expansion of } \left(\frac{5}{6}+\frac{1}{6}\right)^5$$

$$= {}^5C_3\left(\frac{5}{6}\right)^2\left(\frac{1}{6}\right)^3 = \frac{125}{3888}$$

(*ii*) The chance of throwing doublets **at least** three times in five throws

= the sum of the chances when the doublets are thrown three times, four times or five times

= the sum of the 4th, 5th and 6th terms in the expansion of $\left(\frac{5}{6}+\frac{1}{6}\right)^5$

$$= {}^5C_3\left(\frac{5}{6}\right)^2\left(\frac{1}{6}\right)^3 + {}^5C_4\left(\frac{5}{6}\right)^1\left(\frac{1}{6}\right)^4 + {}^5C_5\left(\frac{1}{6}\right)^5$$

$$= 10 \times \frac{25}{6^5} + 5 \times \frac{5}{6^5} + \frac{1}{6^5}$$

$$= \frac{1}{6^5}(250 + 25 + 1) = \frac{276}{6^5} = \frac{23}{648}.$$

Example 3. *In four throws with a pair of dice, what is the chance of throwing doublets twice at least ?*

Sol. Please try yourself. $\left[\textbf{Ans}.\ \frac{19}{144}\right]$

Example 4. *Show that if a coin is tossed n times, the probability of not more than r heads is*

$$({}^nC_0 + {}^nC_1 + \ldots.. + {}^nC_r)\left(\frac{1}{2}\right)^n.$$

Sol. When a coin is tossed once, the probability p of getting a head is $\frac{1}{2}$ and the probability q of not getting a head is

$$1 - \frac{1}{2} = \frac{1}{2}.$$

If the coin is tossed n times, the probability of not more than r heads

= the sum of the probabilities of 0, 1, 2,...., r heads

= the sum of first $(r + 1)$ terms of $\left(\frac{1}{2}+\frac{1}{2}\right)^n$

$$= {}^nC_0\left(\frac{1}{2}\right)^n + {}^nC_1\left(\frac{1}{2}\right)^{n-1}\left(\frac{1}{2}\right) + {}^nC_2\left(\frac{1}{2}\right)^{n-2}\cdot\left(\frac{1}{2}\right)^2 + \ldots + {}^nC_r\left(\frac{1}{2}\right)^{n-r}\left(\frac{1}{2}\right)^r$$

$$= ({}^nC_0 + {}^nC_1 + {}^nC_2 + \ldots.. + {}^nC_r)\left(\frac{1}{2}\right)^n.$$

Example 5. *Ten coins are thrown simultaneously. Find the probability of getting at least seven heads.*

Sol. Required probability

= the sum of the probabilities of getting 7, 8, 9 or 10 heads

= the sum of the 8th, 9th, 10th and 11th terms in the expansion of $\left(\frac{1}{2}+\frac{1}{2}\right)^{10}$

$$= {}^{10}C_7\left(\frac{1}{2}\right)^3\left(\frac{1}{2}\right)^7 + {}^{10}C_8\left(\frac{1}{2}\right)^2\left(\frac{1}{2}\right)^8 + {}^{10}C_9\left(\frac{1}{2}\right)\left(\frac{1}{2}\right)^9 + {}^{10}C_{10}\left(\frac{1}{2}\right)^{10}$$

$$= \left(\frac{1}{2}\right)^{10}[{}^{10}C_3 + {}^{10}C_2 + {}^{10}C_1 + {}^{10}C_0] \qquad | \because \; {}^nC_r = {}^nC_{n-r}$$

$$= \frac{120+45+10+1}{1024} = \frac{176}{1024} = \frac{11}{64}.$$

Example 6. *An experiment succeeds twice as often as it fails. Find the chance that in the next six trials, there will be at least four successes.*

Sol. The probability p that the experiment succeeds $= \frac{2}{3}$

($\because$ odds in favour of it are 2 : 1)

The probability q that the experiment fails

$$= 1 - \frac{2}{3} = \frac{1}{3}.$$

The chance that in the next six trials, there will be at least four successes

= the sum of the chances that there will be 4, 5, 6 successes

= the sum of the 5th, 6th and 7th terms in the expansion of $\left(\frac{1}{3}+\frac{2}{3}\right)^6$

$$= {}^6C_4\left(\frac{1}{3}\right)^2\left(\frac{2}{3}\right)^4 + {}^6C_5\left(\frac{1}{3}\right)\left(\frac{2}{3}\right)^5 + {}^6C_6\left(\frac{2}{3}\right)^6$$

$$= \left(\frac{2}{3}\right)^4\left[\frac{15}{9}+\frac{4}{3}+\frac{4}{9}\right] = \frac{16}{81}\times\frac{31}{9} = \frac{496}{729}.$$

Example 7. *During war, one ship out of every 9 was sunk on an average in making a certain voyage. What was the probability that exactly 3 out of a convoy of 6 ships would arrive safely ?*

Sol. The chance that a ship is sunk $= \frac{1}{9}$.

$\therefore$ The chance that a ship is not sunk $= 1 - \frac{1}{9} = \frac{8}{9}$

The chance that exactly 3 out of 6 ships arrive safely

$$= \text{the fourth term in the expansion of } \left(\frac{1}{9} + \frac{8}{9}\right)^6$$

$$= {}^6C_3 \left(\frac{1}{9}\right)^3 \left(\frac{8}{9}\right)^3 = \frac{10240}{531441}.$$

Example 8. *If on an average, 1 vessel in every 10 is wrecked, find the probability that out of 5 vessels expected to arrive, 4 at least will arrive safely.*

Sol. The chance that a vessel is wrecked $= \frac{1}{10}$.

$\therefore$ The chance that a vessel arrives safely

$$= 1 - \frac{1}{10} = \frac{9}{10}.$$

The chance that at least 4 vessels out of 5 arrive safely

= the sum of the chances that 4 or 5 arrive safely

= the sum of the 5th and 6th terms in the expansion

$$\text{of } \left(\frac{1}{10} + \frac{9}{10}\right)^5$$

$$= {}^5C_4 \left(\frac{1}{10}\right)\left(\frac{9}{10}\right)^4 + {}^5C_5 \left(\frac{9}{10}\right)^5$$

$$= \left(\frac{9}{10}\right)^4 \left[{}^5C_1 \cdot \frac{1}{10} + {}^5C_0 \cdot \frac{9}{10}\right]$$

$$= \left(\frac{9}{10}\right)^4 \left(\frac{1}{2} + \frac{9}{10}\right) = \frac{9^4}{10^4} \times \frac{7}{5} = \frac{45927}{50000}.$$

Example 9. *The incidence of occupational disease in an industry is such that the workers have a 20% chance of suffering from it. What is the probability that out of six workers chosen at random, four or more will suffer from the disease ?*

Sol. The chance of suffering from disease = 20%

$$= \frac{20}{100} = \frac{1}{5}.$$

The chance of not suffering from disease $= 1 - \frac{1}{5} = \frac{4}{5}$.

The probability that out of six workers, four or more will suffer from disease

= the sum of the probabilities that 4, 5 or 6 suffer from disease

= the sum of the 5th, 6th and 7th terms in the expansion of $\left(\frac{4}{5}+\frac{1}{5}\right)^6$

$$= {}^6C_4\left(\frac{4}{5}\right)^2\left(\frac{1}{5}\right)^4 + {}^6C_5\left(\frac{4}{5}\right)\left(\frac{1}{5}\right)^5 + {}^6C_6\left(\frac{1}{5}\right)^6$$

$$= \left(\frac{1}{5}\right)^4\left[{}^6C_2 \cdot \frac{16}{25} + {}^6C_1 \cdot \frac{4}{5} \cdot \frac{1}{5} + {}^6C_0 \cdot \left(\frac{1}{5}\right)^2\right]$$

$$= \frac{1}{625}\left[\frac{240}{25}+\frac{24}{25}+\frac{1}{25}\right] = \frac{1}{625} \times \frac{265}{25} = \frac{53}{3125}.$$

Example 10. *If m things are distributed among 'a' men and 'b' women, show that the probability that the number of things received by men is odd, is*

$$\frac{1}{2}\left[\frac{(b+a)^m-(b-a)^m}{(b+a)^m}\right].$$

Sol. The probability p that a thing is received by a man

$$= \frac{a}{a+b}.$$

The probability q that a thing is received by a women

$$= \frac{b}{a+b}.$$

The probability that out of m things, **exactly** x are received by men and the rest by women

$= (x + 1)$th term in the expansion of $(q + p)^m$

$= {}^mC_x q^{m-x} p^x.$

The probability P that the number of things received by men is odd = the sum of the probabilities of 1, 3, 5,...., things received by men.

$$\therefore \quad P = {}^mC_1 q^{m-1} p + {}^mC_3 q^{m-3} p^3 + {}^mC_5 q^{m-5} p^5 + \ldots\ldots \quad \ldots(1)$$

Now $(q+p)^m = q^m + {}^mC_1 q^{m-1} p + {}^mC_2 q^{m-2} p^2 + {}^mC_3 q^{m-3} p^3 + {}^mC_4 q^{m-4} p^4 + \ldots\ldots$

and $(q-p)^m = q^m - {}^mC_1 q^{m-1} p + {}^mC_2 q^{m-2} p^2 - {}^mC_3 q^{m-3} p^3 + {}^mC_4 q^{m-4} p^4 \ldots\ldots$

Subtracting

$(q+p)^m - (q-p)^m = 2[{}^mC_1 q^{m-1} p + {}^mC_3 q^{m-3} p^3 + \ldots\ldots] = 2P$

[using (1)]

But $\quad q + p = 1 \quad$ and $\quad q - p = \dfrac{b-a}{b+a}.$

$$\therefore \qquad 1-\left(\frac{b-a}{b+a}\right)^m = 2P$$

Hence $\qquad P = \frac{1}{2}\left[\frac{(b+a)^m-(b-a)^m}{(b+a)^m}\right]$

5.20. DE MOIVRE'S PROBLEM. (USE OF MULTINOMIAL EXPANSION)

A die has f faces marked from 1 to f ; such n dice were thrown. To find the chance that the sum of the numbers exhibited on the uppermost faces is equal to p.

Since any one of the f faces may be exposed on any one of the n dice, the total number of ways in which the dice may fall is f^n.

Any of the n numbers from 1, 2, 3,....., f should go to make p as their sum in order that the event may be favourable.

Also in the expansion of $(x^1 + x^2 + x^3 + \ldots + x^f)^n$, the co-efficient of x^p arises out of the different ways in which n of the indices 1, 2, 3,....., f can be taken so as to form p by addition.

$\therefore$ Required chance

$$= \frac{\text{co-eff. of } x^p \text{ in } (x^1+x^2+\ldots.+x^f)^n}{f^n}$$

$$= \frac{\text{co-eff. of } x^p \text{ in } x^n(1+x+x^2+\ldots\ldots+x^{f-1})^n}{f^n}$$

$$= \frac{\text{co-eff. of } x^p \text{ in } x^n\left[\dfrac{(1-x)(1+x+x^2+\ldots+x^{f-1})}{1-x}\right]^n}{f^n}$$

$$= \frac{\text{co-eff. of } x^p \text{ in } x^n \cdot \dfrac{(1-x^f)^n}{(1-x)^n}}{f^n}$$

$$= \frac{\text{co-eff. of } x^{p-n} \text{ in } (1-x^f)^n(1-x)^{-n}}{f^n}.$$

Example 1. *Five coins whose faces are marked 2, 3 are thrown. What is the chance of obtaining a total of 12 ?*

Sol. Total number of ways in which 5 coins can be thrown $= 2^5 = 32$.

Favourable ways of getting a total of 12

$= \text{co-eff. of } x^{12} \text{ in } (x^2+x^3)^5$

$= \text{co-eff. of } x^{12} \text{ in } x^{10}(1+x)^5$

$= \text{co-eff. of } x^{12-10} \text{ in } (1+x)^5$

$= \text{co-eff. of } x^2 \text{ in } (1+x)^5$

$= {}^5C_2 = 10$

$\therefore$ Required chance $= \dfrac{10}{32} = \dfrac{5}{16}$.

Example 2. *Find the chance of throwing 10 with 2 dice.*

Sol. Total number of ways in which 2 dice can be thrown

$$= 6^2 = 36$$

Favourable ways of getting a total of 10

$$= \text{co-eff. of } x^{10} \text{ in } (x^1 + x^2 + x^3 + x^4 + x^5 + x^6)^2$$
$$= \text{co-eff. of } x^{10} \text{ in } x^2(1 + x + x^2 + \ldots\ldots + x^5)^2$$
$$= \text{co-eff. of } x^8 \text{ in } (1 + x + x^2 + \ldots\ldots + x^5)^2$$
$$= \text{co-eff. of } x^8 \text{ in } \left[\frac{(1-x)(1 + x + x^2 \ldots\ldots x^5)}{1-x}\right]^2$$
$$= \text{co-eff. of } x^8 \text{ in } \left(\frac{1-x^6}{1-x}\right)^2$$
$$= \text{co-eff. of } x^8 \text{ in } (1-x^6)^2(1-x)^{-2}$$
$$= \text{co-eff. of } x^8 \text{ in } (1 - 2x^6 + x^{12})(1 + 2x + 3x^2 + \ldots\ldots + 9x^8 + \ldots..)$$
$$= 9 - 6 = 3.$$

$\therefore$ Required chance $= \dfrac{3}{36} = \dfrac{1}{12}$.

Example 3. *Find the chance of throwing 10 with*
(i) three dice *(ii) four dice.*

Sol. Please try yourself. $\left[\textbf{Ans. } (i)\ \dfrac{1}{8} \quad (ii)\ \dfrac{5}{81}\right]$

Example 4. *Four dice are thrown. What is the probability that the sum of the numbers appearing on the dice is 18 ?*

Sol. Total number of ways in which 4 dice can be thrown $= 6^4$

Favourable ways of getting a sum of 18

$$= \text{co-eff. of } x^{18} \text{ in } (x + x^2 + \ldots.. + x^6)^4$$
$$= \text{co-eff. of } x^{18} \text{ in } x^4\,(1 + x + \ldots. + x^5)^4$$
$$= \text{co-eff. of } x^{14} \text{ in } (1 + x \ldots.. + x^5)^4$$
$$= \text{co-eff. of } x^{14} \text{ in } \left[\frac{(1-x)(1 + x + x^2 + \ldots. + x^5)}{1-x}\right]^4$$
$$= \text{co-eff. of } x^{14} \text{ in } \left(\frac{1-x^6}{1-x}\right)^4$$
$$= \text{co-eff. of } x^{14} \text{ in } (1-x^6)^4\,(1-x)^{-4}$$
$$= \text{co-eff. of } x^{14} \text{ in } (1 - 4x^6 + 6x^{12} \ldots..)(1 + 4x + 10x^2 + \ldots\ldots + 165x^8 + \ldots\ldots + 680x^{14} + \ldots..)$$
$$= 680 - 660 + 60 = 80$$

$\therefore$ Required probability $= \dfrac{80}{6^4} = \dfrac{5}{81}$.

Example 5. *Determine the chance of throwing more than 8 with 3 dice.*

Sol. Total number of ways in which 3 dice can be thrown $= 6^3$

Favourable number of ways of getting a sum from 3 to 8

= sum of co-effs. of $x^3, x^4, ..., x^8$ in $(x + x^2 + ... + x^6)^3$

= sum of co-effs. of $x^3, x^4, ..., x^8$ in $x^3 (1 + x + ... + x^5)^3$

= sum of co-effs. of $x^0, x^1, x^2, ..., x^5$ in $(1 + x + ... + x^5)^3$

= sum of co-effs. of $x^0, x^1, ..., x^5$ in $\left[\dfrac{(1-x)(1+x+......+x^5)}{1-x}\right]^3$

= sum of co-effs. of $x^0, x^1, ..., x^5$ in $(1 - x^6)^3 (1 - x)^{-3}$

= sum of co-effs. of $x^0, x^1, ..., x^5$ in $(1 - 3x^6 .....) (1 + 3x + 6x^2 + 10x^3 + 15x^4 + 21x^5 + ...)$

= 1 + 3 + 6 + 10 + 15 + 21 = 56.

$\therefore$ Chance of obtaining a sum from 3 to 8 $= \dfrac{56}{6^3} = \dfrac{7}{27}$

Hence the chance of throwing more than 8 $= 1 - \dfrac{7}{27} = \dfrac{20}{27}$.

Example 6. *Counters marked 1, 2, 3 are placed in a bag, one is withdrawn and replaced three times. What is the chance of obtaining a total of 6 ?*

Sol. Three counters can be withdrawn three times in 3^3 = 27 ways

The favourable no. of ways for obtaining a total of 6

= co-eff. of x^6 in $(x^1 + x^2 + x^3)^3$

= co-eff. of x^6 in $x^3 (1 + x + x^2)^3$

= co-eff. of x^3 in $(1 + x + x^2)^3$

= co-eff. of x^3 in $\left[\dfrac{(1-x)(1+x+x^2)}{1-x}\right]^3$

= co-eff. of x^3 in $(1 - x^3)^3 (1 - x)^{-3}$

= co-eff. of x^3 in $(1 - 3x^3 + ...) (1 + 3x + 6x^2 + 10x^3 + ...)$

= 10 − 3 = 7.

$\therefore$ The reqd. probability $= \dfrac{7}{27}$.

Example 7. *Four tickets marked 00, 01, 10, 11 respectively are placed in a bag. A ticket is drawn at random five times, being replaced each time. Find the probability that the sum of the numbers on tickets thus drawn is 23.*

Sol. Four tickets can be drawn five times in 4^5 ways.

The favourable no. of ways for obtaining a sum 23

= co-eff. of x^{23} in $(x^0 + x^1 + x^{10} + x^{11})^5$

= co-eff. of x^{23} in $(1 + x + x^{10} + x^{11})^5$

= co-eff. of x^{23} in $(1 + x)^5 (1 + x^{10})^5$

$= \text{co-eff. of } x^{23} \text{ in } (1 + 5x + 10x^2 + 10x^3 + 5x^4 + x^5)$

$(1 + 5x^{10} + 10x^{20} + ...)$

$= 10 \times 10 = 100$

$\therefore$ The reqd. probability $= \frac{100}{4^5} = \frac{25}{256}$.

Example 8. *The four faces of a regular tetrahedron are numbered 1, 2, 3, 4. It is thrown five times and the figure on the lowest face is noted down. What is the probability that the sum will be 12 ?*

Sol. Please try yourself. $\left[\textbf{Ans}.\ \frac{155}{1024}\right]$

MISCELLANEOUS EXAMPLES

Example 1. *Six boys and four girls stand in a row at random. What is the probability that 4 girls stand together ?*

Sol. (6 + 4 =) 10 persons can be made to stand in a row in

$^{10}P_{10} = 10\,!$ ways.

$\therefore$ Exhaustive number of cases = 10 !

Let the event '4 girls stand together' be denoted by E.

Regarding the 4 girls as one unit (*i.e.*, group), we have 6 boys and 1 unit of girls *i.e.*, 7 objects which can be arranged among themselves in 7 ! ways.

Corresponding to each such arrangement, the 4 girls can be arranged among themselves in 4 ! ways.

$\therefore$ Number of favourable cases for the event E = 7 ! × 4 !

Hence $P(E) = \frac{7\,! \times 4\,!}{10\,!} = \frac{7\,! \times 4 \times 3 \times 2 \times 1}{10 \times 9 \times 8 \times 7\,!} = \frac{1}{30}$.

Example 2. *The probabilities that a student will receive A, B, C or D grade are 0.30, 0.35, 0.20 and 0.15 respectively. What is the probability that a student will receive at least B grade ?*

Sol. At least B grade $\Rightarrow$ either B grade or A grade.

$\therefore$ Reqd. probability = P(B grade or A grade)

= P(B grade) + P(A grade)

= 0.35 + 0.30 = 0.65.

Example 3. *What is the probability that in a single draw at random from a pack of cards, either a black card or a red ace will be drawn ?*

Sol. Let A = the event of drawing a black card

B = the event of drawing a red ace

Clearly $P(A) = \frac{26}{52}$, $P(B) = \frac{2}{52}$

Since the two events are mutually exclusive

P(A or B) = P(A) + P(B)

$= \frac{26}{52} + \frac{2}{52} = \frac{28}{52} = \frac{7}{13}$.

Example 4. *Find the probability of throwing at least one of the following totals on a single throw of a pair of dice : a total of 5, a total of 6, a total of 7.*

Sol. Exhaustive number of cases = $6^2 = 36$

Let the three events be denoted by A, B, C respectively.

Favourable cases for A are (1, 4), (4, 1), (2, 3), (3, 2) *i.e.*, 4 cases

Favourable cases for B are (1, 5), (5, 1), (2, 4), (4, 2), (3, 3) *i.e.*, 5 cases

Favourable cases for C are (1, 6), (6, 1), (2, 5), (5, 2), (3, 4), (4, 3) *i.e.*, 6 cases

$$\therefore \quad P(A) = \frac{4}{36}, \quad P(B) = \frac{5}{36}, \quad P(C) = \frac{6}{36}$$

The events being mutually exclusive

Reqd. probability = P(A or B or C)

$$= P(A) + P(B) + P(C) = \frac{15}{36} = \frac{5}{12}.$$

Example 5. *Two dice are thrown. Find the probability of getting an odd number on one and a multiple of 3 on the other.*

Sol. Let A = the event of getting an odd number on one die.

B = the event of getting a multiple of 3 on the other die.

Since there are 3 odd numbers from 1 to 6

$$\therefore \quad P(A) = \frac{3}{6} = \frac{1}{2}$$

Also there are 2 multiples of 3 from 1 to 6

$$\therefore \quad P(B) = \frac{2}{6} = \frac{1}{3}$$

A and B being independent events,

$$P(A \text{ and } B) = P(AB) = P(A) \,.\, P(B) = \frac{1}{6}.$$

Example 6. *A pair of dice is rolled. If the sum of 9 has appeared, find the probability that one of the dice shows 3.*

Sol. Let A = the event that the sum is 9

and B = the event that one of the dice shows 3 (on the assumption that A has happened)

Exhaustive cases = $6^2 = 36$

Favourable cases for the event

A = (3, 6), (6, 3), (4, 5), (5, 4) *i.e.*, 4 cases

$$\therefore \quad P(A) = \frac{4}{36} = \frac{1}{9}.$$

Favourable cases for the event AB = (3, 6), (6, 3) *i.e.*, 2 cases

$$\therefore \quad P(AB) = \frac{2}{36} = \frac{1}{18}$$

Since $P(AB) = P(A) \times P(B/A)$

$$\therefore \quad P(B/A) = \frac{P(AB)}{P(A)} = \frac{\frac{1}{18}}{\frac{1}{9}} = \frac{1}{2}.$$

Example 7. *'p' is the probability that a man aged x will die in a year. Find the probability that out of n men $A_1 A_2, \ldots\ldots, A_n$ each aged x, A_1 will die in a year and be the first to die.* (D.U. 1985 ; M.U. 1987)

Sol. The probability that a man dies in a year $= p$

⇒ The probability that a man does not die in a year $= 1 - p$

⇒ The probability that none of the n men die in a year $= (1 - p)^n$

⇒ The probability that *at least* one man dies in a year

$$= 1 - (1 - p)^n$$

Since any man out of n can die first, the probability that A_1 dies first

$$= \frac{1}{n}$$

Hence the probability that A_1 will die in a year and will be the first to die

$$= \frac{1}{n}[1 - (1 - p)^n].$$

Example 8. *The chances of winning 3 out of 5 games and 4 out of 5 games are equal. What is the chance of winning all the 5 games ?*

Sol. Let p, q denote the probabilities of winning and losing a game.

Then $\quad p + q = 1 \quad \ldots(1)$

Probability of winning 3 out of 5 games $= {}^5C_3 q^2 p^3 = 10q^2p^3$

Probability of winning 4 out of 5 games $= {}^5C_4 qp^4 = 5qp^4$

By the given condition $\quad 10q^2p^3 = 5qp^4$

or $\quad 2q = p \quad \ldots(2)$

From (1) and (2),

$$p = \frac{2}{3}, \quad q = \frac{1}{3}$$

∴ Probability of winning 5 out of 5 games

$$= {}^5C_5 p^5 = \left(\frac{2}{3}\right)^5 = \frac{32}{243}.$$

Example 9. *Out of 800 families with 4 children each, how many families would be expected to have (i) 2 boys and 2 girls (ii) at least one boy (iii) no girl (iv) at most 2 girls ? (Assume equal probabilities for boys and girls).* (K.U. 1982 S)

Sol. Here p = probability that a child is a boy = $\frac{1}{2}$.

q = probability that a child is a girl = $\frac{1}{2}$

n = total number of children in a family = 4

(*i*) Probability of having exactly 2 boys (and hence 2 girls)

$$= {}^4C_2 q^2 p^2$$

$$= 6\left(\frac{1}{2}\right)^2 \cdot \left(\frac{1}{2}\right)^2 = \frac{3}{8}$$

(*ii*) Probability of having at least one boy

= 1 – Probability of having no boy

$$= 1 - {}^4C_0 q^4 = 1 - \left(\frac{1}{2}\right)^4 = \frac{15}{16}$$

(*iii*) Probability of having no girl

= Probability of having all the 4 boys

$$= {}^4C_4 p^4 = \left(\frac{1}{2}\right)^4 = \frac{1}{16}$$

(*iv*) Probability of having at most 2 girls

= Probability of having at least 2 boys

= Probability of having (2 boys or 3 boys or 4 boys)

$$= {}^4C_2 q^2 p^2 + {}^4C_3 q p^3 + {}^4C_4 p^4$$

$$= 6\left(\frac{1}{2}\right)^2\left(\frac{1}{2}\right)^2 + 4\left(\frac{1}{2}\right)\left(\frac{1}{2}\right)^3 + \left(\frac{1}{2}\right)^4$$

$$= \frac{6}{16} + \frac{4}{16} + \frac{1}{16} = \frac{11}{16}$$

Total number of families = 800

$\therefore$ The reqd. number of families, in cases (*i*), (*ii*), (*iii*) and (*iv*) respectively are

(*i*) $800 \times \frac{3}{8} = 300$ (*ii*) $800 \times \frac{15}{16} = 750$

(*iii*) $800 \times \frac{1}{16} = 50$ (*iv*) $800 \times \frac{11}{16} = 550.$

Example 10. *A number is chosen at random from the numbers ranging from 1 to 50. What is the probability that the number chosen is a multiple of 2 or 3 or 10 ?* (M.D.U. 1982 S)

Sol. Let A, B, C be the events of chosing a multiple of 2, 3 and 10 respectively.

Exhaustive cases = ${}^{50}C_1 = 50$

The events are not mutually exclusive.

Cases favourable to A are 2, 4, 6,....., 50 *i.e.*, 25 cases

$\therefore \quad P(A) = \frac{25}{50}$

Cases favourable to B are 3, 6, 9, ..., 48 *i.e.,* 16 cases

$\therefore \quad P(B) = \frac{16}{50}$

Cases favourable of C are 10, 20,....., 50, *i.e.,* 5 cases

$\therefore \quad P(C) = \frac{5}{50}$

Cases favourable to AB (the event of chosing a multiple of 2 and 3 *i.e.,* a multiple of 6) are 6, 12, 18,...., 48 *i.e.,* 8 cases

$\therefore \quad P(AB) = \frac{8}{50}$

Cases favourable to AC (the event of chosing a multiple of 2 and 10 *i.e.,* a multiple of 10) are 10, 20,....., 50 *i.e.,* 5 cases

$\therefore \quad P(AC) = \frac{5}{50}$

Cases favourable to BC (the event of chosing a multiple of 3 and 10 *i.e.,* a multiple of 30) are 30 *i.e.,* 1 case

$\therefore \quad P(BC) = \frac{1}{50}$

Cases favourable to ABC (the event of chosing a multiple of 2 and 3 and 10 *i.e.,* a multiple of 30) are 30 *i.e.,* 1 case

$\therefore \quad P(ABC) = \frac{1}{50}.$

Hence $P(A \text{ or } B \text{ or } C) = P(A + B + C)$

$$= P(A) + P(B) + P(C) - P(AB) - P(AC) - P(BC) + P(ABC)$$

$$= \frac{25}{50} + \frac{16}{50} + \frac{5}{50} - \frac{8}{50} - \frac{5}{50} - \frac{1}{50} + \frac{1}{50}$$

$$= \frac{33}{50}.$$

Example 11. *There are four letters and four addressed envelopes. If the letters are put at random in the envelopes, find the probability that*

(*i*) *at least one letter will occupy its correct envelope.*

(*ii*) *no letter will occupy its correct envelope.*

(*iii*) *all letters will occupy correct envelope.*

Sol. (*i*) Let A_i ($i = 1, 2, 3, 4$) denote the event that *i*th letter is placed in the *i*th envelope.

Probability that at least one letter is placed in the right envelope

$$= P(A_1 + A_2 + A_3 + A_4)$$

$$= \Sigma P(A_1) - \Sigma P(A_1 A_2) + \Sigma P(A_1A_2A_3) - P(A_1A_2A_3A_4)$$

$$= {}^4C_1 \,.\, P(A_1) - {}^4C_2 \,.\, P(A_1A_2) + {}^4C_3 \,.\, P(A_1A_2A_3) - P(A_1A_2A_3A_4)$$

$$= 4 \,.\, P(A_1) - 6 \,.\, P(A_1A_2) + 4 \,.\, P(A_1A_2A_3) - P(A_1A_2A_3A_4) \quad ...(1)$$

Now one letter out of four can be placed in right envelope in 1 way.

$\therefore \quad P(A_1) = \dfrac{1}{4}.$

After placing one letter in the right envelope, 3 envelopes are left and the second letter can be placed in the right envelope in 1 way.

$\therefore \quad P(A_1A_2) = P(A_1)\, P(A_2/A_1) = \dfrac{1}{4} \,.\, \dfrac{1}{3} = \dfrac{1}{12}$

After placing two letters in the right envelopes, 2 envelopes are left and the third letter can be placed in the right envelope in 1 way.

$$\therefore \quad P(A_1A_2A_3) = P(A_1)\, P(A_2/A_1)\, P(A_3/A_1A_2)$$

$$= \frac{1}{4} \,.\, \frac{1}{3} \,.\, \frac{1}{2} = \frac{1}{24}$$

Similarly $P(A_1A_2A_3A_4)$

$$= P(A_1)\, P(A_2/A_1)\, P(A_3/A_1A_2)\, P(A_4/A_1A_2A_3)$$

$$= \frac{1}{4} \,.\, \frac{1}{3} \,.\, \frac{1}{2} \,.\, \frac{1}{1} = \frac{1}{24}$$

$\therefore$ From (1), $P(A_1 + A_2 + A_3 + A_4)$

$$= 4 \,.\, \frac{1}{4} - 6 \,.\, \frac{1}{12} + 4 \,.\, \frac{1}{24} - \frac{1}{24}$$

$$= \frac{15}{24} = \frac{5}{8}.$$

(*ii*) Probability that no letter will be placed in the right envelope

$$= 1 - P(A_1 + A_2 + A_3 + A_4)$$

$$= 1 - \frac{5}{8} = \frac{3}{8}.$$

(*iii*) 4 envelopes (each containing a letter) can be arranged among themselves in 4 ! = 24 ways and out of these, there is only one way when each letter is placed in the right envelope.

$\therefore$ The probability that all letters occupy correct envelope

$$= \frac{1}{24}.$$

Example 12. *Four different objects 1, 2, 3, 4 are distributed at random on four places marked 1, 2, 3, 4. What is the probability that none of the objects occupies the place corresponding to its number ?* (D.U. 1982)

Sol. Please try yourself.

[Hint. See Example 11 (*ii*)] **[Ans.** $\frac{3}{8}$]

Example 13. *Find the probability that in a random arrangement of the letters of the word 'UNIVERSITY', two I's do not come together.*

Sol. The word 'UNIVERSITY' has 10 letters of which I occurs twice and the rest are all different.

$\therefore$ Exhaustive cases = the number of arrangements of 10 letters of which I occurs twice

$$= \frac{10\,!}{2\,!} = \frac{10 \times 9\,!}{2} = 5 \times 9\,!$$

Treating the two I's as one letter, we can arrange (8 + 1 =) 9 letters (all different) in 9 ! ways. In all these arrangements, the two I's are together.

$\therefore$ Favourable cases = the number of arrangements in which the two I's donot come together

$$= 5 \times 9\,! - 9\,! = 4 \times 9\,!$$

$$\therefore \text{ Reqd. probability} = \frac{4 \times 9\,!}{5 \times 9\,!} = \frac{4}{5}.$$

Example 14. *From an urn containing 16 balls of the same size of which 8 are red, 5 black and 3 white, 4 balls are drawn at random. What is the probability that (i) all the four balls are red (ii) none of the four balls is black (iii) none of the four balls is white (iv) none of the four balls is red ?* (K.U. 1983)

Sol. 4 balls can be drawn out of 16 in $^{16}C_4$

$$= \frac{16 \times 15 \times 14 \times 13}{4 \times 3 \times 2 \times 1} = 1820$$

$\therefore$ Exhaustive cases = 1820

(*i*) 4 red balls can be drawn out of 8 red balls in $^8C_4$

$$= \frac{8 \times 7 \times 6 \times 5}{4 \times 3 \times 2 \times 1} = 70 \text{ ways}$$

$\therefore$ Favourable cases = 70

$$\text{Reqd. probability} = \frac{70}{1820} = \frac{1}{26}.$$

(*ii*) 4 balls can be drawn out of (8 red + 3 white =) 11 non-black balls in $^{11}C_4$

$$= \frac{11 \times 10 \times 9 \times 8}{4 \times 3 \times 2 \times 1} = 330 \text{ ways}$$

$\therefore$ Favourable cases = 330

$$\text{Reqd. probability} = \frac{330}{1820} = \frac{33}{182}$$

(*iii*) 4 balls can be drawn out of (8 red + 5 black =) 13 non-white balls in $^{13}C_4$

$$= \frac{13 \times 12 \times 11 \times 10}{4 \times 3 \times 2 \times 1} = 715 \text{ ways}$$

$\therefore$ Favourable cases = 715

Reqd. probability $= \frac{715}{1820} = \frac{11}{28}$.

(*iv*) 4 balls can be drawn out of (5 black + 3 white =) 8 non-red balls in ${}^8C_4$

$$= \frac{8 \times 7 \times 6 \times 5}{4 \times 3 \times 2 \times 1} = 70 \text{ ways}$$

Favourable cases = 70

Reqd. probability $= \frac{70}{1820} = \frac{1}{26}$.

Example 15. *A number n is chosen at random from the integers 1, 2, 3,...., n. A and B denote the events that numbers are multiples of 2 and 3 respectively. Show that A and B are independent events when n = 96 and not independent when n = 100.*

Sol. (*i*) *Let n = 96*

Number of exhaustive cases = 96

Cases favourable to A are 2, 4, 6,...., 96 *i.e.,* 48 cases

Cases favourable to B are 3, 6, 9,...., 96 *i.e.,* 32 cases

Cases favourable to AB (the event that the number is a multiple of 2 and 3 *i.e.,* a multiple of 6) are 6, 12, 18,...., 96 *i.e.,* 16 cases

$$\therefore \quad P(A) = \frac{48}{96} = \frac{1}{2}, \quad P(B) = \frac{32}{96} = \frac{1}{3}$$

$$P(AB) = \frac{16}{96} = \frac{1}{6}$$

Since $P(AB) = \frac{1}{6} = \frac{1}{2} \times \frac{1}{3} = P(A) \, . \, P(B)$

the events A and B are independent.

(*ii*) *Let* $n = 100$

Proceeding as in part (*i*),

$$P(A) = \frac{50}{100} = \frac{1}{2},$$

$$P(B) = \frac{33}{100}, \quad P(AB) = \frac{16}{100} = \frac{4}{25}$$

Since $P(AB) \neq P(A) \, . \, P(B)$,

the events A and B are not independent.

Example 16. *An urn contains 10 white and 3 black balls, while another urn contains 3 white and 5 black balls. Two are drawn from the first urn and put into the second urn and then a ball is drawn from the latter. What is the probability that it is a white ball ?* (K.U. 1981)

Sol. The two balls drawn from the first urn may be

(*i*) both white (*ii*) both black

(*iii*) one white and one black

Let these events be denoted by A, B, C respectively.

$$P(A) = \frac{^{10}C_2}{^{13}C_2} = \frac{10 \times 9}{13 \times 12} = \frac{15}{26}$$

$$P(B) = \frac{^{3}C_2}{^{13}C_2} = \frac{3 \times 2}{13 \times 12} = \frac{1}{26}$$

$$P(C) = \frac{^{10}C_1 \times {^{3}C_1}}{^{13}C_2} = \frac{10 \times 3}{\frac{13 \times 12}{2 \times 1}} = \frac{10}{26}$$

When two balls are transferred from first urn to second urn, the second urn will contain

(*i*) 5 white and 5 black balls

(*ii*) 3 white and 7 black balls

(*iii*) 4 white and 6 black balls.

Let W denote the event of drawing a white ball from the second urn in the three cases (*i*), (*ii*) and (*iii*).

Now $$P(W/A) = \frac{5}{10}, \qquad P(W/B) = \frac{3}{10}$$

$$P(W/C) = \frac{4}{10}$$

$\therefore$ Reqd. probability = P(A) . P(W/A) + P(B) . P(W/B) + P(C) . P(W/C)

$$= \frac{15}{26} \cdot \frac{5}{10} + \frac{1}{26} \cdot \frac{3}{10} + \frac{10}{26} \cdot \frac{4}{10}$$

$$= \frac{75 + 3 + 40}{260} = \frac{118}{260} = \frac{59}{130}.$$

Example 17. *Two cards are randomly drawn from a deck of 52 cards and thrown away. What is the probability of drawing an ace in a single draw from the remaining 50 cards ?*

Sol. The two cards drawn randomly may be

(*i*) both aces (*ii*) one ace and one non-ace card

(*iii*) both non-ace cards

Let these events be denoted by A_1, A_2, A_3 respectively.

$$P(A_1) = \frac{^{4}C_2}{^{52}C_2} = \frac{6}{26 \times 51}$$

$$P(A_2) = \frac{^{4}C_1 \times {^{48}C_1}}{^{52}C_2} = \frac{4 \times 48}{26 \times 51} = \frac{192}{26 \times 51}$$

$$P(A_3) = \frac{^{48}C_2}{^{52}C_2} = \frac{24 \times 47}{26 \times 51} = \frac{1128}{26 \times 51}$$

Let A denote the event of drawing an ace in a single draw from the remaining 50 cards in cases (*i*), (*ii*) and (*iii*).

The remaining 50 cards have : two aces in case (*i*), three aces in case (*ii*) and four aces in case (*iii*).

$$P(A/A_1) = \frac{2}{50},\ P(A/A_2) = \frac{3}{50},\ P(A/A_3) = \frac{4}{50}$$

$\therefore$ Reqd. probability $= P(A_1) \,.\, P(A/A_1) + P(A_2) \,.\, P(A/A_2) + P(A_3) \,.\, P(A/A_3)$

$$= \frac{6}{26 \times 51} \times \frac{2}{50} + \frac{192}{26 \times 51} \times \frac{3}{50} + \frac{1128}{26 \times 51} \times \frac{4}{50}$$

$$= \frac{12 + 576 + 4512}{26 \times 51 \times 50} = \frac{5100}{26 \times 51 \times 50} = \frac{1}{13}.$$

6

Correlation and Regression

6.1. UNIVARIATE AND BIVARIATE DISTRIBUTIONS

Distributions involving only one variable are called **univariate distributions** *e.g.*, **marks** obtained by a class of students in a particular subject in an examination ; **wages** earned by a group of workers.

Distributions involving two variables are called bivariate distributions *e.g., heights and weights* of a certain group of persons, ages of husband and wife at the time of marriage.

If (x_i, y_i), $i = 1, 2, 3, ..., n$ is a bivariate distribution and the pair (x_i, y_i) occurs f_i times, then f_i is called the frequency of that pair.

6.2. CORRELATION (*M.D.U. 1982*)

In a bivariate distribution, if the change in one variable affects a change in the other variable, the variables are said to be *correlated.*

If the two variables deviate in the same direction *i.e.,* if the increase (or decrease) in one results in a corresponding increase (or decrease) in the other, correlation is said to be *direct or positive.*

e.g., the correlation between income and expenditure is positive.

If the two variables deviate in opposite directions *i.e.,* if the increase (or decrease) in one results in a corresponding decrease (or increase) in the other, correlation is said to be *inverse or negative.*

e.g., the correlation between volume and the pressure of a perfect gas or the correlation between price and demand is negative.

Correlation is said to be *perfect* if the deviation in one variable is followed by a corresponding and **proportional deviation** in the other.

6.3. SCATTER OR DOT DIAGRAMS

It is the simplest method of the diagrammatic representation of bivariate data. Let (x_i, y_i), $i = 1, 2, 3, ..., n$ be a bivariate distribution. Let the values of the variables x and y be plotted along the x-axis and y-axis on a suitable scale. Then corresponding to every ordered pair, there corresponds a point or dot in the xy-plane. The diagram of dots so obtained is called a *dot or scatter diagram.*

If the dots are very close to each other and the number of observations is not very large, a fairly good correlation is expected. If the dots are widely scattered, a poor correlation is expected.

6.4. CORRELATION TABLE

If the number of observations is large, suitable class-intervals are adopted for each variable. Drawing vertical lines through the end-points of class-intervals along x-axis and horizontal lines through the end-points of class-intervals along y-axis, the co-ordinate plane is divided into rectangles. The number of dots which fall within a rectangle is counted. If any dot falls on the side of a rectangle, it is to be taken in the next interval. Thus, dots can be arranged in a tabular form. A table of this type is called a *correlation table.*

e.g., if the marks obtained by B.A./B.Sc. Part III students of Government College, Gurgaon in the December Test in Mathematics (out of 50) and English (out of 50) are as follows :

Roll Nos.	:	1	2	3	4	5	6	7	8	9	10	11	12
Math.	:	32	37	40	28	26	30	41	46	34	26	48	29
English	:	6	17	21	14	20	19	31	26	23	8	12	18

then denoting the marks in Math. by x and the marks in English by y, the data may be arranged in a tabular form as follows :

Dot Diagram

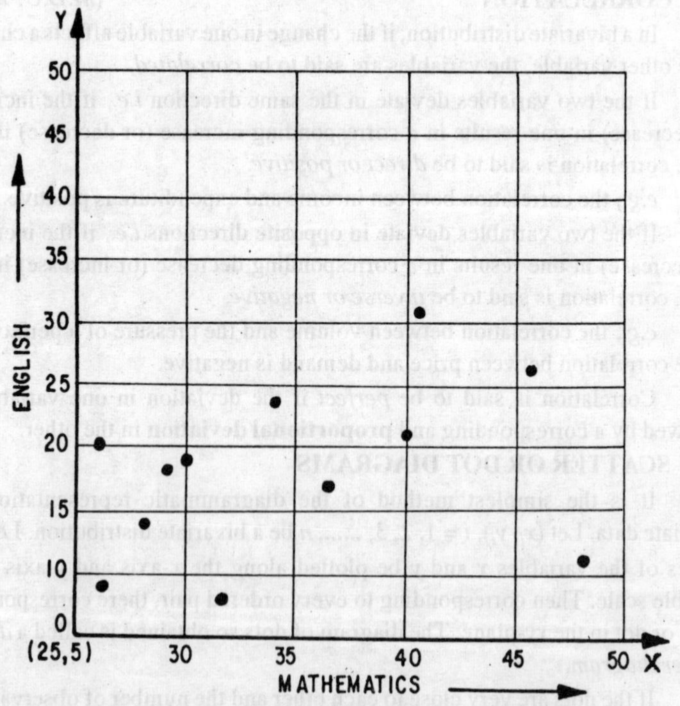

y \ x	25—30	30—35	35—40	40—45	45—50
5—10	1	1			
10—15	1				1
15—20	1	1	1		
20—25	1	1		1	
25—30					1
30—35				1	
35—40					
40—45					
45—50					

6.5. KARL PEARSON'S CO-EFFICIENT OF CORRELATION (OR PRODUCT MOMENT CORRELATION CO-EFFICIENT)

(M.D.U. 1981, 82, 82 S ; D.U. 1987 ; G.N.D.U. 1981 S, 82 ; M.U. 1984, 86)

Correlation co-efficient between two variables x and y, usually denoted by $r(x, y)$ or r_{xy} is a numerical measure of linear relationship between them and is defined as

$$r_{xy} = \frac{\Sigma(x_i - \overline{x})(y_i - \overline{y})}{\sqrt{\Sigma(x_i - \overline{x})^2 \cdot \Sigma(y_i - \overline{y})^2}} = \frac{\frac{1}{n}\Sigma(x_i - \overline{x})(y_i - \overline{y})}{\sqrt{\frac{1}{n}\Sigma(x_i - \overline{x})^2 \cdot \frac{1}{n}\Sigma(y_i - \overline{y})^2}}$$

$$= \frac{\frac{1}{n}\Sigma(x_i - \overline{x})(y_i - \overline{y})}{\sigma_x \sigma_y} = \frac{\text{cov. }(x, y)}{\sigma_x \sigma_y}$$

where $\frac{1}{n}\Sigma(x_i - \overline{x})(y_i - \overline{y})$ is known as co-variance of the variables x and y and is denoted by cov. (x, y).

6.6. LIMITS FOR CORRELATION CO-EFFICIENT

(M.D.U. 1982 ; K.U. 1981 ; D.U. 1987 ; M.U. 1989)

$$r_{xy} = \frac{\Sigma(x_i - \overline{x})(y_i - \overline{y})}{\sqrt{\Sigma(x_i - \overline{x})^2 \cdot \Sigma(y_i - \overline{y})^2}}$$

$$\Rightarrow \qquad r_{xy}^2 = \frac{[\Sigma(x_i - \overline{x})(y_i - \overline{y})]^2}{\Sigma(x_i - \overline{x})^2\, \Sigma(y_i - \overline{y})^2}$$

Putting $a_i = x_i - \bar{x}$ and $b_i = y_i - \bar{y}$

$$r_{xy}^2 = \frac{(\Sigma a_i b_i)^2}{\Sigma a_i^2 \Sigma b_i^2} \quad ...(1)$$

By Schwartz's Inequality, if a_i, b_i are real, then

$$(\Sigma a_i b_i)^2 \le (\Sigma a_i^2)(\Sigma b_i^2)$$

$\therefore$ from (1) $r_{xy}^2 \le 1$

or $\quad |r_{xy}| \le 1 \quad$ or $\quad -1 \le r_{xy} \le 1$

Hence *correlation co-efficient cannot exceed unity numerically.*

If $r = +1$, the correlation is perfect and positive. If $r = -1$, the correlation is perfect and negative.

Note. If $r = 0$, the variables are said to be uncorrelated.

☞6.7. CORRELATION CO-EFFICIENT IS INDEPENDENT OF CHANGE OF ORIGIN AND SCALE

(K.U. 1981, 82 S ; M.D.U. 1981, 82 S ; G.N.D.U. 1981 S, 82 ; D.U. 1986, 87)

Let (x_i, y_i), $i = 1, 2, \ldots\ldots, n$ be the bivariate distribution.

$$r_{xy} = \frac{\frac{1}{n}\Sigma (x_i - \bar{x})(y_i - \bar{y})}{\sigma_x \sigma_y}$$

Let us define two new variables u and v as

$$u = \frac{x-a}{h}, \; v = \frac{y-b}{k}$$

where a, b, h, k are constants, then

$$x = a + hu, \qquad y = b + kv$$

$$\therefore \quad \bar{x} = a + h\bar{u}, \qquad \bar{y} = b + k\bar{v}$$

$$x - \bar{x} = h(u - \bar{u}), \qquad y - \bar{y} = k(v - \bar{v})$$

$$\therefore \quad \sigma_x^2 = \frac{1}{n}\Sigma(x_i - \bar{x})^2 = \frac{h^2}{n}\Sigma(u_i - \bar{u})^2 = h^2\sigma_u^2$$

$$\therefore \quad \sigma_y^2 = \frac{1}{n}\Sigma(y_i - \bar{y})^2 = \frac{k^2}{n}\Sigma(v_i - \bar{v})^2 = k^2\sigma_v^2$$

$$\therefore \quad r_{xy} = \frac{\frac{1}{n}\Sigma (x_i - \bar{x})(y_i - \bar{y})}{\sigma_x \sigma_y} = \frac{\frac{1}{n}\Sigma hk (u_i - \bar{u})(v_i - \bar{v})}{h\sigma_u \cdot k\sigma_v}$$

$$= \frac{\frac{1}{n}\Sigma (u_i - \bar{u})(v_i - \bar{v})}{\sigma_u \sigma_v} = r_{uv}$$

Hence the result.

6.8. COMPUTATION OF CORRELATION CO-EFFICIENT

We know that

$$r_{xy} = \frac{\frac{1}{n}\Sigma (x_i - \bar{x})(y_i - \bar{y})}{\sigma_x \sigma_y}$$

Now $\frac{1}{n}\Sigma(x_i - \bar{x})(y_i - \bar{y}) = \frac{1}{n}\Sigma(x_i y_i - x_i\bar{y} - y_i\bar{x} + \bar{x}\,\bar{y})$

$$= \frac{1}{n}\Sigma x_i y_i - \bar{y}\,.\frac{1}{n}\Sigma x_i - \bar{x}\,.\frac{1}{n}\Sigma y_i + \frac{1}{n}(n\bar{x}\,\bar{y})$$

$$= \frac{1}{n}\Sigma x_i y_i - \bar{y}\,.\,\bar{x} - \bar{x}\,.\,\bar{y} + \bar{x}\,\bar{y} = \frac{1}{n}\Sigma x_i y_i - \bar{x}\,\bar{y}$$

$$\sigma_x^2 = \frac{1}{n}\Sigma(x_i - \bar{x})^2 = \frac{1}{n}\Sigma(x_i^2 - 2x_i\bar{x} + \bar{x}^2)$$

$$= \frac{1}{n}\Sigma x_i^2 - 2\bar{x}\,.\frac{1}{n}\Sigma x_i + \frac{1}{n}(n\bar{x}^2)$$

$$= \frac{1}{n}\Sigma x_i^2 - 2\bar{x}\,.\,\bar{x} + \bar{x}^2 = \frac{1}{n}\Sigma x_i^2 - \bar{x}^2$$

Similarly $\sigma_y^2 = \frac{1}{n}\Sigma y_i^2 - \bar{y}^2$

$$\therefore \quad r_{xy} = \frac{\frac{1}{n}\Sigma x_i y_i - \bar{x}\,\bar{y}}{\sqrt{\left(\frac{1}{n}\Sigma x_i^2 - \bar{x}^2\right)\left(\frac{1}{n}\Sigma y_i^2 - \bar{y}^2\right)}}$$

If $u = \frac{x-a}{h}$ and $v = \frac{y-b}{k}$

then

$$r_{xy} = r_{uv} = \frac{\frac{1}{n}\Sigma u_i v_i - \bar{u}\,\bar{v}}{\sqrt{\left(\frac{1}{n}\Sigma u_i^2 - \bar{u}^2\right)\left(\frac{1}{n}\Sigma v_i^2 - \bar{v}^2\right)}}$$

Note. The students are advised to calculate r_{xy} by arbitrary origin method given above rather than by the direct method.

If $\bar{x}$ and $\bar{y}$ are integers, shift the origin to $(\bar{x}, \bar{y})$.

Example 1. *Calculate the co-efficient of correlation for the following data :*

x :	1	5	4	6	2	3	8	6	2	8
y :	2	10	8	8	5	3	13	10	6	11

Sol.

x	y	$u = x - 6$	$v = y - 10$	u^2	v^2	uv
1	2	− 5	− 8	25	64	40
5	10	− 1	0	1	0	0
4	8	− 2	− 2	4	4	4
6	8	0	− 2	0	4	0
2	5	− 4	− 5	16	25	20
3	3	− 3	− 7	9	49	21
8	13	2	3	4	9	6
6	10	0	0	0	0	0
2	6	− 4	− 4	16	16	16
8	11	2	1	4	1	2
Total		− 15	− 24	79	172	109

$$\bar{u} = \frac{1}{n}\Sigma u_i = \frac{1}{10}(-15) = -1.5$$

$$\bar{v} = \frac{1}{n}\Sigma v_i = \frac{1}{10}(-24) = -2.4$$

$$r_{uv} = \frac{\frac{1}{n}\Sigma u_i v_i - \bar{u}\,\bar{v}}{\sqrt{\left(\frac{1}{n}\Sigma u_i^2 - \bar{u}^2\right)\left(\frac{1}{n}\Sigma v_i^2 - \bar{v}^2\right)}}$$

$$= \frac{\frac{1}{10}(109) - (-1.5)(-2.4)}{\sqrt{\left[\frac{1}{10}(79) - (-1.5)^2\right]\left[\frac{1}{10}(172) - (-2.4)^2\right]}}$$

$$= \frac{10.9 - 3.6}{\sqrt{(7.9 - 2.25)(17.2 - 5.76)}} = \frac{7.3}{\sqrt{5.65 \times 11.44}}$$

$$= \frac{7.3}{8} = 0.91.$$

Hence $r_{xy} = r_{uv} = 0.91$

Example 2. *Calculate the correlation co-efficient for the following heights in inches of fathers (X) and their sons (Y).*

$X:$	65	66	67	67	68	69	70	72
$Y:$	67	68	65	68	72	72	69	71

Sol.

X	Y	U = X – 68	V = Y – 69	U^2	V^2	UV
65	67	– 3	– 2	9	4	6
66	68	– 2	– 1	4	1	2
67	65	– 1	– 4	1	16	4
67	68	– 1	– 1	1	1	1
68	72	0	3	0	9	0
69	72	1	3	1	9	3
70	69	2	0	4	0	0
72	71	4	2	16	4	8
Total		0	0	36	44	24

$$\overline{U} = \frac{1}{n}\Sigma U_i = 0,\ V = \frac{1}{n}\Sigma V_i = 0$$

$$r_{UV} = \frac{\frac{1}{n}\Sigma U_iV_i - \overline{U}\,\overline{V}}{\sqrt{\left(\frac{1}{n}\Sigma U_i^{\,2} - \overline{U}^{\,2}\right)\left(\frac{1}{n}\Sigma V_i^{\,2} - \overline{V}^{\,2}\right)}}$$

$$= \frac{\frac{1}{8}(24) - 0}{\sqrt{\frac{1}{8}(36)\,.\,\frac{1}{8}(44)}} = \frac{24}{12\sqrt{11}} = \frac{2\sqrt{11}}{11} = 0.603.$$

Hence $r_{XY} = r_{UV} = 0.603$.

Example 3. *Ten students got the following percentage of marks in Principles of Economics and Statistics.*

Roll Nos.	:	*1*	*2*	*3*	*4*	*5*	*6*	*7*	*8*	*9*	*10*
Marks in Economics	:	*78*	*36*	*98*	*25*	*75*	*82*	*90*	*62*	*65*	*39*
Marks in Statistics	:	*84*	*51*	*91*	*60*	*68*	*62*	*86*	*58*	*53*	*47*

Calculate the co-efficient of correlation.

Sol. Let the marks in the two subjects be denoted by x and y respectively.

x	y	$u = x - 65$	$v = y - 66$	u^2	v^2	uv
78	84	13	18	169	324	234
36	51	– 29	– 15	841	225	435
98	91	33	25	1089	625	825
25	60	– 40	– 6	1600	36	240
75	68	10	2	100	4	20
82	62	17	– 4	289	16	– 68
90	86	25	20	625	400	500
62	58	– 3	– 8	9	64	24
65	53	0	– 13	0	169	0
39	47	– 26	– 19	676	361	494
Total		0	0	5398	2224	2734

$$\bar{u} = \frac{1}{n}\Sigma u_i = 0, \bar{v} = \frac{1}{n}\Sigma v_i = 0$$

$$r_{uv} = \frac{\frac{1}{n}\Sigma u_i v_i - \bar{u}\,\bar{v}}{\sqrt{\left(\frac{1}{n}\Sigma u_i^2 - \bar{u}^2\right)\left(\frac{1}{n}\Sigma v_i^2 - \bar{v}^2\right)}}$$

$$= \frac{\frac{1}{10}(2734)}{\sqrt{\frac{1}{10}(5398) \cdot \frac{1}{10}(2224)}} = \frac{2734}{\sqrt{5398 \times 2224}} = 0.787$$

Hence $r_{xy} = r_{uv} = 0.787$.

Example 4. *Find the co-efficient of correlation for the following table :*

x :	10	14	18	22	26	30
y :	18	12	24	6	30	36

Sol. Let $u = \frac{x-22}{4}$, $v = \frac{y-24}{6}$

x	y	u	v	u^2	v^2	uv
10	18	−3	−1	9	1	3
14	12	−2	−2	4	4	4
18	24	−1	0	1	0	0
22	6	0	−3	0	9	0
26	30	1	1	1	1	1
30	36	2	2	4	4	4
Total		−3	−3	19	19	12

$$\bar{u} = \frac{1}{n}\Sigma u_i = \tfrac{1}{6}(-3) = -\tfrac{1}{2}$$

$$\bar{v} = \frac{1}{n}\Sigma v_i = \tfrac{1}{6}(-3) = -\tfrac{1}{2}$$

$$r_{uv} = \frac{\frac{1}{n}\Sigma u_i v_i - \bar{u}\,\bar{v}}{\sqrt{\left(\frac{1}{n}\Sigma u_i^2 - \bar{u}^2\right)\left(\frac{1}{n}\Sigma v_i^2 - \bar{v}^2\right)}}$$

$$= \frac{\frac{1}{6}(12) - \frac{1}{4}}{\sqrt{\left[\frac{1}{6}(19) - \frac{1}{4}\right]\left[\frac{1}{6}(19) - \frac{1}{4}\right]}}$$

$$= \frac{\frac{7}{4}}{\frac{35}{12}} = \frac{3}{5} = 0.6.$$

Hence $r_{xy} = r_{uv} = 0.6.$

Example 5. *Calculate the co-efficient of correlation for the following ages of husbands and wives :*

Husband's age x	:	23	27	28	28	29	30	31	33	35	36
Wife's age y	:	18	20	22	27	21	29	27	29	28	29

(M.D.U. 1982)

Sol. Let $u = x - 30, \quad v = y - 25$

x	y	u	v	u^2	v^2	uv
23	18	−7	−7	49	49	49
27	20	−3	−5	9	25	15
28	22	−2	−3	4	9	6
28	27	−2	2	4	4	−4
29	21	−1	−4	1	16	4
30	29	0	4	0	16	0
31	27	1	2	1	4	2
33	29	3	4	9	16	12
35	28	5	3	25	9	15
36	29	6	4	36	16	24
Total		0	0	138	164	123

$$\bar{u} = \frac{1}{n} \Sigma u_i = \frac{1}{10} (0) = 0$$

$$\bar{v} = \frac{1}{n} \Sigma v_i = \frac{1}{10} (0) = 0$$

$$r_{uv} = \frac{\frac{1}{n} \Sigma u_i v_i - \bar{u}\,\bar{v}}{\sqrt{\left(\frac{1}{n} \Sigma u_i^2 - \bar{u}^2\right)\left(\frac{1}{n} \Sigma v_i^2 - \bar{v}^2\right)}}$$

$$= \frac{\frac{1}{10}(123) - 0}{\sqrt{\left[\frac{1}{10}(138) - 0\right]\left[\frac{1}{10}(164) - 0\right]}} = \frac{123}{\sqrt{138 \times 164}}$$

$$= .82$$

Hence $r_{xy} = r_{uv} = 0.82.$

Example 6. *By affecting suitable change of origin and scale, compute the product moment correlation co-efficient for the following set of 5 observations on (X, Y) :*

X :	*– 10*	*– 5*	*0*	*5*	*10*
Y :	*5*	*9*	*7*	*11*	*13*

Sol. Please try yourself. **[Ans.** 0.9]

Example 7. *A computer while calculating correlation co-efficient between two variables X and Y from 25 pairs of observations obtained the following results :*

$$n = 25, \quad \Sigma X = 125, \quad \Sigma X^2 = 650,$$
$$\Sigma Y = 100, \quad \Sigma Y^2 = 460, \quad \Sigma XY = 508.$$

It was, however, later discovered at the time of checking that he had copied down two pairs as

X	Y
6	14
8	6

while the correct values were

X	Y
8	12
6	8

Obtain the correct value of correlation co-efficient.

(D.U. 1988 ; Agra 1987)

Sol.

$$\left.\begin{aligned}
\text{Corrected } \Sigma X &= 125 - 6 - 8 + 8 + 6 = 125\\
\text{Corrected } \Sigma Y &= 100 - 14 - 6 + 12 + 8 = 100\\
\text{Corrected } \Sigma X^2 &= 650 - 6^2 - 8^2 + 8^2 + 6^2 = 650\\
\text{Corrected } \Sigma Y^2 &= 460 - 14^2 - 6^2 + 12^2 + 8^2 = 436\\
\text{Corrected } \Sigma XY &= 508 - 6 \times 14 - 8 \times 6 + 8 \times 12 + 6 \times 8 = 520
\end{aligned}\right\}$$

(Subtract the incorrect values and add the corresponding correct values)

$$\overline{X} = \frac{1}{n}\Sigma X = \frac{1}{25} \times 125 = 5$$

$$\overline{Y} = \frac{1}{n}\Sigma Y = \frac{1}{25} \times 100 = 4$$

$$\text{Corrected } r_{XY} = \frac{\frac{1}{n}\Sigma XY - \overline{X}\,\overline{Y}}{\sqrt{\left(\frac{1}{n}\Sigma X^2 - \overline{X}^2\right)\left(\frac{1}{n}\Sigma Y^2 - \overline{Y}^2\right)}}$$

$$= \frac{\frac{1}{25} \times 520 - 5 \times 4}{\sqrt{\left(\frac{1}{25} \times 650 - 25\right)\left(\frac{1}{25} \times 436 - 16\right)}}$$

$$= \frac{\frac{4}{5}}{\sqrt{(1)\left(\frac{36}{25}\right)}} = \frac{4}{5} \times \frac{5}{6} = \frac{2}{3} = 0.67.$$

Example 8. *Show that if x′, y′ are the deviations of the variables x and y from their respective means, then*

$$(i)\ r = 1 - \frac{1}{2N} \sum_i \left(\frac{x_i'}{\sigma_x} - \frac{y_i'}{\sigma_y} \right)^2$$

$$(ii)\ r = -1 + \frac{1}{2N} \sum_i \left(\frac{x_i'}{\sigma_x} + \frac{y_i'}{\sigma_y} \right)^2$$

Deduce that $\quad -1 \le r \le +1.$

Sol. Here $x_i' = x_i - \bar{x},\ y_i' = y_i - \bar{y}$

$$(i)\quad 1 - \frac{1}{2N} \sum_i \left(\frac{x_i'}{\sigma_x} - \frac{y_i'}{\sigma_y} \right)^2$$

$$= 1 - \frac{1}{2N} \sum_i \left[\frac{x_i'^2}{\sigma_x^2} + \frac{y_i'^2}{\sigma_y^2} - \frac{2x_i'y_i'}{\sigma_x\sigma_y} \right]$$

$$= 1 - \frac{1}{2N} \left[\frac{1}{\sigma_x^2} \sum_i x_i'^2 + \frac{1}{\sigma_y^2} \sum_i y_i'^2 - \frac{2}{\sigma_x\sigma_y} \sum_i x_i'y_i' \right]$$

$$= 1 - \frac{1}{2N} \left[\frac{1}{\sigma_x^2} \sum_i (x_i - \bar{x})^2 + \frac{1}{\sigma_y^2} \sum_i (y_i - \bar{y})^2 - \frac{2}{\sigma_x\sigma_y} \sum_i (x_i - \bar{x})(y_i - \bar{y}) \right]$$

$$= 1 - \frac{1}{2\sigma_x^2} \cdot \frac{1}{N} \sum (x_i - \bar{x})^2 - \frac{1}{2\,\sigma_y^2} \cdot \frac{1}{N} \sum_i (y_i - \bar{y})^2 + \frac{\frac{1}{N} \sum_i (x_i - \bar{x})(y_i - \bar{y})}{\sigma_x\sigma_y}$$

$$= 1 - \frac{1}{2\sigma_x^2} \cdot \sigma_x^2 - \frac{1}{2\sigma_y^2} \cdot \sigma_y^2 + r$$

$$= 1 - \tfrac{1}{2} - \tfrac{1}{2} + r = r.$$

$$(ii)\quad -1 + \frac{1}{2N} \sum_i \left(\frac{x_i'}{\sigma_x} + \frac{y_i'}{\sigma_y} \right)^2$$

$$= -1 + \frac{1}{2N} \sum_i \left(\frac{x_i'^2}{\sigma_x^2} + \frac{y_i'^2}{\sigma_y^2} + \frac{2x_i' y_i'}{\sigma_x \sigma_y} \right)$$

$$= -1 + \frac{1}{2N} \left[\frac{1}{\sigma_x^2} \sum_i x_i'^2 + \frac{1}{\sigma_y^2} \sum_i y_i'^2 + \frac{2}{\sigma_x \sigma_y} \sum_i x_i' y_i' \right]$$

$$= -1 + \frac{1}{2N} \left[\frac{1}{\sigma_x^2} \sum_i (x_i - \overline{x})^2 + \frac{1}{\sigma_y^2} \sum_i (y_i - \overline{y})^2 + \frac{2}{\sigma_x \sigma_y} \sum_i (x_i - \overline{x})(y_i - \overline{y}) \right]$$

$$= -1 + \frac{1}{2\sigma_x^2} \cdot \frac{1}{N} \sum_i (x_i - \overline{x})^2 + \frac{1}{2\sigma_y^2} \cdot \frac{1}{N} \sum_i (y_i - \overline{y})^2 + \frac{\frac{1}{N} \sum_i (x_i - \overline{x})(y_i - \overline{y})}{\sigma_x \sigma_y}$$

$$= -1 + \frac{1}{2\sigma_x^2} \cdot \sigma_x^2 + \frac{1}{2\sigma_y^2} \cdot \sigma_x^2 + r$$

$$= -1 + \tfrac{1}{2} + \tfrac{1}{2} + r = r.$$

Deduction. Since $\left(\frac{x_i'}{\sigma_x} \pm \frac{y_i'}{\sigma_y} \right)^2$ being the square of a real quantity is always non-negative (*i.e.*, ≥ 0), so is

$$\sum_i \left(\frac{x_i'}{\sigma_x} \pm \frac{y_i'}{\sigma_y} \right)^2$$

$$\therefore \qquad r = 1 - \frac{1}{2N} \sum_i \left(\frac{x_i'}{\sigma_x} - \frac{y_i'}{\sigma_y} \right)^2 \leq 1$$

$$r = -1 + \frac{1}{2N} \sum_i \left(\frac{x_i'}{\sigma_x} + \frac{y_i'}{\sigma_y} \right)^2 \geq -1$$

Hence $\qquad -1 \leq r \leq 1.$

Example 9. *The variables x and y are connected by the equation $ax + by + c = 0$. Show that the correlation co-efficient between them is – 1 if the signs of a and b are alike and + 1 if they are different.* (G.N.D.U. 1982)

Sol. $\qquad ax + by + c = 0 \qquad \Rightarrow \qquad y = -\frac{a}{b} x - \frac{c}{b}$

$$\therefore \qquad y_i = -\frac{a}{b} x_i - \frac{c}{b} \qquad \text{and} \qquad \overline{y} = -\frac{a}{b} \overline{x} - \frac{c}{b}$$

$$y_i - \overline{y} = -\frac{a}{b}(x_i - \overline{x})$$

$$r_{xy} = \frac{\Sigma (x_i - \overline{x})(y_i - \overline{y})}{\sqrt{\Sigma (x_i - \overline{x})^2 \, \Sigma (y_i - \overline{y})^2}}$$

$$= \frac{-\frac{a}{b}\Sigma (x_i - \overline{x})^2}{\sqrt{\frac{a^2}{b^2}[\Sigma (x_i - \overline{x})^2]^2}} = \frac{-\frac{a}{b}\Sigma (x_i - \overline{x})^2}{\sqrt{\frac{a^2}{b^2}} \, . \, \Sigma (x_i - \overline{x})^2}$$

$$= -\frac{a}{b} . \sqrt{\frac{b^2}{a^2}} = -\frac{a}{b}\left|\frac{b}{a}\right|$$

If a and b have opposite signs,

$$\frac{b}{a} \text{ is } - \text{ve and } \therefore \quad \left|\frac{b}{a}\right| = -\frac{b}{a}$$

$$\therefore \quad r_{xy} = -\frac{a}{b}\left(-\frac{b}{a}\right) = +1$$

If a and b have same sign,

$$\frac{b}{a} \text{ is } + \text{ve and } \therefore \quad \left|\frac{b}{a}\right| = \frac{b}{a}$$

$$\therefore \quad r_{xy} = -\frac{a}{b}\left(\frac{b}{a}\right) = -1.$$

Example 10. *If $z = ax + by$ and r is the correlation co-efficient between x and y, show that*

$$\sigma_z^2 = a^2\sigma_x^2 + b^2\sigma_y^2 + 2abr\,\sigma_x\sigma_y$$

(M.D.U. 1981 S ; D.U. 1992)

Sol. $z = ax + by$

$\Rightarrow \quad \overline{z} = a\,\overline{x} + b\,\overline{y}, \quad z_i = ax_i + by_i$

$$z_i - \overline{z} = a(x_i - \overline{x}) + b(y_i - \overline{y})$$

Now

$$\sigma_z^2 = \frac{1}{n}\Sigma(z_i - \overline{z})^2$$

$$= \frac{1}{n}\Sigma[a(x_i - \overline{x}) + b(y_i - \overline{y})]^2$$

$$= \frac{1}{n} . \Sigma[a^2(x_i - \overline{x})^2 + b^2(y_i - \overline{y})^2 + 2ab(x_i - \overline{x})(y_i - \overline{y})]$$

$$= a^2 . \frac{1}{n}\Sigma(x_i - \overline{x})^2 + b^2 . \frac{1}{n}\Sigma(y_i - \overline{y})^2 + 2ab . \frac{1}{n}\Sigma(x_i - \overline{x})(y_i - \overline{y})$$

$$= a^2\sigma_x{}^2 + b^2\sigma_y{}^2 + 2abr\sigma_x\sigma_y.$$

$$\because \quad r = \frac{\frac{1}{n}\Sigma (x_i - \bar{x})(y_i - \bar{y})}{\sigma_x\sigma_y}$$

Example 11. *Show that the correlation co-efficient r between two variables x and y is given by*

$$r = \frac{\sigma_x{}^2 + \sigma_y{}^2 - \sigma_{x-y}^2}{2\sigma_x\sigma_y}$$

where σ_x, σ_y and σ_{x-y} are the standard deviations of x, y and x – y respectively.

(D.U. 1983, 89, 91, 93)

Sol. Let $\quad z = x - y$ then $z_i = x_i - y_i$ and $\bar{z} = \bar{x} - \bar{y}$

$$\sigma^2_{x-y} = \sigma_z{}^2 = \frac{1}{n}\Sigma(z - \bar{z})^2$$

$$= \frac{1}{n}\Sigma[(x_i - y_i) - (\bar{x} - \bar{y})]^2$$

$$= \frac{1}{n}\Sigma\,[(x_i - \bar{x}) - (y_i - \bar{y})]^2$$

$$= \frac{1}{n}\Sigma\,(x_i - \bar{x})^2 + \frac{1}{n}\Sigma\,(y_i - \bar{y})^2 - 2\,.\,\frac{1}{n}\Sigma(x_i - \bar{x})(y_i - \bar{y})$$

$$= \sigma_x{}^2 + \sigma_y{}^2 - 2r\sigma_x\sigma_y$$

$$\because \quad r = \frac{\frac{1}{n}\Sigma(x_i - \bar{x})(y_i - \bar{y})}{\sigma_x\sigma_y}$$

$$\Rightarrow \quad 2r\,\sigma_x\sigma_y = \sigma_x{}^2 + \sigma_y{}^2 - \sigma^2_{x-y}$$

$$\therefore \quad r = \frac{\sigma_x{}^2 + \sigma_y{}^2 - \sigma^2_{x-y}}{2\,\sigma_x\sigma_y}.$$

Example 12. *Show that if a and b are constants and r is the correlation co-efficient between x and y, then the correlation co-efficient between ax and by is equal to r if the signs of a and b are alike and – r if they are different.*

(D.U. 1990)

Sol. Let $\quad u = ax, \quad v = by$

then $\quad u_i = ax_i, \quad v_i = by_i$

and $\quad \bar{u} = a\,\bar{x}, \quad \bar{v} = b\,\bar{y}$

$$r_{uv} = \frac{\Sigma\,(u_i - \bar{u})(v_i - \bar{v})}{\sqrt{\Sigma\,(u_i - \bar{u})^2\,\Sigma\,(v_i - \bar{v})^2}}$$

$$= \frac{\Sigma a\,(x_i - \bar{x})b(y_i - \bar{y})}{\sqrt{\Sigma a^2\,(x_i - \bar{x})^2\,\Sigma b^2\,(y_i - \bar{y})^2}}$$

$$= \frac{ab\Sigma\,(x_i - \overline{x})(y_i - \overline{y})}{\sqrt{a^2b^2\Sigma\,(x_i - \overline{x})^2\,\Sigma\,(y_i - \overline{y})^2}}$$

$$= \frac{ab\,[illegible]\,(y_i - \overline{y})}{\sqrt{a^2\,[illegible]\,-\overline{y})^2}} = \frac{ab}{|\,ab\,|}\,r$$

If a and b have same si[illegible] $\therefore$ $|\,ab\,| = ab$

$\therefore$ $r_{uv} = \frac{ab}{ab}\,r = r.$

If a and b have opposite signs, [illegible] ve

and $\therefore$ $|\,ab\,| = -\,ab$

$\therefore$ $r_{uv} = \frac{ab}{-\,ab}\,r = -\,r.$

Example 13. *Prove that correlation co-efficient between x and y is positive or negative according as* $\sigma_{x+y} >$ *or* $< \sigma_{x-y}$.

Sol. Proceeding as in Ex. 10.

$$\sigma^2_{ax+by} = a^2\sigma_x^{\,2} + b^2\sigma_y^{\,2} + 2abr\,\sigma_x\sigma_y$$

where r is the correlation co-efficient between x and y.

Putting $a = b = 1$

$$\sigma^2_{x+y} = \sigma_x^{\,2} + \sigma_y^{\,2} + 2r\sigma_x\sigma_y \quad \text{...(1)}$$

Putting $a = 1, b = -1$

$$\sigma^2_{x-y} = \sigma_x^{\,2} + \sigma_y^{\,2} - 2r\,\sigma_x\sigma_y \quad \text{...(2)}$$

Subtracting (2) from (1)

$$\sigma^2_{x+y} - \sigma^2_{x-y} = 4r\sigma_x\sigma_y$$

$$\therefore \quad r = \frac{\sigma^2_{x+y} - \sigma^2_{x-y}}{4\sigma_x\sigma_y}$$

Since standard deviation is always positive ; σ_x, σ_y are positive.

$\therefore$ If $\sigma_{x+y} > \sigma_{x-y}$, r is positive.

If $\sigma_{x+y} < \sigma_{x-y}$, r is negative.

Hence the result.

Example 14. *Prove that for two independent variables, r = 0. Show by an example that the converse is not true.* (M.U. 1984)

Sol. If x and y are independent variables, then, cov $(x, y) = 0$

$$\therefore \quad r_{xy} = \frac{\text{cov}\,(x, y)}{\sigma_x\sigma_y} = 0$$

Hence two independent variables are uncorrelated.

But the converse is not true, *i.e.*, two uncorrelated variables may not be independent as is shown in the following example :

x :	– 3	– 2	– 1	1	2	3
y :	9	4	1	1	4	9
xy :	– 27	– 8	– 1	1	8	27

Here $\Sigma x = 0, \quad \Sigma y = 28, \quad \Sigma xy = 0$

$$\therefore \quad \bar{x} = \frac{1}{n} \Sigma x = 0$$

$$\text{cov}(x, y) = \frac{1}{n} \Sigma xy - \bar{x}\,\bar{y} = 0$$

$$\therefore \quad r_{xy} = \frac{\text{cov}(x, y)}{\sigma_x \sigma_y} = 0$$

Thus the variables x and y are uncorrelated. But a careful examination shows that $y = x^2$ so that x and y are not independent.

Example 15. *Two independent random variables x and y have the following variances :*

$$\sigma_x^{\,2} = 36, \quad \sigma_y^{\,2} = 16.$$

Find the co-efficient of correlation between u = x + y and v = x − y

Sol. $u = x + y \quad \Rightarrow \quad \bar{u} = \bar{x} + \bar{y}$

$v = x - y \quad \Rightarrow \quad \bar{v} = \bar{x} - \bar{y}$

Since x and y are independent variables, $\quad r_{xy} = 0$

$$\Rightarrow \quad \text{cov}(x, y) = 0$$

or

$$\frac{1}{n} \Sigma (x - \bar{x})(y - \bar{y}) = 0 \qquad \text{...(1)}$$

Now

$$\text{cov}(u, v) = \frac{1}{n} \Sigma (u - \bar{u})(v - \bar{v})$$

$$= \frac{1}{n} \Sigma [(x - \bar{x}) + (y - \bar{y})][(x - \bar{x}) - (y - \bar{y})]$$

$$= \frac{1}{n} \Sigma [(x - \bar{x})^2 - (y - \bar{y})^2]$$

$$= \frac{1}{n} \Sigma (x - \bar{x})^2 - \frac{1}{n} \Sigma (y - \bar{y})^2$$

$$= \sigma_x^{\,2} - \sigma_y^{\,2} = 36 - 16 = 20$$

Also

$$\sigma_u^{\,2} = \frac{1}{n} \Sigma (u - \bar{u})^2 = \frac{1}{n} \Sigma [(x - \bar{x}) + (y - \bar{y})]^2$$

$$= \frac{1}{n} \Sigma (x - \bar{x})^2 + \frac{1}{n} \Sigma (y - \bar{y})^2 + 2 \cdot \frac{1}{n} \Sigma (x - \bar{x})(y - \bar{y})$$

$$= \sigma_x^{\,2} + \sigma_y^{\,2} + 0 \qquad \text{[using (1)]}$$

$$= 36 + 16 = 52$$

$$\sigma_v^{\,2} = \frac{1}{n} \Sigma (v - \bar{v})^2 = \frac{1}{n} \Sigma [(x - \bar{x}) - (y - \bar{y})]^2.$$

$$= \frac{1}{n} \Sigma (x - \bar{x})^2 + \frac{1}{n} \Sigma (y - \bar{y})^2 - 2 \cdot \frac{1}{n} \Sigma (x - \bar{x})(y - \bar{y})$$

$$= \sigma_x^{\,2} + \sigma_y^{\,2} - 0 \qquad \text{[using (1)]}$$

$$= 36 + 16 = 52$$

$$\therefore \quad r_{uv} = \frac{\text{cov}(u, v)}{\sigma_u \sigma_v} = \frac{20}{52} = \frac{5}{13}.$$

Example 16. *If x and y are two uncorrelated variables and if u = x + y, v = x – y, find the co-efficient of correlation between u and v in terms of* σ_x *and* σ_y*, the standard deviations of x and y respectively.*

Sol. Proceeding as in example 15.

$$\text{cov}(u, v) = \sigma_x^2 - \sigma_y^2$$

$$\sigma_u^2 = \sigma_x^2 + \sigma_y^2, \sigma_v^2 = \sigma_x^2 + \sigma_y^2$$

$$\therefore \quad r_{uv} = \frac{\text{cov}(u, v)}{\sigma_u \sigma_v} = \frac{\sigma_x^2 - \sigma_y^2}{\sigma_x^2 + \sigma_y^2}.$$

Example 17. *If u = ax + by and v = ax – by where x and y are measured from their respective means and* ρ *is the co-efficient of correlation between x and y ; if u and v are uncorrelated, then show that*

$$\sigma_u \sigma_v = 2ab\, \sigma_x \sigma_y \sqrt{1 - \rho^2}.$$

Sol. Since x and y are measured from their respective means

$$\therefore \quad \bar{x} = 0, \quad \bar{y} = 0$$

($\because$ Σx = sum of deviation from mean = 0, similarly $\Sigma y = 0$)

Now $u = ax + by \Rightarrow \bar{u} = a\bar{x} + b\bar{y} = 0$

$v = ax - by \Rightarrow \bar{v} = a\bar{x} - b\bar{y} = 0$

$\because$ u and v are uncorrelated, $\text{cov}(u, v) = 0$

$$\Rightarrow \quad \frac{1}{n} \Sigma(u - \bar{u})(v - \bar{v}) = 0 \quad \text{or} \quad \frac{1}{n} \Sigma uv = 0$$

or $$\frac{1}{n} \Sigma(ax + by)(ax - by) = 0$$

or $$\frac{1}{n} (a^2 x^2 - b^2 y^2) = 0$$

or $$a^2 . \frac{1}{n} \Sigma x^2 - b^2 . \frac{1}{n} \Sigma y^2 = 0$$

or $$a^2 \sigma_x^2 - b^2 \sigma_y^2 = 0 \qquad ...(1)$$

$$\left[\because \quad \sigma_x^2 = \frac{1}{n} \Sigma(x - \bar{x})^2 = \frac{1}{n} \Sigma x^2, \text{etc.} \right]$$

Also $$\sigma_u^2 = \frac{1}{n} \Sigma(u - \bar{u})^2 = \frac{1}{n} \Sigma u^2$$

$$= \frac{1}{n} \Sigma(ax + by)^2 = \frac{1}{n} \Sigma(a^2x^2 + b^2y^2 + 2abxy)$$

$$= a^2 . \frac{1}{n} \Sigma x^2 + b^2 . \frac{1}{n} \Sigma y^2 + 2ab . \frac{1}{n} \Sigma xy$$

$$= a^2 \sigma_x^2 + b^2 \sigma_y^2 + 2ab\, \rho\, \sigma_x \sigma_y \qquad ...(2)$$

$$\left[\because \quad \rho = \frac{\frac{1}{n}\Sigma(x-\bar{x})(y-\bar{y})}{\sigma_x\sigma_y} = \frac{\frac{1}{n}\Sigma xy}{\sigma_x\sigma_y}\right]$$

Similarly $\sigma_v{}^2 = \frac{1}{n}\Sigma(v-\bar{v})^2 = \frac{1}{n}\Sigma v^2 = \frac{1}{n}\Sigma(ax-by)^2$

$$= \frac{1}{n}\Sigma(a^2x^2 + b^2y^2 - 2abxy)$$

$$= a^2 . \frac{1}{n}\Sigma x^2 + b^2 . \frac{1}{n}\Sigma y^2 - 2ab . \frac{1}{n}\Sigma xy$$

$$= a^2\sigma_x{}^2 + b^2\sigma_y{}^2 - 2ab\,\rho\sigma_x\sigma_y \quad ...(3)$$

Multiplying (2) and (3)

$$\sigma_u{}^2\sigma_v{}^2 = (a^2\sigma_x{}^2 + b^2\sigma_y{}^2 + 2ab\,\rho\,\sigma_x\sigma_y)(a^2\sigma_x{}^2 + b^2\sigma_y{}^2 - 2ab\rho\,\sigma_x\sigma_y)$$

$$= (a^2\sigma_x{}^2 + b^2\sigma_y{}^2)^2 - 4a^2b^2\rho^2\,\sigma_x{}^2\sigma_y{}^2$$

$$= (a^2\sigma_x{}^2 - b^2\sigma_y)^2 + 4a^2\sigma_x{}^2 . b^2\sigma_y{}^2 - 4a^2b^2\rho^2\,\sigma_x{}^2\sigma_y{}^2$$

$$= 0 + 4a^2b^2\,\sigma_x{}^2\sigma_y{}^2(1-\rho^2) \quad \text{[using (1)]}$$

$$\therefore \quad \sigma_u\sigma_v = 2ab\,\sigma_x\sigma_y\sqrt{1-\rho^2}.$$

Example 18. *If $u = ax + by$ and $v = bx - ay$, where x and y represent deviations from the respective means, and if the co-efficient of correlation between x and y is r but u and v are uncorrelated, show that*

(*i*) $\sigma_u{}^2 + \sigma_v{}^2 = (a^2+b^2)(\sigma_x{}^2 + \sigma_y{}^2)$

(*ii*) $ab(\sigma_x{}^2 - \sigma_y{}^2) = r\sigma_x\sigma_y(a^2 - b^2)$

(*iii*) $\sigma_u\sigma_v = (a^2+b^2)\,\sigma_x\sigma_y\sqrt{1-r^2}.$ (M.U. 1985)

Sol. Since x and y are measured from their respective means

$\therefore \quad \bar{x} = 0, \quad \bar{y} = 0$

($\because$ Σx = sum of deviations from mean = 0, similarly $\Sigma y = 0$)

Now $u = ax + by \Rightarrow \bar{u} = a\bar{x} + b\bar{y} = 0$

$v = bx - ay \Rightarrow \bar{v} = b\bar{x} - a\bar{y} = 0$

Also $\sigma_u{}^2 = \frac{1}{n}\Sigma(u-\bar{u})^2 = \frac{1}{n}\Sigma u^2 = \frac{1}{n}\Sigma(ax+by)^2$

$$= a^2 . \frac{1}{n}\Sigma x^2 + b^2 . \frac{1}{n}\Sigma y^2 + 2ab . \frac{1}{n}\Sigma xy$$

$$= a^2\sigma_x{}^2 + b^2\sigma_y{}^2 + 2abr\,\sigma_x\sigma_y \quad ...(1)$$

$$\left[\because \quad r = \frac{\frac{1}{n}\Sigma(x-\bar{x})(y-\bar{y})}{\sigma_x\sigma_y} = \frac{\frac{1}{n}\Sigma xy}{\sigma_x\,\sigma_y}\right]$$

Similarly $\sigma_v{}^2 = b^2\sigma_x{}^2 + a^2\sigma_y{}^2 - 2abr\,\sigma_x\sigma_y$ ...(2)

Adding (1) and (2)

$$\sigma_u{}^2 + \sigma_v{}^2 = (a^2+b^2)(\sigma_x{}^2 + \sigma_y{}^2) \quad ...(I)$$

Since u and v are uncorrelated, cov $(u, v) = 0$

$$\Rightarrow \quad \frac{1}{n}\Sigma(u-\bar{u})(v-\bar{v}) = 0 \quad \text{or} \quad \frac{1}{n}\Sigma uv = 0$$

or $$\frac{1}{n}\Sigma(ax+by)(bx-ay) = 0$$

or $$\frac{1}{n}\Sigma[abx^2 - aby^2 - (a^2-b^2)xy] = 0$$

or $$ab\,.\frac{1}{n}\Sigma x^2 - ab\,.\frac{1}{n}\Sigma y^2 - (a^2-b^2)\,.\frac{1}{n}\Sigma xy = 0$$

or $$ab\,\sigma_x^{\,2} - ab\,\sigma_y^{\,2} - (a^2-b^2)\,r\,\sigma_x\sigma_y = 0$$

$$\therefore \quad ab(\sigma_x^{\,2} - \sigma_y^{\,2}) = r\sigma_x\sigma_y(a^2-b^2) \qquad \text{...(II)}$$

Multiplying (1) and (2)

$$\begin{aligned}
\sigma_u^{\,2}\,\sigma_v^{\,2} &= (a^2\sigma_x^{\,2} + b^2\sigma_y^{\,2} + 2abr\,\sigma_x\sigma_y)(b^2\sigma_x^{\,2} + a^2\sigma_y^{\,2} - 2abr\,\sigma_x\sigma_y) \\
&= (a^4+b^4)\,\sigma_x^{\,2}\sigma_y^{\,2} + a^2b^2(\sigma_x^{\,4} + \sigma_y^{\,4}) \\
&\qquad + 2abr\,\sigma_x\sigma_y\,(b^2\sigma_x^{\,2} + a^2\sigma_y^{\,2} - a^2\sigma_x^{\,2} - b^2\sigma_y^{\,2}) - 4a^2b^2r^2\,\sigma_x^{\,2}\sigma_y^{\,2} \\
&= (a^4+b^4)\,\sigma_x^{\,2}\sigma_y^{\,2} + a^2b^2\,(\sigma_x^{\,4} + \sigma_y^{\,4}) \\
&\qquad + 2abr\,\sigma_x\sigma_y\,(a^2-b^2)(\sigma_y^{\,2} - \sigma_x^{\,2}) - 4a^2b^2r^2\,\sigma_x^{\,2}\sigma_y^{\,2} \\
&= (a^2+b^2)^2\sigma_x^{\,2}\sigma_y^{\,2} - 2a^2b^2\sigma_x^{\,2}\sigma_y^{\,2} + a^2b^2(\sigma_y^{\,2} - \sigma_x^{\,2})^2 + 2a^2b^2\sigma_x^{\,2}\sigma_y^{\,2} \\
&\qquad + 2abr\,\sigma_x\sigma_y(a^2-b^2)(\sigma_y^{\,2} - \sigma_x^{\,2}) - 4a^2b^2r^2\,\sigma_x^{\,2}\sigma_y^{\,2} \\
&= (a^2+b^2)^2\,\sigma_x^{\,2}\sigma_y^{\,2} + a^2b^2\,(\sigma_y^{\,2} - \sigma_x^{\,2}) \\
&\qquad + 2abr\,\sigma_x\sigma_y\,(a^2-b^2)\,(\sigma_y^{\,2} - \sigma_x^{\,2}) - 4a^2b^2r^2\,\sigma_x^{\,2}\sigma_y^{\,2} \qquad \text{...(3)}
\end{aligned}$$

From II $0 = ab(\sigma_y^{\,2} - \sigma_x^{\,2}) + r\,\sigma_x\sigma_y\,(a^2-b^2)$

Squaring $0 = a^2\,b^2\,(\sigma_y^{\,2} - \sigma_x^{\,2})^2 + 2abr\,\sigma_x\sigma_y(a^2-b^2)(\sigma_y^{\,2} - \sigma_x^{\,2})$
$+ r^2\,\sigma_x^{\,2}\sigma_y^{\,2}(a^2-b^2)^2$ ...(4)

Subtracting (4) from (3)

$$\begin{aligned}
\sigma_u^{\,2}\sigma_v^{\,2} &= (a^2+b^2)^2\sigma_x^{\,2}\sigma_y^{\,2} - 4a^2b^2r^2\,\sigma_x^{\,2}\sigma_y^{\,2} - r^2\,\sigma_x^{\,2}\sigma_y^{\,2}(a^2-b^2)^2 \\
&= (a^2+b^2)^2\,\sigma_x^{\,2}\sigma_y^{\,2} - r^2\,\sigma_x^{\,2}\sigma_y^{\,2}[(a^2-b^2)^2 + 4a^2b^2] \\
&= (a^2+b^2)^2\,\sigma_x^{\,2}\sigma_y^{\,2} - r^2\,\sigma_x^{\,2}\sigma_y^{\,2}(a^2+b^2)^2 \\
&= (a^2+b^2)^2\,\sigma_x^{\,2}\sigma_y^{\,2}\,(1-r^2)
\end{aligned}$$

$$\therefore \quad \sigma_u\sigma_v = (a^2+b^2)\,\sigma_x\sigma_y\,\sqrt{1-r^2}. \qquad \text{...(III)}$$

Example 19. *x and y are two variates with variances* $\sigma_x^{\,2}$ *and* $\sigma_y^{\,2}$ *respectively and k is the co-efficient of correlation between them.*

If $u = x + ky$, $v = x + \dfrac{\sigma_x}{\sigma_y}y$, *find the value of k so that u and v are uncorrelated.* (D.U. 1987, 89, 92)

Sol.
$$k = \frac{\frac{1}{n}\Sigma\,(x-\bar{x})(y-\bar{y})}{\sigma_x\sigma_y} \qquad \text{...(1)}$$

$$u = x + ky \quad \Rightarrow \quad \bar{u} = \bar{x} + k\bar{y}$$

$$v = x + \frac{\sigma_x}{\sigma_y} y \quad \Rightarrow \quad \bar{v} = \bar{x} + \frac{\sigma_x}{\sigma_y} \bar{y}$$

Since u and v are uncorrelated, cov $(u, v) = 0$

$$\Rightarrow \quad \frac{1}{n} \Sigma(u - \bar{u})(v - \bar{v}) = 0$$

or $$\frac{1}{n} \Sigma[(x - \bar{x}) + k(y - \bar{y})] \left[(x - \bar{x}) + \frac{\sigma_x}{\sigma_y} (y - \bar{y}) \right] = 0$$

or $$\frac{1}{n} \Sigma(x - \bar{x})^2 + k \frac{\sigma_x}{\sigma_y} . \frac{1}{n} \Sigma(y - \bar{y})^2 + \left(k + \frac{\sigma_x}{\sigma_y} \right) . \frac{1}{n} \Sigma (x - \bar{x})(y - \bar{y}) = 0$$

or $$\sigma_x^2 + k . \frac{\sigma_x}{\sigma_y} . \sigma_y^2 + \left(k + \frac{\sigma_x}{\sigma_y} \right) . k \sigma_x \sigma_y = 0 \qquad [using\ (1)]$$

or $$\sigma_x^2 + k\sigma_x\sigma_y + k^2\sigma_x\sigma_y + k\sigma_x^2 = 0$$

or $$\sigma_x(\sigma_x + k\sigma_y) + k\sigma_x(\sigma_x + k\sigma_y) = 0$$

or $$(\sigma_x + k\sigma_y)(1 + k)\sigma_x = 0$$

or $$\sigma_x + k\sigma_y = 0 \qquad [\because \ k \neq -1, \sigma_x \neq 0]$$

$$\therefore \quad k = -\frac{\sigma_x}{\sigma_y}.$$

Example 20. *Two variates x and y have zero means, the same variance σ^2 and zero correlation co-efficient. Show that*

$$u = x \cos \alpha + y \sin \alpha \quad and \quad v = x \sin \alpha - y \cos \alpha$$

have the same variance σ^2 and zero correlation. (D.U. 1987, 91)

Sol. Here $\bar{x} = 0 = \bar{y}, \quad \sigma_x^2 = \sigma_y^2 = \sigma^2$

Correlation co-efficient = 0 $\Rightarrow$ cov $(x, y) = 0$

or $$\frac{1}{n} \Sigma(x - \bar{x})(y - \bar{y}) = 0 \quad \text{or} \quad \frac{1}{n} \Sigma xy = 0$$

Now $$u = x \cos \alpha + y \sin \alpha \quad \Rightarrow \quad \bar{u} = \bar{x} \cos \alpha + \bar{y} \sin \alpha = 0$$

$$v = x \sin \alpha - y \cos \alpha \quad \Rightarrow \quad \bar{v} = \bar{x} \sin \alpha - \bar{y} \cos \alpha = 0$$

$$\sigma_u^2 = \frac{1}{n} \Sigma(u - \bar{u})^2 = \frac{1}{n} \Sigma u^2 = \frac{1}{n} \Sigma (x \cos \alpha + y \sin \alpha)^2$$

$$= \cos^2 \alpha . \frac{1}{n} \Sigma x^2 + 2 \cos \alpha \sin \alpha . \frac{1}{n} \Sigma xy + \sin^2 \alpha . \frac{1}{n} \Sigma y^2$$

$$= \cos^2 \alpha . \sigma_x^2 + 2 \cos \alpha \sin \alpha . 0 + \sin^2 \alpha . \sigma_y^2$$

$$= \sigma^2 (\cos^2 \alpha + \sin^2 \alpha) = \sigma^2$$

Similarly $\sigma_v^2 = \sigma^2$

Also cov $$(u, v) = \frac{1}{n} \Sigma(u - \bar{u})(v - \bar{v})$$

$$= \frac{1}{n}\Sigma uv = \frac{1}{n}\Sigma(x\cos\alpha + y\sin\alpha)(x\sin\alpha - y\cos\alpha)$$

$$= \frac{1}{n}\Sigma[x^2\cos\alpha\sin\alpha - y^2\sin\alpha\cos\alpha - xy(\cos^2\alpha - \sin^2 y)]$$

$$= \cos\alpha\sin\alpha \,.\, \frac{1}{n}\Sigma x^2 - \cos\alpha\sin\alpha \,.\, \frac{1}{n}\Sigma y^2 - \cos 2\alpha \,.\, \frac{1}{n}\Sigma xy$$

$$= \cos\alpha\sin\alpha \,.\, \sigma_x^2 - \cos\alpha\sin\alpha \,.\, \sigma_y^2 - \cos 2\alpha \,.\, 0$$

$$= \sigma^2\cos\alpha\sin\alpha - \sigma^2\cos\alpha\sin\alpha = 0$$

$\Rightarrow$ u and v are uncorrelated.

Example 21. *Two variates x and y are normally correlated with co-efficient r and u, v are defined by $u = x\cos\alpha + y\sin\alpha$, $v = y\cos\alpha - x\sin\alpha$. Show that u, v will be uncorrelated if*

$$\tan 2\alpha = \frac{2r\,\sigma_x\sigma_y}{\sigma_x^2 - \sigma_y^2}.$$ (Rohilkhand 1987)

Sol. Since x and y are normal variates, $\bar{x} = 0, \bar{y} = 0$

$$r = \frac{\frac{1}{n}\Sigma(x - \bar{x})(y - \bar{y})}{\sigma_x\sigma_y} = \frac{\frac{1}{n}\Sigma xy}{\sigma_x\sigma_y} \quad ...(1)$$

Now $u = x\cos\alpha + y\sin\alpha \Rightarrow \bar{u} = \bar{x}\cos\alpha + \bar{y}\sin\alpha = 0$

$v = y\cos\alpha - x\sin\alpha \Rightarrow \bar{v} = \bar{y}\cos\alpha - \bar{x}\sin\alpha = 0$

Since u and v are uncorrelated, cov $(u, v) = 0$

$$\Rightarrow \frac{1}{n}\Sigma(u - \bar{u})(v - \bar{v}) = 0 \quad \text{or} \quad \frac{1}{n}\Sigma uv = 0$$

or $$\frac{1}{n}\Sigma(x\cos\alpha + y\sin\alpha)(y\cos\alpha - x\sin\alpha) = 0$$

or $$\frac{1}{n}\Sigma[xy(\cos^2\alpha - \sin^2\alpha) - x^2\cos\alpha\sin\alpha + y^2\cos\alpha\sin\alpha] = 0$$

or $$\cos 2\alpha \,.\, \frac{1}{n}\Sigma xy - \cos\alpha\sin\alpha \,.\, \frac{1}{n}\Sigma x^2 + \cos\alpha\sin\alpha \,.\, \frac{1}{n}\Sigma y^2 = 0$$

or $$r\,\sigma_x\sigma_y\cos 2\alpha - \cos\alpha\sin\alpha\,\sigma_x^2 + \cos\alpha\sin\alpha\,\sigma_y^2 = 0$$ *[using (1)]*

or $$r\,\sigma_x\sigma_y\cos 2\alpha = (\sigma_x^2 - \sigma_y^2)\cos\alpha\sin\alpha$$

or $$2r\,\sigma_x\sigma_y\cos 2\alpha = (\sigma_x^2 - \sigma_y^2)\sin 2\alpha$$

$$\therefore \quad \tan 2\alpha = \frac{2r\,\sigma_x\sigma_y}{\sigma_x^2 - \sigma_y^2}.$$

Example 22. *If x and y are uncorrelated random variables with means zero and variances σ_1^2 and σ_2^2 respectively, show that*

$$u = x\cos\alpha + y\sin\alpha, \; v = x\sin\alpha - y\cos\alpha$$

have correlation co-efficient ρ given by

$$\rho = \frac{\sigma_1^2 - \sigma_2^2}{\sqrt{(\sigma_1^2 - \sigma_2^2)^2 + 4\sigma_1^2\sigma_2^2 \operatorname{cosec}^2 2\alpha}}\,. \qquad \text{(D.U. 1982)}$$

Sol. Here $\sigma_x^2 = \sigma_1^2, \quad \sigma_y^2 = \sigma_2^2, \quad \bar{x} = 0 = \bar{y}$

$$r_{xy} = 0 \quad \Rightarrow \quad \operatorname{cov}(x, y) = 0$$

or $$\frac{1}{n}\Sigma(x - \bar{x})(y - \bar{y}) = 0 \quad \text{or} \quad \frac{1}{n}\Sigma xy = 0 \qquad \text{...(1)}$$

$$u = x\cos\alpha + y\sin\alpha \quad \Rightarrow \quad \bar{u} = \bar{x}\cos\alpha + \bar{y}\sin\alpha = 0$$

$$v = x\sin\alpha - y\cos\alpha \quad \Rightarrow \quad \bar{v} = \bar{x}\sin\alpha - \bar{y}\cos\alpha = 0$$

Now $$\sigma_u^2 = \frac{1}{n}\Sigma(u - \bar{u})^2 = \frac{1}{n}\Sigma u^2$$

$$= \frac{1}{n}\Sigma(x\cos\alpha + y\sin\alpha)^2$$

$$= \cos^2\alpha \,.\, \frac{1}{n}\Sigma x^2 + \sin^2\alpha \,.\, \frac{1}{n}\Sigma y^2 + 2\cos\alpha\sin\alpha \,.\, \frac{1}{n}\Sigma xy$$

$$= \sigma_1^2 \cos^2\alpha + \sigma_2^2 \sin^2\alpha \qquad \text{[using (1)]}$$

Similarly $$\sigma_v^2 = \sigma_1^2 \sin^2\alpha + \sigma_2^2\cos^2\alpha$$

$$\therefore \quad \sigma_u^2\sigma_v^2 = (\sigma_1^2\cos^2\alpha + \sigma_2^2\sin^2\alpha)(\sigma_1^2\sin^2\alpha + \sigma_2^2\cos^2\alpha)$$

$$= (\sigma_1^4 + \sigma_2^4)\cos^2\alpha\sin^2\alpha + \sigma_1^2\sigma_2^2(\cos^4\alpha + \sin^4\alpha)$$

$$= (\sigma_1^2 - \sigma_2^2)^2\cos^2\alpha\sin^2\alpha + 2\sigma_1^2\sigma_2^2\cos^2\alpha\sin^2\alpha$$

$$+ \sigma_1^2\sigma_2^2(\cos^2\alpha + \sin^2\alpha)^2 - 2\sigma_1^2\sigma_2^2\cos^2\alpha\sin^2\alpha$$

$$= (\sigma_1^2 - \sigma_2^2)^2\cos^2\alpha\sin^2\alpha + \sigma_1^2\sigma_2^2$$

$$= \cos^2\alpha\sin^2\alpha\left[(\sigma_1^2 - \sigma_2^2)^2 + \frac{4\sigma_1^2\sigma_2^2}{4\cos^2\alpha\sin^2\alpha}\right]$$

$$= \cos^2\alpha\sin^2\alpha\,[(\sigma_1^2 - \sigma_2^2)^2 + 4\sigma_1^2\sigma_2^2 \operatorname{cosec}^2 2\alpha]$$

or $$\sigma_u\sigma_v = \cos\alpha\sin\alpha\sqrt{(\sigma_1^2 - \sigma_2^2)^2 + 4\sigma_1^2\sigma_2^2\operatorname{cosec}^2 2\alpha}$$

Also $$\operatorname{cov}(u, v) = \frac{1}{n}(u - \bar{u})(v - \bar{v}) = \frac{1}{n}\Sigma uv$$

$$= \frac{1}{n}\Sigma(x\cos\alpha + y\sin\alpha)(x\sin\alpha - y\cos\alpha)$$

$$= \frac{1}{n}\Sigma[x^2\cos\alpha\sin\alpha - y^2\cos\alpha\sin\alpha - xy(\cos^2\alpha - \sin^2\alpha)]$$

$$= \cos\alpha\sin\alpha \,.\, \frac{1}{n}\Sigma x^2 - \cos\alpha\sin\alpha \,.\, \frac{1}{n}\Sigma y^2 - \cos 2\alpha \,.\, \frac{1}{n}\Sigma xy$$

$$= \cos\alpha \sin\alpha\, \sigma_1^2 - \cos\alpha \sin\alpha\, \sigma_2^2 - 0$$
$$= (\sigma_1^2 - \sigma_2^2) \cos\alpha \sin\alpha$$

$$\therefore \qquad \rho = r_{uv} = \frac{\text{cov}(u, v)}{\sigma_u \sigma_v}$$
$$= \frac{\sigma_1^2 - \sigma_2^2}{\sqrt{(\sigma_1^2 - \sigma_2^2)^2 + 4\sigma_1^2\sigma_2^2 \text{cosec}^2\, 2\alpha}}.$$

Example 23. *Two independent variates x_1 and x_2 have means 5 and 10 and variances 4 and 9 respectively. Obtain the correlation co-efficient between $y_1 = 3x_1 + 4x_2$ and $y_2 = 3x_1 - x_2$.* (D.U. 1985)

Sol. Here $\bar{x}_1 = 5$, $\bar{x}_2 = 10$, $\sigma_{x_1}^2 = 4$, $\sigma_{x_2}^2 = 9$

Since x_1 and x_2 are independent variates, cov $(x_1, x_2) = 0$

$$\Rightarrow \qquad \frac{1}{n} \Sigma(x_1 - \bar{x}_1)(x_2 - \bar{x}_2) = 0 \qquad ...(1)$$

Now $y_1 = 3x_1 + 4x_2 \Rightarrow \bar{y}_1 = 3\bar{x}_1 + 4\bar{x}_2$

$y_2 = 3x_1 - x_2 \Rightarrow \bar{y}_2 = 3\bar{x}_1 - \bar{x}_2$

$$\text{Cov}(y_1, y_2) = \frac{1}{n} \Sigma(y_1 - \bar{y}_1)(y_2 - \bar{y}_2)$$
$$= \frac{1}{n} \Sigma[3(x_1 - \bar{x}_1) + 4(x_2 - \bar{x}_2)][3(x_1 - \bar{x}_1) - (x_2 - \bar{x}_2)]$$
$$= 9 \cdot \frac{1}{n} \Sigma(x_1 - \bar{x}_1)^2 - 4 \cdot \frac{1}{n} \Sigma(x_2 - \bar{x}_2)^2 + 9 \cdot \frac{1}{n} \Sigma(x_1 - \bar{x}_1)(x_2 - \bar{x}_2)$$
$$= 9\sigma_{x_1}^2 - 4\sigma_{x_2}^2 + (0) \qquad \text{[using (1)]}$$
$$= 9(4) - 4(9) = 0$$

$\Rightarrow$ y_1 and y_2 are uncorrelated.

Example 24. *If x_1, x_2 and x_3 are three uncorrelated variables having standard deviations σ_1, σ_2 and σ_3 respectively, obtain the co-efficient of correlation between $(x_1 + x_2)$ and $(x_2 + x_3)$.*

Sol. $\because$ x_1, x_2, x_3 are uncorrelated in pairs

$\therefore$ cov $(x_1, x_2) = 0$, cov $(x_1, x_3) = 0$, cov $(x_2, x_3) = 0$

Let $u = x_1 + x_2$ and $v = x_2 + x_3$

then $\bar{u} = \bar{x}_1 + \bar{x}_2$ and $\bar{v} = \bar{x}_2 + \bar{x}_3$

$$\text{cov}(u, v) = \frac{1}{n} \Sigma(u - \bar{u})(v - \bar{v})$$
$$= \frac{1}{n} \Sigma[(x_1 - \bar{x}_1) + (x_2 - \bar{x}_2)][(x_2 - \bar{x}_2) + (x_3 - \bar{x}_3)]$$

$$= \frac{1}{n} \Sigma(x_1 - \bar{x}_1)(x_2 - \bar{x}_2) + \frac{1}{n} \Sigma(x_1 - \bar{x}_1)(x_3 - \bar{x}_3)$$

$$+ \frac{1}{n} \Sigma(x_2 - \bar{x}_2)(x_3 - \bar{x}_3) + \frac{1}{n} \Sigma(x_2 - \bar{x}_2)^2$$

$$= \text{cov}(x_1, x_2) + \text{cov}(x_1, x_3) + \text{cov}(x_2, x_3) + \sigma_{x_2}^2$$

$$= 0 + 0 + 0 + \sigma_2^2 = \sigma_2^2$$

Also $\quad \sigma_u^2 = \frac{1}{n} \Sigma(u - \bar{u})^2 = \frac{1}{n} \Sigma[(x_1 - \bar{x}_1) + (x_2 - \bar{x}_2)]^2$

$$= \frac{1}{n} \Sigma(x_1 - \bar{x}_1)^2 + \frac{1}{n} \Sigma(x_2 - \bar{x}_2)^2$$

$$+ 2 \cdot \frac{1}{n} \Sigma(x_1 - \bar{x}_1)(x_2 - \bar{x}_2)$$

$$= \sigma_{x_1}^2 + \sigma_{x_2}^2 + 2 \text{cov}(x_1, x_2)$$

$$= \sigma_1^2 + \sigma_2^2 \qquad [\because \text{cov}(x_1, x_1) = 0]$$

Similarly $\quad \sigma_v^2 = \sigma_2^2 + \sigma_3^2$

$$\therefore \qquad r_{uv} = \frac{\text{cov}(u, v)}{\sigma_u \sigma_v} = \frac{\sigma_2^2}{\sqrt{(\sigma_1^2 + \sigma_2^2)(\sigma_2^2 + \sigma_3^2)}}.$$

Example 25. *If x_1, x_2 and x_3 are three uncorrelated variables having the same standard deviation, obtain the co-efficient of correlation between $(x_1 + x_2)$ and $(x_2 + x_3)$.* (D.U. 1988)

Sol. Proceed as in example 24.

Here $\quad \sigma_1 = \sigma_2 = \sigma_3 = \sigma$ (*say*)

$$\therefore \qquad r_{uv} = \frac{\sigma^2}{\sqrt{2\sigma^2 \cdot 2\sigma^2}} = \frac{1}{2}.$$

Example 26. *If x and y are two correlated variables with the same standard deviation and the correlation co-efficient r, show that the correlation co-efficient between x and x + y is* $\sqrt{\frac{1+r}{2}}$. (D.U. 1993 ; Meerut 1991)

Sol. Please try yourself.

Example 27. *If x and y are two independent random variables, show that $r_{x+y,\, x-y} = r^2_{x,\, x+y} - r^2_{y,\, x+y}$, where $r_{x+y,\, x-y}$ denotes the co-efficient of correlation between x + y and x – y.* (D.U. 1986 ; Kanpur 1986)

Sol. x and y are independent variables $\Rightarrow$ cov $(x, y) = 0$

Let $\qquad u = x + y \quad$ and $\quad v = x - y$

then $\qquad \bar{u} = \bar{x} + \bar{y} \quad$ and $\quad \bar{v} = \bar{x} - \bar{y}$

$$\text{cov}(u, v) = \frac{1}{n} \Sigma(u - \bar{u})(v - \bar{v})$$

$$= \frac{1}{n} \Sigma[(x - \bar{x}) + (y - \bar{y})][(x - \bar{x}) - (y - \bar{y})]$$

$$= \frac{1}{n} \Sigma[(x - \bar{x})^2 - (y - \bar{y})^2]$$

$$= \frac{1}{n} \Sigma(x - \bar{x})^2 - \frac{1}{n} \Sigma(y - \bar{y})^2$$

$$= \sigma_x^2 - \sigma_y^2$$

$$\sigma_u^2 = \frac{1}{n} \Sigma(u - \bar{u})^2 = \frac{1}{n} \Sigma[(x - \bar{x}) + (y - \bar{y})]^2$$

$$= \frac{1}{n} \Sigma(x - \bar{x})^2 + \frac{1}{n} \Sigma(y - \bar{y})^2 + 2 \cdot \frac{1}{n} \Sigma(x - \bar{x})(y - \bar{y})$$

$$= \sigma_x^2 + \sigma_y^2 + 2 \text{ cov } (x, y) = \sigma_x^2 + \sigma_y^2$$

Similarly $\sigma_v^2 = \sigma_x^2 + \sigma_y^2$

$$\therefore \quad r_{uv} = \frac{\text{cov } (u, v)}{\sigma_u \sigma_v} = \frac{\sigma_x^2 - \sigma_y^2}{\sigma_x^2 + \sigma_y^2} \qquad \text{...(1)}$$

Now $\text{cov } (x, u) = \frac{1}{n} \Sigma(x - \bar{x})(u - \bar{u})$

$$= \frac{1}{n} \Sigma(x - \bar{x})[(x - \bar{x}) + (y - \bar{y})]$$

$$= \frac{1}{n} \Sigma(x - \bar{x})^2 + \frac{1}{n} \Sigma(x - \bar{x})(y - \bar{y})$$

$$= \sigma_x^2 + \text{cov } (x, y) = \sigma_x^2$$

$$r_{x, u} = \frac{\text{cov } (x, u)}{\sigma_x \sigma_u} = \frac{\sigma_x^2}{\sigma_x \sqrt{\sigma_x^2 + \sigma_y^2}} = \frac{\sigma_x}{\sqrt{\sigma_x^2 + \sigma_y^2}}$$

Also $\text{cov } (y, u) = \frac{1}{n} \Sigma(y - \bar{y})(u - \bar{u})$

$$= \frac{1}{n} \Sigma(y - \bar{y})[(x - \bar{x}) + (y - \bar{y})]$$

$$= \frac{1}{n} \Sigma(x - \bar{x})(y - \bar{y}) + \frac{1}{n} \Sigma(y - \bar{y})^2$$

$$= \text{cov } (x, y) + \sigma_y^2 = \sigma_y^2$$

$$r_{y, u} = \frac{\text{cov } (y, u)}{\sigma_y \sigma_u} = \frac{\sigma_y^2}{\sigma_y \sqrt{\sigma_x^2 + \sigma_y^2}} = \frac{\sigma_y}{\sqrt{\sigma_x^2 + \sigma_y^2}}$$

$$r^2_{x, u} - r^2_{y, u} = \frac{\sigma_x^2 - \sigma_y^2}{\sigma_y^2 + \sigma_y^2} \qquad \text{...(2)}$$

From (1) and (2)

$$r_{u, v} = r^2_{x, u} - r^2_{y, u}$$

r $$r_{x + y, x - y} = r^2_{x, x + y} - r^2_{y, x + y}$$

Example 28. *If x_1, x_2 and x_3 are three variates measured from their espective means as origin and of equal variances, find the co-efficient of* *correlation between $x_1 + x_3$ and $x_2 + x_3$ in terms of r_{12}, r_{13} and r_{23} and show that it is equal to*

(i) $\dfrac{r_{12}+1}{2}$, *if* $r_{13} = r_{23} = 0$

(ii) $\dfrac{r_{12}+3}{4}$, *if* $r_{13} = r_{23} = 1$.

Sol. Since x_1, x_2, x_3 are measured from their respective means

$$\Sigma x_1 = 0, \quad \Sigma x_2 = 0, \quad \Sigma x_3 = 0$$

$$\Rightarrow \quad \bar{x}_1 = \bar{x}_2 = \bar{x}_3 = 0$$

Also $\quad \sigma_1^2 = \sigma_2^2 = \sigma_3^2 = \sigma^2$ *(say)*

$$\text{cov}(x_1, x_2) = \frac{1}{n}\Sigma(x_1 - \bar{x}_1)(x_2 - \bar{x}_2) = \frac{1}{n}\Sigma x_1 x_2$$

$$\left.\begin{array}{l} \therefore \quad r_{12} = \dfrac{\text{cov}(x_1, x_2)}{\sigma_1\sigma_2} = \dfrac{\frac{1}{n}\Sigma x_1 x_2}{\sigma^2} \\ \text{Similarly } r_{13} = \dfrac{\frac{1}{n}\Sigma x_1 x_3}{\sigma^2}, \; r_{23} = \dfrac{\frac{1}{n}\Sigma x_2 x_3}{\sigma^2} \end{array}\right] \quad \text{...(A)}$$

Now, let $\quad u = x_1 + x_3 \quad$ and $\quad v = x_2 + x_3$

then $\quad \bar{u} = \bar{x}_1 + \bar{x}_3 = 0, \quad \bar{v} = \bar{x}_2 + \bar{x}_3 = 0$

$$\text{cov}(u, v) = \frac{1}{n}\Sigma(u - \bar{u})(v - \bar{v}) = \frac{1}{n}\Sigma uv$$

$$= \frac{1}{n}\Sigma(x_1 + x_3)(x_2 + x_3)$$

$$= \frac{1}{n}\Sigma x_1 x_2 + \frac{1}{n}\Sigma x_1 x_3 + \frac{1}{n}\Sigma x_2 x_3 + \frac{1}{n}\Sigma x_3^2$$

$$= \sigma^2 r_{12} + \sigma^2 r_{13} + \sigma^2 r_{23} + \sigma_3^2 \qquad \text{[using (A)]}$$

$$= \sigma^2(r_{12} + r_{13} + r_{23} + 1) \qquad [\because \; \sigma_3^2 = \sigma^2]$$

$$\sigma_u^2 = \frac{1}{n}\Sigma(u - \bar{u})^2 = \frac{1}{n}\Sigma u^2 = \frac{1}{n}\Sigma(x_1 + x_3)^2$$

$$= \frac{1}{n}\Sigma x_1^2 + \frac{1}{n}\Sigma x_3^2 + 2 \cdot \frac{1}{n}\Sigma x_1 x_3$$

$$= \sigma_1^2 + \sigma_3^2 + 2\sigma^2 r_{13} \qquad \text{[using (A)]}$$

$$= 2\sigma^2(1 + r_{13})$$

Similarly $\quad \sigma_v^2 = 2\sigma^2(1 + r_{23})$

$$\therefore \quad r_{u,v} = \frac{\text{cov}(u, v)}{\sigma_u \sigma_v} = \frac{\sigma^2(r_{12} + r_{13} + r_{23} + 1)}{2\sigma^2\sqrt{(1 + r_{13})(1 + r_{23})}}$$

$$= \frac{r_{12} + r_{13} + r_{23} + 1}{2\sqrt{(1 + r_{13})(1 + r_{23})}}$$

(*i*) If $r_{13} = r_{23} = 0$, $r_{u,\,v} = \dfrac{r_{12} + 1}{2}$

(*ii*) If $r_{13} = r_{23} = 1$, $r_{u,\,v} = \dfrac{r_{12} + 3}{4}$.

Example 29. *If X and Y are standardized random variables, and*

$$r(aX + bY,\ bX + aY) = \frac{1 + 2ab}{a^2 + b^2}$$

find r (X, Y), the co-efficient of correlation between X and Y.

Sol. (*Standardized variables have mean zero and variance unity.*)

Here $\overline{X} = \overline{Y} = 0, \quad \sigma_X^2 = \sigma_Y^2 = 1$

$$\text{cov}(X, Y) = \frac{1}{n}\Sigma(X - \overline{X})(Y - \overline{Y}) = \frac{1}{n}\Sigma XY$$

$$\therefore \quad r(X, Y) = \frac{\text{cov}(X, Y)}{\sigma_X \sigma_Y} = \frac{1}{n}\Sigma XY \qquad ...(1)$$

Let $u = aX + bY$ and $v = bX + aY$

then $\overline{u} = a\overline{X} + b\,\overline{Y} = 0, \qquad \overline{v} = b\,\overline{X} + a\,\overline{Y} = 0$

$$\text{cov}(u, v) = \frac{1}{n}\Sigma(u - \overline{u})(v - \overline{v}) = \frac{1}{n}\Sigma uv$$

$$= \frac{1}{n}\Sigma(aX + bY)(bX + aY)$$

$$= \frac{1}{n}\Sigma[(abX^2 + abY^2 + (a^2 + b^2)XY]$$

$$= ab\,.\frac{1}{n}\Sigma X^2 + ab\,.\frac{1}{n}\Sigma Y^2 + (a^2 + b^2)\,.\frac{1}{n}\Sigma XY$$

$$= ab\sigma_X^2 + ab\sigma_Y^2 + (a^2 + b^2)\,r(X, Y) \qquad \text{[using (1)]}$$

$$= 2ab + (a^2 + b^2)\,r(X, Y) \qquad (\because \ \sigma_X^2 = \sigma_Y^2 = 1)$$

$$\sigma_u^2 = \frac{1}{n}\Sigma(u - \overline{u})^2 = \frac{1}{n}\Sigma u^2$$

$$= \frac{1}{n}\Sigma(aX + bY)^2 = a^2\,.\frac{1}{n}\Sigma X^2 + b^2\,.\frac{1}{n}\Sigma Y^2 + 2ab\,.\frac{1}{n}\Sigma XY$$

$$= a^2\sigma_X^2 + b^2\sigma_Y^2 + 2ab\,r(X, Y)$$

Similarly $\sigma_v^2 = a^2 + b^2 + 2ab\,r(X, Y)$

$$\therefore \quad r(u, v) = \frac{\text{cov}(u, v)}{\sigma_u \sigma_v} = \frac{2ab + (a^2 + b^2)\,r(X, Y)}{a^2 + b^2 + 2ab\,r(X, Y)}$$

But $r(u, v) = r(aX + bY, bX + aY) = \frac{1 + 2ab}{a^2 + b^2}$ (given)

$$\therefore \quad \frac{1 + 2ab}{a^2 + b^2} = \frac{2ab + (a^2 + b^2)\, r(X, Y)}{a^2 + b^2 + 2ab\, r(X, Y)}$$

$$\Rightarrow \quad (a^2 + b^2) + 2ab\,(a^2 + b^2) + 2ab\, r(X, Y) + 4a^2b^2\, r(X, Y)$$
$$= 2ab\,(a^2 + b^2) + (a^2 + b^2)^2\, r(X, Y)$$

or $\quad [(a^2 + b^2)^2 - 4a^2b^2 - 2ab]\, r(X, Y) = a^2 + b^2$

$$\therefore \quad r(X, Y) = \frac{a^2 + b^2}{(a^2 - b^2)^2 - 2ab}.$$

6.9. CALCULATION OF CO-EFFICIENT OF CORRELATION FOR A BIVARIATE FREQUENCY DISTRIBUTION

If the bivariate data on x and y is presented on a two way correlation table and f is the frequency of a particular rectangle in the correlation table, then

$$r_{xy} = \frac{\Sigma fxy - \frac{1}{n}\Sigma fx\, \Sigma fy}{\sqrt{\left[\Sigma fx^2 - \frac{1}{n}(\Sigma fx)^2\right]\left[\Sigma fy^2 - \frac{1}{n}(\Sigma fy)^2\right]}}$$

Since change of origin and scale do not affect the co-efficient of correlation,

$$\therefore \quad r_{xy} = r_{uv}$$

where the new variables u, v are properly chosen.

Also, by symmetry, $r_{xy} = r_{yx} = r_{uv} = r_{vu}$.

Example 1. *The following table gives according to age the frequency of marks obtained by 100 students in an intelligence test :*

Age (in years) / *Marks*	*18*	*19*	*20*	*21*	*Total*
10—20	*4*	*2*	*2*		*8*
20—30	*5*	*4*	*6*	*4*	*19*
30—40	*6*	*8*	*10*	*11*	*35*
40—50	*4*	*4*	*6*	*8*	*22*
50—60		*2*	*4*	*4*	*10*
60—70		*2*	*3*	*1*	*6*
Total	*19*	*22*	*31*	*28*	*100*

Calculate the coefficient of correlation between age and intelligence.

Sol. Let age and intelligence be denoted by x and y respectively.

Let us define two new variables u and v as

$$u = \frac{y - 45}{10}, \qquad v = x - 20$$

Mid value	$y \backslash x$	*18*	*19*	*20*	*21*	f	u	fu	fu^2	fuv
15	10—20	4 24	2 6	2 0		8	– 3	– 24	72	30
25	20—30	5 20	4 8	6 0	4 – 8	19	– 2	– 38	76	20
35	30—40	6 12	8 8	10 0	11 – 11	35	– 1	– 35	35	9
45	40—50	4 0	4 0	6 0	8 0	22	0	0	0	0
55	50—60		2 – 2	4 0	4 4	10	1	10	10	2
65	60—70		2 – 4	3 0	1 2	6	2	12	24	– 2
	f	19	22	31	28	100	Totals	– 75	**217**	**59**
	v	– 2	– 1	0	1	Totals				
	fv	– 38	– 22	0	28	**– 32**				
	fv^2	76	22	0	28	**126**				
	fuv	56	16	0	– 13	**59**				

$$r_{xy} = r_{vu} = r_{uv} = \frac{\Sigma fuv - \frac{1}{n}\Sigma fu \Sigma fv}{\sqrt{\left[\Sigma fu^2 - \frac{1}{n}(\Sigma fu)^2\right]\left[\Sigma fv^2 - \frac{1}{n}(\Sigma fv)^2\right]}}$$

$$= \frac{59 - \frac{1}{100}(-75)(-32)}{\sqrt{\left[217 - \frac{1}{100}(-75)^2\right]\left[126 - \frac{1}{100}(-32)^2\right]}}$$

$$= \frac{59 - 24}{\sqrt{\frac{643}{4} \times \frac{2894}{25}}} = \frac{35 \times 2 \times 5}{\sqrt{643 \times 2894}} = 0.25.$$

Example 2. *Find the co-efficient of correlation for the following table :*

y \ *x*	*67*	*72*	*77*	*82*	*87*	*92*	*97*
92				*1*	*2*	*3*	*1*
87			*1*	*3*	*8*	*1*	*5*
82	*4*	*4*	*6*	*4*	*9*	*1*	
77	*3*	*3*	*7*	*6*	*4*		
72	*2*	*3*	*5*	*6*	*1*	*1*	
67	*3*	*2*					
62	*1*						

Sol. Let $u = \frac{y - 77}{5}$, $v = \frac{x - 82}{5}$

$y \backslash x$	67	72	77	82	87	92	97	f	u	fu	fu^2	fuv
92				1 0	2 6	3 18	1 9	7	3	21	63	33
87			1 – 2	3 0	8 16	1 4	5 30	18	2	36	72	48
82	4 – 12	4 – 8	6 – 6	4 0	9 9	1 2		28	1	28	28	– 15
77	3 0	3 0	7 0	6 0	4 0			23	0	0	0	0
72	2 6	3 6	5 5	6 0	1 – 1	1 – 2		18	– 1	– 18	18	14
67	3 18	2 8						5	– 2	– 10	20	26
62	1 9							1	– 3	– 3	9	9
f	13	12	19	20	24	6	6	100	Totals	**54**	210	**115**
v	– 3	– 2	– 1	0	1	2	3	Totals				
fv	– 39	–24	–19	0	24	12	18	**– 28**				
fv^2	117	48	19	0	24	24	54	286				
fuv	21	6	– 3	0	30	22	39	**115**				

$$r_{xy} = r_{vu} = r_{uv} = \frac{\Sigma fuv - \frac{1}{n}\Sigma fu\,\Sigma fv}{\sqrt{\left[\Sigma fu^2 - \frac{1}{n}(\Sigma fu)^2\right]\left[\Sigma fv^2 - \frac{1}{n}(\Sigma fv)^2\right]}}$$

$$= \frac{115 - \frac{1}{100}(54)(-28)}{\sqrt{\left[210 - \frac{1}{100}(54)^2\right]\left[286 - \frac{1}{100}(-28)^2\right]}}$$

$$= \frac{115 + 15.12}{\sqrt{(210 - 29.16).(286 - 7.84)}} = \frac{130.12}{\sqrt{180.84 \times 278.16}}$$

$= 0.58.$

Example 3. *Calculate the co-efficient of correlation for the following table :*

y \ x	*16—18*	*18—20*	*20—22*	*22—24*
10—20	*2*	*1*	*1*	
20—30	*3*	*2*	*3*	*2*
30—40	*3*	*4*	*5*	*6*
40—50	*2*	*2*	*3*	*4*
50—60		*1*	*2*	*2*
60—70		*1*	*2*	*1*

Sol. Let $u = \frac{y - 35}{10}$ and $v = \frac{x - 19}{2}$

[See table on next page]

$$r_{xy} = r_{vu} = r_{uv} = \frac{\Sigma fuv - \frac{1}{n}\Sigma fu\,\Sigma fv}{\sqrt{\left[\Sigma fu^2 - \frac{1}{n}(\Sigma fu)^2\right]\left[\Sigma fv^2 - \frac{1}{n}(\Sigma fv)^2\right]}}$$

$$= \frac{31 - \frac{1}{52}(15)(36)}{\sqrt{\left[93 - \frac{1}{52}(15)^2\right]\left[86 - \frac{1}{52}(36)^2\right]}}$$

$$= \frac{52 \times 31 - 15 \times 36}{\sqrt{(93 \times 52 - 225)(86 \times 52 - 1296)}}$$

$$= \frac{1612 - 540}{\sqrt{(4836 - 225)\,(4472 - 1296)}}$$

$$= \frac{1072}{\sqrt{4611 \times 3176}} = 0.28.$$

	Mid values	*17*	*19*	*21*	*23*					
Mid values	*x* \ *y*	*16—18*	*18—20*	*20—22*	*22—24*	*f*	*u*	*fu*	fu^2	*fuv*
15	10—20	2 4	1 0	1 – 2		4	– 2	– 8	16	2
25	20—30	3 3	2 0	3 – 3	2 – 4	10	– 1	–10	10	– 4
35	30—40	3 0	4 0	5 0	6 0	18	0	0	0	0
45	40—50	2 – 2	2 0	3 3	4 8	11	1	11	11	9
55	50—60		1 0	2 4	2 8	5	2	10	20	12
65	60—70		1 0	2 6	1 6	4	3	12	36	12
	f	10	11	16	15	52	Totals	**15**	**93**	**31**
	v	– 1	0	1	2	Totals				
	fv	– 10	0	16	30	**36**				
	fv^2	10	0	16	60	**86**				
	fuv	5	0	8	18	**31**				

Example 4. *Find the co-efficient of correlation for the following tables :*

(*a*)

y \ *x*	*0—5*	*5—10*	*10—15*	*15—20*	*20—25*
0—4	*1*	*2*			
4—8		*4*	*5*	*8*	
8—12			*3*	*4*	
12—16				*2*	*1*

(*b*)

y \ *x*	*0—4*	*4—8*	*8—12*	*12—16*
0—5	*7*			
5—10	*6*	*8*		
10—15		*5*	*3*	
15—20		*7*	*2*	
20—25				*9*

Sol. Please try yourself. **[Ans.** (*a*) 0.6, (*b*) 0.8]

Example 5. *From the following bivariate frequency table calculate the value of correlation co-efficient.*

Wife's age (yrs.) \ *Husband's age (yrs.)*	*20—25*	*25—30*	*30—35*	*35—40*
15—20	*20*	*10*	*3*	*2*
20—25	*4*	*28*	*6*	*4*
25—30		*5*	*11*	
30—35			*2*	
35—40				*5*

Sol. Please try yourself. **[Ans.** 0.613]

Example 6. *Calculate Karl Pearson's co-efficient of correlation from the following data :*

Age in years

Marks	*18*	*19*	*20*	*21*	*22*	*Total*
20—25	3	2	—	—	—	5
15—20	—	5	4	—	—	9
10—15	—	—	7	10	—	17
5—10	—	—	—	3	2	5
0—5	—	—	—	3	1	4
Total	3	7	11	16	3	40

(K.U. 1982 S)

Sol. Please try yourself. **[Ans.** -0.84]

6.10. RANK CORRELATION

Sometimes we have to deal with problems in which data cannot be quantitatively measured but qualitative assessment is possible.

Let a group of n individuals be arranged in order of merit or proficiency in possession of two characteristics A and B. The ranks in the two characteristics are, in general, different. For example, if A stands for intelligence and B for beauty, it is not necessary that the most intelligent individual may be the most beautiful and *vice versa*. Thus an individual who is ranked at the top for the characteristic A *may be* ranked at the bottom for the characteristic B. Let (x_i, y_i), $i = 1, 2, ..., n$ be the ranks of the n individuals in the group for the characteristics A and B respectively. Pearsonian co-efficient of correlation between the ranks x_i's and y_i's is called the *rank correlation co-efficient* between the characteristics A and B for that group of individuals.

Thus rank correlation co-efficient

$$r = \frac{\Sigma(x_i - \bar{x})(y_i - \bar{y})}{\sqrt{\Sigma(x_i - \bar{x})^2 \, \Sigma(y_i - \bar{y})^2}}$$

$$= \frac{\frac{1}{n}\Sigma(x_i - \bar{x})(y_i - \bar{y})}{\sigma_x \sigma_y} \quad ...(1)$$

Now x_i's and y_i's are merely the permutations of n numbers from 1 to n. *Assuming that no two individuals are bracketed or tied in either classification i.e.,* $(x_i, y_i) \neq (x_j, y_j)$ *for* $i \neq j$, *both x and y take all integral values from 1 to n.*

$$\therefore \quad \bar{x} = \bar{y} = \frac{1}{n}(1 + 2 + 3 + ... + n) = \frac{1}{n} \cdot \frac{n(n+1)}{2} = \frac{n+1}{2}$$

$$\Sigma x_i = 1 + 2 + 3 + ... + n = \frac{n(n+1)}{2} = \Sigma y_i$$

$$\Sigma x_i^2 = 1^2 + 2^2 + \ldots + n^2 = \frac{n(n+1)(2n+1)}{6} = \Sigma y_i^2$$

If d_i denotes the difference in ranks of the ith individual, then

$$d_i = x_i - y_i = (x_i - \bar{x}) - (y_i - \bar{y}) \qquad [\because \quad \bar{x} = \bar{y}]$$

$$\frac{1}{n}\Sigma d_i^2 = \frac{1}{n}\Sigma[(x_i - \bar{x}) - (y_i - \bar{y})]^2$$

$$= \frac{1}{n}\Sigma(x_i - \bar{x})^2 + \frac{1}{n}\Sigma(y_i - \bar{y})^2 - 2 \cdot \frac{1}{n}\Sigma(x_i - \bar{x})(y_i - \bar{y})$$

$$= \sigma_x^2 + \sigma_y^2 - 2r\sigma_x\sigma_y \qquad ...(2) \qquad | \textit{using } (1)$$

But $\sigma_x^2 = \frac{1}{n}\Sigma x_i^2 - \bar{x}^2 = \frac{1}{n}\Sigma y_i^2 - \bar{y}^2 = \sigma_y^2$

$\therefore$ From (2) $\frac{1}{n}\Sigma d_i^2 = 2\sigma_x^2 - 2r\sigma_x^2 = 2(1-r)\sigma_x^2$

$$= 2(1-r)\left[\frac{1}{n}\Sigma x_i^2 - \bar{x}^2\right]$$

$$= 2(1-r)\left[\frac{1}{n} \cdot \frac{n(n+1)(2n+1)}{6} - \frac{(n+1)^2}{4}\right]$$

$$= (1-r)(n+1)\left[\frac{4n+2-3n-3}{6}\right]$$

$$= \frac{(1-r)(n^2-1)}{6}$$

or $$1 - r = \frac{6\Sigma d_i^2}{n(n^2-1)}$$

Hence $$r = 1 - \frac{6\Sigma d_i^2}{n(n^2-1)}$$ *(D.U. 1985 ; M.U. 1989)*

Note. This is called *Spearman's Formula for* Rank Correlation.

$$\Sigma d_i = \Sigma(x_i - y_i) = \Sigma x_i - \Sigma y_i = 0$$

always. This serves as a ckeck on calculations.

Example 1. *If $d_i = 0$ for all i, show that $r = 1$ and conversely, if $r = 1$ then $d_i = 0$ for all i.*

Sol. $d_i = 0$ for all $i \Rightarrow \Sigma d_i^2 = 0$

$$\therefore \qquad r = 1 - \frac{6\Sigma d_i^2}{n(n^2-1)} = 1$$

Conversely, if $r = 1$, then $\Sigma d_i^2 = 0$

$\Rightarrow \quad d_1^2 + d_2^2 + \ldots + d_n^2 = 0$

$\Rightarrow \quad d_1 = d_2 = \ldots = d_n = 0$

$\Rightarrow \quad d_i = 0$ for all i.

Example 2. *If $x_i + y_i = n + 1$, show that $r = -1$.*

Sol. $x_i - y_i = d_i$

$x_i + y_i = n + 1$

Adding $2x_i = d_i + (n + 1)$

$\therefore$ $d_i = 2x_i - (n + 1)$

$\therefore$ $\Sigma d_i^{\,2} = 4\Sigma x_i^{\,2} - 4(n + 1)\,\Sigma x_i + (n + 1)^2 \,.\, n$

$$= 4 \,.\, \frac{n(n+1)(2n+1)}{6} - 4\,(n + 1)\,.\, \frac{n(n+1)}{2} + n\,(n + 1)^2$$

$$= \tfrac{1}{3}\, n(n + 1)\,[2(2n + 1) - 6(n + 1) + 3(n + 1)]$$

$$= \frac{1}{3}\, n(n + 1)(n - 1) = \frac{n(n^2 - 1)}{3}$$

$$\therefore \quad r = 1 - \frac{6\Sigma d_i^{\,2}}{n(n^2 - 1)} = 1 - \frac{2n(n^2 - 1)}{n(n^2 - 1)} = 1 - 2 = -1.$$

Example 3. *Show that the rank correlation co-efficient varies between – 1 and + 1.* (K.U. 1982)

Sol. Spearman's Rank Correlation Co-efficient is given by

$$r = 1 - \frac{6\Sigma d_i^{\,2}}{n(n^2 - 1)} \qquad ...(1)$$

(*i*) r is maximum when $\Sigma d_i^{\,2}$ is minimum

i.e., when $\Sigma d_i^{\,2} = 0$

i.e., when $d_i = x_i - y_i = 0$ for each i

i.e., when $x_i = y_i$ for each i.

This happens when every individual gets the same rank in the two characteristics.

$\therefore$ Maximum value of $r = 1 - 0 = 1$.

(*ii*) r is minimum when $\Sigma d_i^{\,2}$ is maximum.

Σd^2 is maximum when $|\,d_i\,| = |\,x_i - y_i\,|$ is maximum.

This happens when an individual getting the highest rank in one characteristic gets the lowest rank in the second characteristic ; *i.e.,* when the ranks (x_i, y_i) are $(1, n)$, $(2, n - 1)$, $(3, n - 2)$, ..., $(n, 1)$.

$\therefore$ Σd^2 is maximum when $x_i + y_i = n + 1$ for each i.

Proceeding as in example 2, $r = -1$.

Hence $-1 \le r \le +1$

i.e., r varies between – 1 and + 1.

Example 4. *The ranking of ten students in two subjects A and B are as follows :*

A :	*3*	*5*	*8*	*4*	*7*	*10*	*2*	*1*	*6*	*9*
B :	*6*	*4*	*9*	*8*	*1*	*2*	*3*	*10*	*5*	*7*

What is the co-efficient of rank correlation ?

Sol.

Ranks in $A(x)$	3	5	8	4	7	10	2	1	6	9	Total
Ranks in $B(y)$	6	4	9	8	1	2	3	10	5	7	
$d = x - y$	−3	1	−1	−4	6	8	−1	−9	1	2	0
d^2	9	1	1	16	36	64	1	81	1	4	214

$$\therefore \quad r = 1 - \frac{6\Sigma d^2}{n(n^2 - 1)} \quad \text{here } n = 10$$

$$= 1 - \frac{6 \times 214}{10 \times 99} = -\frac{147}{495} = -0.3.$$

Example 5. *The marks secured by recruits in the selection test (X) and in the proficiency test Y are given below :*

Serial No :	*1*	*2*	*3*	*4*	*5*	*6*	*7*	*8*	*9*
x :	*10*	*15*	*12*	*17*	*13*	*16*	*24*	*14*	*22*
y :	*30*	*42*	*45*	*46*	*33*	*34*	*40*	*35*	*39*

Calculate the rank correlation co-efficient.

Sol. In example 3, ranks of the students in subjects A and B were given. *Here the marks are given. Therefore, first of all, write down ranks.* In each series, the item with the largest size is ranked 1, next largest 2 and so on.

X	10	15	12	17	13	16	24	14	22	Totals
Y	30	42	45	46	33	34	40	35	39	
Ranks in $X(x)$	9	5	8	3	7	4	1	6	2	
Ranks in $Y(y)$	9	3	2	1	8	7	4	6	5	
$d = x - y$	0	2	6	2	−1	−3	−3	0	−3	0
d^2	0	4	36	4	1	9	9	0	9	72

$$\therefore \quad r = 1 - \frac{6\Sigma d^2}{n(n^2 - 1)}, \quad \text{here } n = 9$$

$$= 1 - \frac{6 \times 72}{9 \times 80} = 1 - 0.6 = 0.4.$$

Example 6. *The following data provides the ranks of 10 students in Statistics and Economics :*

Statistics	:	*4*	*5*	*7*	*8*	*10*	*1*	*3*	*6*	*2*	*9*
Economics	:	*3*	*4*	*7*	*9*	*10*	*8*	*6*	*5*	*2*	*1*

Calculate the rank correlation co-efficient.

Sol. Please try yourself. **[Ans.** 0.24]

Example 7. *Ten students got the following percentage of marks in Principles of Economics and Statistics :*

Students	:	*1*	*2*	*3*	*4*	*5*	*6*	*7*	*8*	*9*	*10*
Marks in Economics	:	*78*	*36*	*98*	*25*	*75*	*82*	*90*	*62*	*65*	*39*
Marks in Statistics	:	*84*	*51*	*91*	*60*	*68*	*62*	*86*	*58*	*63*	*47*

Calculate the rank correlation co-efficient.

Sol. Please try yourself. **[Ans.** 84]

Example 8. *The ranks of same 16 students in Mathematics and Physics are as follows. Two numbers within bracket denote the ranks of the students in Mathematics and Physics :*

(1, 1), (2, 10), (3, 3), (4, 4), (5, 5), (6, 7), (7, 2), (8, 6), (9, 8), (10, 11), (11, 15), (12, 9), (13, 14), (14, 12), (15, 16), (16, 13).

Calculate the rank correlation co-efficient for proficiencies of this group in Mathematics and Physics.

Sol.

Ranks in Math. (x)	:	1	2	3	4	5	6	7	8	9	10	11	12	13	14	15	16
Ranks in Physics (y)	:	1	10	3	4	5	7	2	6	8	11	15	9	14	12	16	13
$d = x - y$	:	0	−8	0	0	0	−1	5	2	1	1	−4	3	−1	2	−1	3
d^2	:	0	64	0	0	0	1	25	4	1	1	16	9	1	4	1	9

$$\Sigma d = 0, \qquad \Sigma d^2 = 136$$

$$r = 1 - \frac{6\Sigma d^2}{n(n^2 - 1)}, \quad \text{here } n = 16$$

$$= 1 - \frac{6 \times 136}{16 \times 255} = 1 - \frac{1}{5} = \frac{4}{5} = 0.8.$$

Example 9. *Ten competitors in a musical test were ranked by the three judges x, y and z in the following order :*

Ranks by x :	*1*	*6*	*5*	*10*	*3*	*2*	*4*	*9*	*7*	*8*
Ranks by y :	*3*	*5*	*8*	*4*	*7*	*10*	*2*	*1*	*6*	*9*
Ranks by z :	*6*	*4*	*9*	*8*	*1*	*2*	*3*	*10*	*5*	*7*

Using rank correlation method, discuss which pair of judges has the nearest approach to common likings in music.

Sol. Here $n = 10$

Ranks by x (X)	Ranks by y (Y)	Ranks by z (Z)	$d_1 = X - Y$	$d_2 = X - Z$	$d_3 = Y - Z$	d_1^2	d_2^2	d_3^2
1	3	6	−2	−5	−3	4	25	9
6	5	4	1	2	1	1	4	1
5	8	9	−3	−4	−1	9	16	1
10	4	8	6	2	−4	36	4	16
3	7	1	−4	2	6	16	4	36
2	10	2	−8	0	8	64	0	64
4	2	3	2	1	−1	4	1	1
9	1	10	8	−1	−9	64	1	81
7	6	5	1	2	1	1	4	1
8	9	7	−1	1	2	1	1	4
Total			0	0	0	200	60	214

$$r_{XY} = 1 - \frac{6\Sigma d_1^2}{n(n^2 - 1)} = 1 - \frac{6 \times 200}{10 \times 99} = 1 - \frac{40}{33} = -\frac{7}{33}$$

$$r_{XZ} = 1 - \frac{6\Sigma d_2^2}{n(n^2 - 1)} = 1 - \frac{6 \times 60}{10 \times 99} = 1 - \frac{4}{11} = \frac{7}{11}$$

$$r_{YZ} = 1 - \frac{6\Sigma d_3^2}{n(n^2 - 1)} = 1 - \frac{6 \times 214}{10 \times 99} = 1 - \frac{214}{165} = -\frac{49}{165}$$

Since r_{XZ} is positive and maximum, we arrive at the conclusion that the pair of judges x and z has the nearest approach to common likings in music.

Example 10. *The co-efficient of rank correlation between marks in Statistics and marks in Mathematics by a certain group of students is 0.8. If the sum of the squares of the difference in ranks is given to be 33, find the number of students in the group.*

Sol. Here $r = 0.8,\quad \Sigma d^2 = 33,\quad n = ?$

Using $r = 1 - \dfrac{6\Sigma d^2}{n(n^2 - 1)}$, we have

$$0.8 = 1 - \frac{6 \times 33}{n(n^2 - 1)}$$

$$\Rightarrow \quad \frac{198}{n(n^2 - 1)} = 1 - 0.8 = 0.2 = \frac{1}{5}$$

$$\Rightarrow \quad n(n^2 - 1) = 990$$

$$\Rightarrow \quad n^3 - n - 990 = 0$$

By inspection $n = 10.$

Example 12. *The co-efficient of rank correlation of the marks obtained by 10 students in Mathematics and Statistics was found to be 0.5. It was later discovered that the difference in ranks in two subjects obtained by one of the students was wrongly taken as 3 instead of 7. Find the correct co-efficient of rank correlation.*

Sol. Here $n = 10$ Incorrect $r = 0.5$

Using $r = 1 - \dfrac{6\Sigma d_i^{\,2}}{n(n^2-1)}$, we have

$$0.5 = 1 - \frac{6\Sigma d_i^{\,2}}{10 \times 99}$$

$$\Rightarrow \quad 6\Sigma d_i^{\,2} = 990 \times 0.5 = 495$$

$$\therefore \quad \text{Incorrect } \Sigma d_i^{\,2} = \frac{495}{6} = 82.5$$

Corrected $\Sigma d_i^{\,2} = 82.5 - 3^2 + 7^2 = 122.5$

$\therefore$ Correct value of r

$$= 1 - \frac{6 \times \text{corrected } \Sigma d_i^{\,2}}{n(n^2-1)} = 1 - \frac{6 \times 122.5}{10 \times 99} = 0.2576.$$

6.11. REPEATED RANKS

If any two or more individuals have same rank or the same value in the series of marks, then the above formula fails and requires an adjustment. In such cases, each individual is given an average rank. This common average rank is the average of the ranks which these individuals would have assumed if they were slightly different from each other. Thus, if two individuals are ranked equal at the sixth place, they would have assumed the 6th and 7th ranks if they were ranked slightly different. Their common rank $= \dfrac{6+7}{2} = 6.5$. If three individuals are ranked equal at fourth place, they would have assumed the 4th, 5th and 6th ranks if they were ranked slightly different. Their common rank $= \dfrac{4+5+6}{3} = 5$.

Adjustment. Add $\dfrac{1}{12}\, m(m^2-1)$ to Σd^2, where m stands for the number of times an item is repeated.

☞ This adjustment factor is to be added for each repeated item.

$$\text{Thus } r = 1 - \frac{6\left\{\Sigma d^2 + \frac{1}{12} m(m^2-1) + \frac{1}{12} m(m^2-1) + \ldots\right\}}{n(n^2-1)}$$

Example 1. *Obtain the rank correlation co-efficient for the following data :*

X :	68	64	75	50	64	80	75	40	55	64
Y :	62	58	68	45	81	60	68	48	50	70

Sol. Here, marks are given, so write down the ranks.

X	68	64	75	50	64	80	75	40	55	64	Totals
Y	62	58	68	45	81	60	68	48	50	70	
Ranks in X (x)	4	6	2.5	9	6	1	2.5	10	8	6	
Ranks in Y (y)	5	7	3.5	10	1	6	3.5	9	8	2	
$d = x - y$	−1	−1	−1	−1	5	−5	−1	1	0	4	0
d^2	1	1	1	1	25	25	1	1	0	16	72

In the X-series, the value 75 occurs twice. Had these values been slightly different, they would have been given the ranks 2 and 3. Therefore, the common rank given to them is $\frac{2+3}{2} = 2.5$. The value 64 occurs thrice. Had these values been slightly different, they would have been given the ranks 5, 6 and 7. Therefore, the common rank given to them is $\frac{5+6+7}{3} = 6$. Similarly in the Y-series, the value 68 occurs twice. Had these values been slightly different, they would have been given the ranks 3 and 4. Therefore, the common rank given to them is $\frac{3+4}{2} = 3.5$.

Thus, m has the values 2, 3, 2.

$$\therefore \quad r = 1 - \frac{6\left\{\Sigma d^2 + \frac{1}{12}m(m^2-1) + \frac{1}{12}m(m^2-1) + \ldots\right\}}{n(n^2-1)}$$

$$= 1 - \frac{6\left[72 + \frac{1}{12}\cdot 2(2^2-1) + \frac{1}{12}\cdot 3(3^2-1) + \frac{1}{12}\cdot 2(2^2-1)\right]}{10(10^2-1)}$$

$$= 1 - \frac{6[72 + \frac{1}{2} + 2 + \frac{1}{2}]}{990} = 1 - \frac{6 \times 75}{990} = \frac{6}{11} = 0.545.$$

Example 2. *A sample of 12 fathers and their eldest sons gave the following data about their height in inches :*

Father :	65	63	67	64	68	62	70	66	68	67	69	71
Son :	68	66	68	65	69	66	68	65	71	67	68	70

Calculate the co-efficient of rank correlation.

Sol. Please try yourself. **[Ans.** 0.722]

Example 3. *From the following data, calculate the co-efficient of rank correlation between X and Y :*

$X:$	33	56	50	65	44	38	44	50	15	26
$Y:$	50	35	70	25	35	58	75	60	55	26

Sol. Please try yourself. **[Ans.** – 0.076]

6.12. REGRESSION

Regression is the estimation or prediction of unknown values of one variable from known values of another variable.

After establishing the fact of correlation between two variables, it is natural curiosity to know the extent to which one variable varies in response to a given variation in the other variable *i.e.,* one is interested to know the nature of relationship between the two variables.

Regression measures the nature and extent of correlation.

6.13. LINEAR REGRESSION

If two variates x and y are correlated *i.e.,* there exists an association or relationship between them, then the scatter diagram will be more or less concentrated round a curve. This curve is called the *curve of regression* and the relationship is said to be expressed by means of *curvilinear regression.* In the particular case, when the curve is a straight line, it is called a *line of regression* and the regression is said to be *linear.*

A line of regression is the straight line which gives the best fit in the least square sense to the given frequency. (*K.U. 1981 S; D.U. 1981*)

If the line of regression is so chosen that the sum of squares of deviations parallel to the axis of y is minimised [*See. Fig.* (*a*)], it is called *the line of regression of y on x* and it gives the *best estimate of y for any given value of x.*

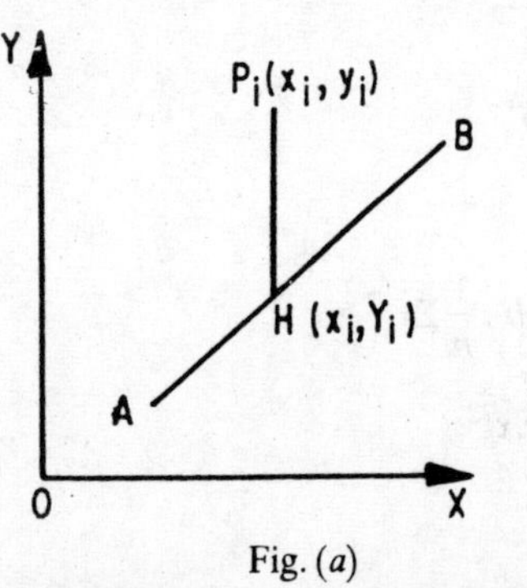

Fig. (*a*)

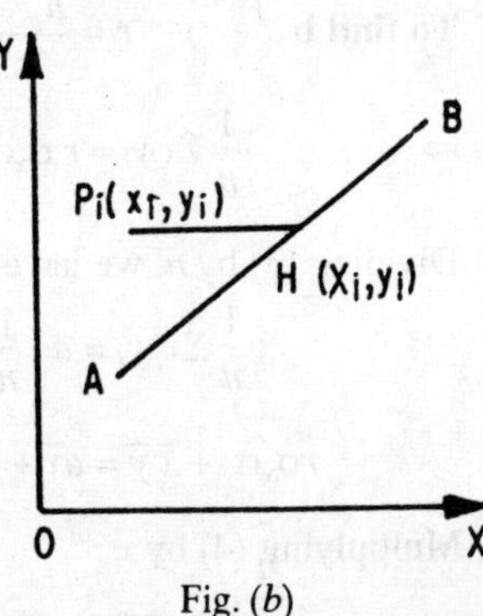

Fig. (*b*)

If the line of regression is so chosen that the sum of squares of deviations parallel to the axis of x is minimised [*See Fig.* (*b*)], it is called *the line of regression of x on y* and it gives the *best estimate of x for any given value of y.*

☞ 6.14. EQUATIONS TO THE LINES OF REGRESSION.

(D.U. 1981, 89 ; M.U. 1984)

Let $y = a + bx$ ...(1)

be the equation of the line of regression of y on x so that corresponding to $x = x_i$, the expected value of y is $Y_i\ (= a + bx_i)$ and S, the sum of the squares of deviations of observed value y_i from expected value Y_i is given by

$$S = \sum_{i=1}^{n} (y_i - Y_i)^2 = \sum_{i=1}^{n} (y_i - a - bx_i)^2$$

According to the principle of least squares, we have to choose a and b so that S is minimum. The normal equations are

$$\frac{\partial S}{\partial a} = 0, \quad \frac{\partial S}{\partial b} = 0$$

or

$$\begin{cases} -2\Sigma(y_i - a - bx_i) = 0 \\ \text{and} \quad -2\Sigma x_i(y_i - a - bx_i) = 0 \end{cases}$$

or $\Sigma y_i = na + b\,\Sigma x_i$ $[\because \quad \Sigma a = na]$ ...(2)

and $\Sigma x_i\, y_i = a\Sigma x_i + b\Sigma x_i^2$ ...(3)

Dividing (2) by n, we have

$$\frac{1}{n}\Sigma y_i = a + b \cdot \frac{1}{n}\Sigma x_i$$

or $\bar{y} = a + b\bar{x}$ ...(4)

Subtracting (4) from (1) [to eliminate a]

$$y - \bar{y} = b(x - \bar{x}) \qquad ...(5)$$

To find b

$$r = \frac{\frac{1}{n}\Sigma x_i\, y_i - \bar{x}\,\bar{y}}{\sigma_x \sigma_y}$$

$$\Rightarrow \quad \frac{1}{n}\Sigma x_i y_i = r\,\sigma_x\sigma_y + \bar{x}\,\bar{y}$$

Dividing (3) by n, we have

$$\frac{1}{n}\Sigma x_i\, y_i = a \cdot \frac{1}{n}\Sigma x_i + b \cdot \frac{1}{n}\Sigma x_i^2$$

or

$$r\sigma_x\sigma_y + \bar{x}\,\bar{y} = a\bar{x} + b \cdot \frac{1}{n}\Sigma x_i^2 \qquad ...(6)$$

Multiplying (4) by $\bar{x}$

$$\bar{x}\,\bar{y} = a\,\bar{x} + b\,\bar{x}\,\bar{y} \qquad ...(7)$$

Subtracting (7) from (6)

$$r\sigma_x\sigma_y = b\left(\frac{1}{n}\Sigma x_i^2 - \bar{x}\,\bar{y}\right)$$

or $\quad r\sigma_x\sigma_y = b\sigma_x^2 \qquad \therefore \quad b = \dfrac{r\sigma_y}{\sigma_x}$

$\therefore$ From (5), the line of regression of y on x is

$$y - \bar{y} = \frac{r\sigma_y}{\sigma_x}(x - \bar{x})$$

Similarly, the equation to the line of regression of x on y is

$$x - \bar{x} = \frac{r\sigma_x}{\sigma_y}(y - \bar{y}).$$

☞**Note.** From the equations of the two lines of regression, it is clear that both the lines pass through the point $(\bar{x}, \bar{y})$.

Thus to find the mean of x and y series, solve the equations of the two lines of regression.

Note.
$$\frac{r\sigma_x}{\sigma_y} = r \cdot \frac{\sigma_x}{\sigma_y} = \frac{\frac{1}{n}\Sigma xy - \bar{x}\,\bar{y}}{\sigma_x\sigma_y} \cdot \frac{\sigma_x}{\sigma_y}$$

$$= \frac{\frac{1}{n}\Sigma xy - \bar{x}\,\bar{y}}{\sigma_y^2} = \frac{\frac{1}{n}\Sigma xy - \bar{x}\,\bar{y}}{\frac{1}{n}\Sigma y^2 - \bar{y}^2}$$

Similarly,
$$\frac{r\sigma_y}{\sigma_x} = \frac{\frac{1}{n}\Sigma xy - \bar{x}\,\bar{y}}{\frac{1}{n}\Sigma x^2 - \bar{x}^2}.$$

6.15. REGRESSION CO-EFFICIENTS

(M.D.U. 1981 S ; G.N.D.U. 1981, 82 S)

The equation to the line of regression of y on x is

$$y - \bar{y} = \frac{r\sigma_y}{\sigma_x}(x - \bar{x})$$

and the equation to the line of regression of x on y is

$$x - \bar{x} = \frac{r\sigma_x}{\sigma_y}(y - \bar{y})$$

$\dfrac{r\sigma_y}{\sigma_x}$ is called the regression co-efficient of y on x and is denoted by b_{yx}.

$\dfrac{r\sigma_x}{\sigma_y}$ is called the regression co-efficient of x on y and is denoted by b_{xy}.

Note. If $r = 0$, the two lines of regression become $y = \bar{y}$ an $x = \bar{x}$ which are two straight lines parallel to X and Y-axes respectively and passing through their means $\bar{y}$ and $\bar{x}$. They are mutually perpendicular.

If $r = \pm 1$, the two lines of regression will coincide

6.16. PROPERTIES OF REGRESSION CO-EFFICIENTS

Property I. Correlation co-efficient is the geometric mean between the regression co-efficients. *(M.D.U. 1981, S ; G.N.D.U. 1981)*

Proof. The regression co-efficients are $\frac{r\sigma_y}{\sigma_x}$ and $\frac{r\sigma_x}{\sigma_y}$

G.M., between them = $\sqrt{\frac{r\sigma_y}{\sigma_x} \times \frac{r\sigma_x}{\sigma_y}} = \sqrt{r^2} = r$

= correlation co-efficient.

Property II. If one of the regression co-efficients is greater than unity numerically, the other must be less than unity numerically.

(G.N.D.U. 1982 S)

Proof. The two regression co-efficients are

$$b_{yx} = \frac{r\sigma_y}{\sigma_x} \quad \text{and} \quad b_{xy} = \frac{r\sigma_x}{\sigma_y}.$$

Let $b_{yx} > 1$, then $\frac{1}{b_{yx}} < 1$ ...(1)

Since $b_{yx} \, . \, b_{xy} = r^2 \le 1$ $(\because \; -1 \le r \le 1)$

$\therefore$ $b_{xy} \le \frac{1}{b_{yx}} < 1$ | *using* (1)

Similarly, if $b_{xy} > 1$, then $b_{yx} < 1$.

Property III. Arithmetic mean of regression co-efficients is greater than the correlation co-efficient.

Proof. We have to prove that $\frac{b_{yx} + b_{xy}}{2} > r$

or $$\frac{\frac{r\sigma_y}{\sigma_x} + \frac{r\sigma_x}{\sigma_y}}{2} > r$$

or $$\sigma_y^{\,2} + \sigma_x^{\,2} > 2\sigma_x\sigma_y$$

or $$\sigma_y^{\,2} + \sigma_x^{\,2} - 2\sigma_x\sigma_y > 0$$

or $$(\sigma_y - \sigma_x)^2 > 0 \text{ which is true.}$$

Hence the result.

Second Method. We know that A.M. > G.M.

$\therefore$ $$\frac{b_{yx} + b_{xy}}{2} > \sqrt{b_{yx} \, . \, b_{xy}}$$

or $$\frac{b_{yx} + b_{xy}}{2} > r \qquad \left[\because \; r = \sqrt{b_{yx} \, . \, b_{xy}}\,\right]$$

$\therefore$ A.M. of regression co-efficients > correlation co-efficient.

Property IV. Regression co-efficients are independent of the origin but not of scale. *(M.D.U. 1981 S)*

Proof. Let $u = \dfrac{x-a}{h}, \quad v = \dfrac{y-b}{k}$

where a, b, h and k are constants

$$b_{yx} = \frac{r\sigma_y}{\sigma_x} = r \cdot \frac{k\sigma_v}{h\sigma_u} = \frac{k}{h}\left(\frac{r\sigma_v}{\sigma_u}\right) = \frac{k}{h} b_{vu}$$

Similarly $b_{xy} = \dfrac{h}{k} b_{uv}$.

Thus b_{yx} and b_{xy} are both independent of a and b but not of h and k.

Property V. The correlation co-efficient and the two regression co-efficients have same sign.

Proof. Regression co-efficient of y on $x = b_{yx} = r\dfrac{\sigma_y}{\sigma_x}$

Regression co-efficient of x on $y = b_{xy} = r\dfrac{\sigma_x}{\sigma_y}$

Since σ_x and σ_y are both positive, b_{yx}, b_{xy} and r have same sign

6.17. ANGLE BETWEEN TWO LINES OF REGRESSION

If θ is the acute angle between the two regression lines in the case of two variables x and y, show that

$$\tan\theta = \frac{1-r^2}{|r|} \cdot \frac{\sigma_x\sigma_y}{\sigma_x^2+\sigma_y^2}$$

where r, σ_x, σ_y have their usual meanings.

Explain the significance of the formula when $r = 0$ and $r = \pm 1$.

(D.U. 1984, 86, 93 ; M.U. 1987, 90)

Proof. Equations to the lines of regression of y on x and x on y are

$$y - \bar{y} = \frac{r\sigma_y}{\sigma_x}(x - \bar{x}) \quad \text{and} \quad x - \bar{x} = \frac{r\sigma_x}{\sigma_y}(y - \bar{y})$$

Their slopes are $m_1 = \dfrac{r\sigma_y}{\sigma_x}$ and $m_2 = \dfrac{\sigma_y}{r\sigma_x}$.

$$\therefore \quad \tan\theta = \left|\frac{m_2 - m_1}{1 + m_2 m_1}\right|$$

$$= \left|\frac{\dfrac{\sigma_y}{r\sigma_x} - \dfrac{r\sigma_y}{\sigma_x}}{1 + \dfrac{\sigma_y^2}{\sigma_x^2}}\right|$$

$$= \left|\frac{1-r^2}{r} \cdot \frac{\sigma_y}{\sigma_x} \cdot \frac{\sigma_x^2}{\sigma_x^2+\sigma_y^2}\right| = \frac{1-r^2}{|r|} \cdot \frac{\sigma_x\sigma_y}{\sigma_x^2+\sigma_y^2}$$

since $r^2 \le 1$ and σ_x, σ_y are positive.

Hence $\tan\theta = \frac{1-r^2}{|r|} \cdot \frac{\sigma_x\sigma_y}{\sigma_x^2+\sigma_y^2}$

when r = 0, $\theta = \frac{\pi}{2}$

∴ The two lines of regression are perpendicular to each other.

Hence the estimated value of y is the same for all values of x and *vice versa.*

when r = ± 1, $\tan\theta = 0$ so that, $\theta = 0$ or π.

Hence the lines of regression coincide and there is perfect correlation between the two variates x and y.

Example 1. *If r = 0, show that the two lines of regression are parallel to the axes.*

Sol. Equations to the two lines of regression are

$$y - \bar{y} = \frac{r\sigma_y}{\sigma_x}(x - \bar{x}) \quad \text{and} \quad x - \bar{x} = \frac{r\sigma_x}{\sigma_y}(y - \bar{y})$$

When $r = 0$, they reduce to $y = \bar{y}$ and $x = \bar{x}$ which are equations of lines parallel to x-axis and y-axis respectively.

Example 2. *If the two regression co-efficients are 0.8 and 0.2, what would be the value of co-efficient of correlation.*

Sol. Since the co-efficient of correlation is the G.M. between the regreesion co-efficients.

∴ $r^2 = 0.8 \times 0.2 = 0.16, \quad r = \pm 0.4$

Since r has same sign as the two regression co-efficients

∴ r must be + ve. Hence $r = 0.4$.

Example 3. *Is the following statement correct ? Give reasons. The regression co-efficient of x on y is 3.2 and that of y on x is 0.8.*

Sol. $b_{xy} = 3.2, b_{yx} = 0.8$

Now $r^2 = b_{xy} \times b_{yx} = 3.2 \times 0.8 = 2.56 > 1$

Since $r^2 \le 1$ always, the statement is false.

Example 4. *Is the following statement true ? Give reasons.*

$$40x - 18y = 5 \quad \textit{and} \quad 8x - 10y + 6 = 0$$

are respectively the regression equations of y on x and x on y.

Sol. Regression equation of y on x is

$$40x - 18y = 5$$

or
$$18y = 40x - 5$$

or
$$y = \frac{20}{9}x - \frac{5}{18}$$

⇒ Regression co-eff. of y on $x = b_{yx} = \frac{20}{9}$

Regression equation of x on y is $8x - 10y + 6 = 0$

or $$8x = 10y - 6$$

or $$x = \frac{5}{4}y - \frac{3}{4}$$

$\Rightarrow$ Regression co-eff. of x on $y = b_{xy} = \frac{5}{4}$

Now $$r^2 = b_{yx} \cdot b_{xy} = \left(\frac{20}{9}\right)\left(\frac{5}{4}\right) = \frac{25}{9} > 1$$

But r^2 is always ≤ 1.

$\therefore$ The statement is false.

Example 5. *Find the correlation co-efficient and the equations of regression lines for the following values of x and y :*

x :	1	2	3	4	5
y :	2	5	3	8	7

Sol.

x	y	$u = x - 3$	$v = y - 5$	u^2	v^2	uv
1	2	−2	−3	4	9	6
2	5	−1	0	1	0	0
3	3	0	−2	0	4	0
4	8	1	3	1	9	3
5	7	2	2	4	4	4
Total		0	0	10	26	13

$$\bar{u} = \tfrac{1}{5}\Sigma u = 0, \qquad \bar{v} = \tfrac{1}{5}\Sigma v = 0$$

$$r_{xy} = r_{uv} = \frac{\frac{1}{n}\Sigma uv - \bar{u}\,\bar{v}}{\sqrt{\left(\frac{1}{n}\Sigma u^2 - \bar{u}^2\right)\left(\frac{1}{n}\Sigma v^2 - \bar{v}^2\right)}}$$

$$= \frac{\frac{1}{5}(13) - 0}{\sqrt{[\frac{1}{5}(10) - 0][\frac{1}{5}(26) - 0]}} = \frac{\frac{1}{5}(13)}{\frac{1}{5}\sqrt{260}} = \frac{13}{16.1} = 0.8$$

Now $$\frac{r\sigma_y}{\sigma_x} = \frac{r\sigma_v}{\sigma_u} = \frac{\frac{1}{n}\Sigma uv - \bar{u}\,\bar{v}}{\frac{1}{n}\Sigma u^2 - \bar{u}^2} = \frac{\frac{1}{5}(13) - 0}{\frac{1}{5}(10) - 0} = 1.3$$

$$\bar{x} = 3 + \tfrac{1}{5}\Sigma u = 3, \qquad \bar{y} = 5 + \tfrac{1}{5}\Sigma v = 5$$

Equation to the line of regression of y on x is

$$y - \bar{y} = \frac{r\sigma_y}{\sigma_x}(x - \bar{x})$$

or $\quad y - 5 = 1.3\,(x - 3)$

or $\quad y = 1.3x + 1.1$

Also
$$\frac{r\sigma_x}{\sigma_y} = \frac{r\sigma_u}{\sigma_v} = \frac{\frac{1}{n}\Sigma uv - \overline{u}\,\overline{v}}{\frac{1}{n}\Sigma v^2 - \overline{v}^2} = \frac{\frac{1}{5}(13) - 0}{\frac{1}{5}(26) - 0} = .5$$

Equation to the line of regression of x on y is

$$x - \overline{x} = \frac{r\sigma_x}{\sigma_y}(y - \overline{y})$$

or $\quad x - 3 = .5(y - 5) \quad$ or $\quad x = .5y + .5.$

Example 6. *Calculate the co-efficient of correlation and obtain the least square regression lines for the following data :*

x :	*1*	*2*	*3*	*4*	*5*	*6*	*7*	*8*	*9*
y :	*9*	*8*	*10*	*12*	*11*	*13*	*14*	*16*	*15*

Also obtain an estimate of y which should correspond on the average to x = 6.2. (G.N.D.U. 1982 S ; K.U. 1981 ; D.U. 1989)

Sol.

x	y	$u = x - 5$	$v = y - 12$	u^2	v^2	uv
1	9	−4	−3	16	9	12
2	8	−3	−4	9	16	12
3	10	−2	−2	4	4	4
4	12	−1	0	1	0	0
5	11	0	−1	0	1	0
6	13	1	1	1	1	1
7	14	2	2	4	4	4
8	16	3	4	9	16	12
9	15	4	3	16	9	12
Total		0	0	60	60	57

$$r_{xy} = r_{uv} = \frac{\frac{1}{n}\Sigma uv - \overline{u}\,\overline{v}}{\sqrt{\left(\frac{1}{n}\Sigma u^2 - \overline{u}^2\right)\left(\frac{1}{n}\Sigma v^2 - \overline{v}^2\right)}}$$

$$= \frac{\frac{1}{9}(57) - 0}{\sqrt{[\frac{1}{9}(60) - 0][\frac{1}{9}(60) - 0]}} = \frac{\frac{1}{9}(57)}{\frac{1}{9}(60)} = \frac{19}{20} = 0.95$$

$$\frac{r\sigma_y}{\sigma_x} = \frac{r\sigma_v}{\sigma_u} = \frac{\frac{1}{n}\Sigma uv - \overline{u}\,\overline{v}}{\frac{1}{n}\Sigma u^2 - \overline{u}^2} = \frac{\frac{1}{9}(57) - 0}{\frac{1}{9}(60) - 0} = \frac{19}{20} = 0.95$$

Also $\bar{x} = 5 + \frac{1}{9}\Sigma u = 5, \qquad \bar{y} = 12 + \frac{1}{9}\Sigma v = 12$

Equation to the line of regression of y on x is

$$y - \bar{y} = \frac{r\sigma_y}{\sigma_x}(x - \bar{x})$$

or $\qquad y - 12 = 0.95(x - 5) \qquad$ or $\qquad y = 0.95x + 7.25$

When $\quad x = 6.2$, estimated value of $y = 0.95 \times 6.2 + 7.25$

$= 5.89 + 7.25 = 13.14$

$$\text{Now} \quad r\frac{\sigma_x}{\sigma_y} = r\frac{\sigma_u}{\sigma_v} = \frac{\frac{1}{n}\Sigma uv - \bar{u}\,\bar{v}}{\frac{1}{n}\Sigma v^2 - \bar{v}^2} = \frac{\frac{1}{9}(57) - 0}{\frac{1}{9}(60) - 0} = \frac{19}{20} = 0.95$$

Equation to the line of regression of x on y is

$$x - \bar{x} = \frac{r\sigma_x}{\sigma_y}(y - \bar{y})$$

or $\qquad x - 5 = 0.95(y - 12) \qquad$ or $\qquad x = 0.95y - 6.4.$

Example 7. *From the following data, obtain the equations to the two regression lines :*

x :	*6*	*2*	*10*	*4*	*8*
y :	*9*	*11*	*5*	*8*	*7*

Sol. Please try yourself.

[Ans. $y = -.65x + 11.9$; $x = -1.3y + 16.4$]

Example 8. *Calculate the co-efficient of correlation and obtain the equations to the lines of regression for the following data :*

x :	*1*	*2*	*3*	*4*	*5*	*6*	*7*
y :	*7*	*8*	*9*	*11*	*10*	*13*	*12*

Also obtain an estimated value of y which should correspond to x = 9.

Sol. Please try yourself.

[Ans. $r = .93, y = .93x + 6.28, x = .93y - 5.3, 14.65$]

Example 9. *From the following table, form the two regression equations :*

Height of father (in inches) x :	*65*	*66*	*67*	*67*	*68*	*69*	*71*	*73*
Height of son (in inches) y :	*67*	*68*	*64*	*68*	*72*	*70*	*69*	*70*

(K.U. 1981 S)

Sol. Please try yourself.

[Ans. $y = .424x + 3.956$; $x = .525y + 3.229$]

Example 10. *In a partially destroyed laboratory record of an analysis of a correlation data, the following resutls only are eligible :*

Variance of x = 9

Regression equations : $8x - 10y + 66 = 0, 40x - 18y = 214$

What were (a) the mean values of x and y, (b) the standard deviation of y, and (c) the co-efficient of correlation between x and y.

(Meerut 1987, 90 ; D.U. 1986 ; Kanpur 1988)

Sol. (*i*) **Since both the lines of regression pass through the point** $(\overline{x}, \overline{y})$, therefore, we have

$$8\overline{x} - 10\overline{y} + 66 = 0 \quad \text{...(1)}$$

$$40\overline{x} - 18\overline{y} - 214 = 0 \quad \text{...(2)}$$

Multiplying (1) by 5

$$40\overline{x} - 50\overline{y} + 330 = 0 \quad \text{...(3)}$$

Subtracting (3) from (2) $32\overline{y} - 544 = 0$

$$\therefore \quad \overline{y} = \frac{544}{32} = 17$$

$\therefore$ From (1) $8\overline{x} - 170 + 66 = 0$

or $8\overline{x} = 104 \quad \therefore \quad \overline{x} = 13$

Hence $\overline{x} = 13, \quad \overline{y} = 17$ ...(*a*)

(*ii*) Variance of $x = \sigma_x^2 = 9$ (given)

$$\therefore \quad \sigma_x = 3$$

The equations of lines of regression can be written as

$$y = .8x + 6.6 \quad \text{and} \quad x = .45y + 5.35$$

$\therefore$ The regression co-efficient of y on x is

$$\frac{r\sigma_y}{\sigma_x} = .8 \quad \text{...(4)}$$

The regression co-efficient of x on y is

$$\frac{r\sigma_x}{\sigma_y} = .45 \quad \text{...(5)}$$

Multiplting (4) and (5)

$$r^2 = .8 \times .45 = .360$$

$$\therefore \quad r = 0.6 \quad \text{...(c)}$$

(+ ve sign with sq. root is taken because regression co-efficients are + ve).

From (4) $\sigma_y = \dfrac{.8\sigma_x}{r} = \dfrac{.8 \times 3}{0.6} = 4.$ ...(*b*)

Example 11. *The equations of two regression lines obtained in a correlation analysis are as follows :*

$$3x + 2y = 26, \; 6x + y = 31.$$

Obtain (i) the value of correlation co-efficient,

(ii) the mean values of x and y.

(D.U. 1981, 90 ; M.D.U. 1982 S)

Sol. Please try yourself. $\left[\textbf{Ans. } (i) -\frac{1}{2}, (ii)\ \overline{x} = 4, \overline{y} = 7 \right]$

Example 12. *Given that $x = 4y + 5$, $y = kx + 4$ are the regression lines, show that $0 < 4k < 1$. If $k = \frac{1}{16}$; find the means of the two variables and the co-efficient of correlation.*

Sol. The given equations of lines of regression are

$$x = 4y + 5, \qquad y = kx + 4$$

∴ The regression co-efficient of x on y is

$$\frac{r\sigma_x}{\sigma_y} = 4 \qquad \text{...(i)}$$

The regression co-efficient of y on x is

$$\frac{r\sigma_y}{\sigma_x} = k \qquad \text{...(ii)}$$

Multiplying (*i*) and (*ii*) $\quad r^2 = 4k$

Since the lines of regression are not coincident, $r \neq \pm 1$

Since the lines of regression are not parallel to axes $r \neq 0$

$$\therefore \qquad 0 < r^2 < 1 \qquad (\because \ |r| \le 1 \text{ or } -1 \le r \le 1)$$

$$\Rightarrow \qquad 0 < 4k < 1$$

When $\quad k = \dfrac{1}{16}$, the two regression lines are

$$x = 4y + 5, \qquad y = \frac{1}{16}x + 4$$

Since both the lines of regression pass through the point $(\bar{x}, \bar{y})$

$$\therefore \qquad \bar{x} - 4\bar{y} - 5 = 0$$

$$\bar{x} - 16\bar{y} + 64 = 0$$

Solving these equations $\bar{x} = 28, \quad \bar{y} = \dfrac{23}{4} = 5.75.$

Example 13. *If θ is the acute angle between the two regression lines, prove that $\sin\theta \le 1 - r^2$.*

Sol. We know that $\quad \tan\theta = \dfrac{1 - r^2}{r} \cdot \dfrac{\sigma_x\sigma_y}{\sigma_x^2 + \sigma_y^2} \qquad$ ...(1)

Since A.M. ≥ G.M.

$$\therefore \qquad \frac{\sigma_x^2 + \sigma_y^2}{2} \ge \sqrt{\sigma_x^2 \cdot \sigma_y^2} \quad \Rightarrow \quad \frac{\sigma_x^2 + \sigma_y^2}{2} \ge \sigma_x\sigma_y$$

$$\Rightarrow \qquad \frac{\sigma_x\sigma_y}{\sigma_x^2 + \sigma_y^2} \le \frac{1}{2}$$

$$\therefore \qquad \text{From (1) } \tan\theta \le \frac{1 - r^2}{2r} \qquad \text{...(2)}$$

Let $\qquad \tan\alpha = \dfrac{1 - r^2}{2r} \quad$ then $\quad \sin\alpha = \dfrac{1 - r^2}{1 + r^2}$

From (2) $\tan\theta \le \tan\alpha$

$\Rightarrow$ $\theta \le \alpha$ $\quad \therefore \quad \sin\theta \le \sin\alpha$

or $\sin\theta \le \dfrac{1-r^2}{1+r^2}$

or $\sin\theta \le 1 - r^2$. $\quad \left[\begin{array}{l} \because \quad 1 + r^2 \ge 1 \\ \therefore \quad \dfrac{1}{1+r^2} \le 1 \end{array}\right]$

Example 14. *For two variables x and y with the same mean, the two regresson equations are* $y = ax + b$ *and* $x = \alpha y + \beta$. *Show that* $\dfrac{b}{\beta} = \dfrac{1-a}{1-\alpha}$. *Find also the common mean.* (Rajasthan 1985)

Sol. Let the common mean be m. Then the two regression equations are

$$y - m = a(x - m) \quad \text{and} \quad x - m = \alpha(y - m)$$

or

$$y = ax + m(1 - a) \quad \text{and} \quad x = \alpha y + m(1 - \alpha)$$

Comparing with the given regression equations

$$b = m(1 - a), \qquad \beta = m(1 - \alpha)$$

$$\therefore \quad \frac{b}{\beta} = \frac{1-a}{1-\alpha}$$

Also $$m = \frac{b}{1-a} = \frac{\beta}{1-\alpha}.$$

7

Theoretical Distributions

7.1. Frequency distributions can be classified under two heads :

(*i*) Observed Frequency Distributions.

(*ii*) Theoretical or Expected Frequency Distributions.

Observed frequency distributions are based on actual observation and experimentation. If certain hypothesis is assumed, it is sometimes possible to derive mathematically what the frequency distributions of certain universe should be. Such distributions are called **Theoretical Distributions.**

There are many types of theoretical frequency distributions but we shall consider only three which are of great importance :

(*i*) Binomial Distribution (or Bernoulli's Distribution)

(*ii*) Poisson's Distribution, (*iii*) Normal Distribution.

I. BINOMIAL DISTRIBUTION

7.2. Suppose that n trials constitute an experiment and let the experiment be repeated N times. Let us call the occurrence of an event E a 'success' and its non-occurrence a 'failure'. Let p be the probability of a success and q be the probability of a failure in a single trial so that $p + q = 1$.

Let us assume that the trials are independent and the probability p of success is the same (*i.e.,* constant) for each trial.

The number of successes in n trials may be 0, 1, 2, ..., r ...... , n. The probability that r trials are successes and the remaining $(n - r)$ are failures is, by the compound probability theorem,

$$p.p.q.p.q.q.q.......pqp \qquad \text{(written at random)}$$

$$= \underbrace{p.p.p.............p}_{r \text{ times}}.\underbrace{q.q.q..........q}_{(n-r) \text{ times}} = p^r q^{n-r}$$

But we are to consider all the cases where r trials are successes. r successes in n trials can occur in ${}^nC_r$ ways and the probability for each of these ways is $p^r q^{n-r}$. The probability of r successes in n trials **in any order,** whatsoever, is given by ${}^nC_r p^r q^{n-r}$.

Hence in all the N sets, the number of sets with r successes is $N. {}^nC_r p^r q^{n-r}$. Putting $r = 0, 1, 2, ..., n$, the number of sets corresponding to the number of successes 0, 1, 2, ..., r, ..., n are respectively.

$$Nq^n, N\,{}^nC_1pq^{n-1}, N\,{}^nC_2p^2q^{n-2}, ..., N\,{}^nC_rp^rq^{n-r}, ... , Np^n$$

Hence for N sets of n trials, the frequencies of 0, 1, 2, ..., r, ..., n successes are given by the successive terms in the expression

$$N\,[q^n + {}^nC_1q^{n-1}p + {}^nC_2q^{n-2}p^2 + ... + {}^nC_rq^{n-r}p^r + ... + p^n]$$

which is the binomial expansion of

$$N(q+p)^n.$$

This is called **Binomial Frequency Distribution.**

(M.D.U. 1983 ; K.U. 1981 S)

Note 1. The probability that the event will happen **not more than r times**

$$= N[q^n + {}^nC_1q^{n-1}p + {}^nC_2q^{n-2}p^2 + ... + {}^nC_rq^{n-r}p^r]$$

$$= N\sum_{s=0}^{r} {}^nC_sq^{n-s}p^s.$$

Note 2. The probability that the event will happen at **least r times**

$$= N[\,{}^nC_rq^{n-r}p^r + {}^nC_{r+1}q^{n-r-1}p^{r+1} + ... + p^n]$$

$$= N\sum_{s=r}^{n} {}^nC_sq^{n-s}p^s.$$

Note 3. $r\,{}^nC_r = r\,.\,\frac{n}{r}\,.\,{}^{n-1}C_{r-1} = n\,{}^{n-1}C_{r-1}$

$$r(r-1)\,{}^nC_r = r(r-1)\,.\,\frac{n(n-1)}{r(r-1)}\,.\,{}^{n-2}C_{r-2} = n\,(n-1)\,{}^{n-2}C_{r-2}$$

Similarly $r(r-1)(r-2)\,{}^nC_r = n\,(n-1)(n-2)\,{}^{n-3}C_{r-3}$

$$r(r-1)(r-2)\,......\,(r-10) = n(n-1)(n-2)\,......\,(n-10)\,{}^{n-11}C_{r-11}.$$

7.3. FOR THE BINOMIAL DISTRIBUTION N(q + p)^n, q + p = 1 TO COMPUTE THE MEAN, THE STANDARD DEVIATION AND PEARSON'S β AND γ CO-EFFICIENTS.

(*K.U. 1982 S ; M.D.U. 1982 S ; D.U. 1984*)

In the binomial distribution $N(q+p)^n$, the theoretical fequencies of 0, 1, 2, ..., r, ... n successes are the successive terms in the binomial expansion of $N(q+p)^n$.

$\Rightarrow x$: 0 1 2 .... r .... n

f : Nq^n $N\,.\,{}^nC_1q^{n-1}p$ $N\,.\,{}^nC_2q^{n-2}p^2$ ...... $N\,.\,{}^nC_rq^{n-r}p^r$ ...... $N\,.\,p^n$

Now, the nth moment about any point A is given by

$$\mu_n' = \frac{1}{N}\,\Sigma f(x-A)^n$$

Taking an arbitrary origin at 0 success *i.e.* taking A = 0

$$\mu_n' = \frac{1}{N}\,\Sigma f x^n$$

$$\therefore \qquad \mu_1' = \frac{1}{N}\Sigma fx = \frac{1}{N}\sum_{r=0}^{n} N \,.\, {}^nC_r q^{n-r} p^r \,.\, r$$

$$= \sum_{r=0}^{n} r \,.\, {}^nC_r q^{n-r} p^r = \sum_{r=1}^{n} n \,.\, {}^{n-1}C_{r-1}\, q^{n-r} \,.\, p.p^{r-1}$$

$$= np \sum_{r=1}^{n} {}^{n-1}C_{r-1} q^{n-r} p^{r-1} = np(q+p)^{n-1}$$

$$= np \qquad\qquad (\because \quad q + p = 1)$$

Since $\overline{x} = A + \mu_1' = 0 + np = np.$

Hence **mean** $\mathbf{\overline{x} = np.}$ *(M.D.U.1983 ; D.U. 1982)*

$$\mu_2' = \frac{1}{N}\Sigma fx^2 = \frac{1}{N}\sum_{r=0}^{n} N \,.\, {}^nC_r q^{n-r} p^r \,.\, r^2$$

$$= \sum_{r=0}^{n} {}^nC_r q^{n-r} p^r [r + r(r-1)] \text{ writing } \mathbf{r^2 = r + r(r-1)}$$

$$= \sum_{r=0}^{n} r \,.\, {}^nC_r q^{n-r} p^r + \sum_{r=0}^{n} r(r-1) \,.\, {}^nC_r q^{n-r} p^r$$

$$= \sum_{r=1}^{n} n.{}^{n-1}C_{r-1} q^{n-r}.p.p^{r-1} + \sum_{r=2}^{n} n(n-1) \,.\, {}^{n-2}C_{r-2} q^{n-r}.p^2.p^{r-2}.$$

$$= np \sum_{r=1}^{n} {}^{n-1}C_{r-1} q^{n-r} p^{r-1} + n(n-1)p^2 \sum_{r=2}^{n} {}^{n-2}C_{r-2} q^{n-r} p^{r-2}$$

$$= np(q+p)^{n-1} + n(n-1)p^2(q+p)^{n-2}$$

$$= np + n(n-1)p^2 \qquad\qquad (\because \quad q + p = 1)$$

$$= np[1 + (n-1)p] = np(1 - p + np)$$

$$= np(q + np) = npq + n^2p^2$$

$$\therefore \qquad \sigma^2 = \mu_2 = \mu_2' - {\mu'_1}^2 = npq + n^2p^2 - n^2p^2 = npq.$$

Hence **Standard Deviation** $\sigma = \sqrt{\mathbf{npq}}$. (M.D.U. 1983)

$$\mu_3' = \frac{1}{N}\Sigma fx^3 = \frac{1}{N}\sum_{r=0}^{n} N\, {}^nC_r q^{n-r} p^r \,.\, r^3$$

$$= \sum_{r=0}^{n} [r(r-1)(r-2) + 3r(r-1) + r]\,{}^nC_r q^{n-r} p^r$$

$$| \text{ writing } \mathbf{r^3 = r(r-1)(r-2) + 3r(r-1) + r}$$

$$= \sum_{r=0}^{n} r(r-1)(r-2) \,.\, {}^nC_r q^{n-1} p^r + 3 \sum_{r=0}^{n} r(r-1) {}^nC_r q^{n-r} p^r + \sum_{r=0}^{n} r \, {}^nC_r q^{n-r} p^r$$

$$= \sum_{r=3}^{n} n(n-1)(n-2) \, {}^{n-3}C_{r-3} q^{n-r} p^3 p^{r-3} + 3 \sum_{r=2}^{n} n(n-1) \,.\, {}^{n-2}C_{r-2} q^{n-r} p^2 \,.\, p^{r-2} + \sum_{r=1}^{n} n \,.\, {}^{n-1}C_{r-1} q^{n-r} \,.\, p \,.\, p^{r-1}$$

$$= n(n-1)(n-2) \, p^3 \sum_{r=3}^{n} {}^{n-3}C_{r-3} q^{n-r} p^{r-3} + 3n(n-1)p^2 \sum_{r=2}^{n} {}^{n-2}C_{r-2} q^{n-r} p^{r-2} + np \sum_{r=1}^{n} {}^{n-1}C_{r-1} q^{n-r} p^{r-1}$$

$$= n(n-1)(n-2)p^3(q+p)^{n-3} + 3n(n-1)p^2(q+p)^{n-2} + np(q+p)^{n-1}$$

$$= n(n-1)(n-2)p^3 + 3n(n-1)p^2 + np$$

$$\mu_3 = \mu_3' - 3\mu'_2\mu_1' + 2\mu_1'^{\,3}$$

$$= n(n-1)(n-2)p^3 + 3n(n-1)p^2 + np - 3(npq + n^2p^2)\, np + 2n^3p^3$$

$$= np[(n^2 - 3n + 2)p^2 + 3(n-1)p + 1 - 3(npq + n^2p^2) + 2n^2p^2]$$

$$= np(1 - 3p + 2p^2 - 3np^2 + 3np - 3npq)$$

$$= np[1 - 3p + 2p^2 - 3np(p + q - 1)]$$

$$= np(1 - 3p + 2p^2) \qquad [\because \; p + q = 1]$$

$$= np(1-p)(1-2p)$$

$$= npq(1-2p) = npq(1-p-p) = npq\,(q-p)$$

$$\mu_4' = \frac{1}{N} \Sigma f x^4 = \frac{1}{N} \sum_{r=0}^{n} N \,.\, {}^nC_r q^{n-r} p^r \,.\, r^4$$

Writing $r^4 = r(r-1)(r-2)(r-3) + 6r(r-1)(r-2) + 7r(r-1) + r$

$$= \sum_{r=0}^{n} r(r-1)(r-2)(r-3) \, {}^nC_r q^{n-r} p^r + 6 \sum_{r=0}^{n} r(r-1)(r-2) \, {}^nC_r q^{n-r} p^r + 7 \sum_{r=0}^{n} r(r-1) {}^nC_r q^{n-r} p^r + \sum_{r=0}^{n} r \, {}^nC_r q^{n-r} p^r$$

$$= n(n-1)(n-2)(n-3)\, p^4 \sum_{r=4}^{n} {}^{n-4}C_{r-4} q^{n-r} p^{r-4}$$

$$+ 6n(n-1)(n-2)p^3 \sum_{r=3}^{n} {}^{n-3}C_{r-3} q^{n-r} p^{r-3}$$

$$+ 7n(n-1)p^2 \sum_{r=2}^{n} {}^{n-2}C_{r-2} q^{n-r} p^{r-2} + np \sum_{r=1}^{n} {}^{n-1}C_{r-1} q^{n-r} p^{r-1}$$

$$= n(n-1)(n-2)(n-3)p^4 (q+p)^{n-4} + 6n(n-1)(n-2)p^3 (q+p)^{n-3} + 7n(n-1)p^2 (q+p)^{n-2} + np(q+p)^{n-1}$$

$$= n(n-1)(n-2)(n-3)p^4 + 6n(n-1)(n-2)p^3 + 7n(n-1)p^2 + np \qquad \Big|\because\ q+p=1$$

$$\mu_4 = \mu_4' - 4\mu_3'\mu_1' + 6\mu_2'\mu_1'^2 - 3\mu_1'^4$$

$$= n(n-1)(n-2)(n-3)\, p^4 + 6n(n-1)(n-2)p^3 + 7n(n-1)p^2 + np - 4np[n(n-1)(n-2)p^3 + 3n(n-1)p^2 + np] + 6n^2p^2[np + n(n-1)p^2] - 3n^4p^4$$

$$= np[(n^3 - 6n^2 + 11n - 6)p^3 + 6(n^2 - 3n + 2)p^2 + 7(n-1)p + 1 - 4(n^3 - 3n^2 + 2n)\, p^3 - 12(n^2 - n)p^2 - 4np + 6n^2p^2 + 6n^2p^3(n-1) - 3n^3p^3]$$

$$= np[(p^3 - 4p^3 + 6p^3 - 3p^3)n^3 + (-6p^3 + 6p^2 + 12p^3 - 12p^2 + 6p^2 - 6p^3)n^2 + (11p^3 - 18p^2 + 7p - 8p^3 + 12p^3 - 4p)n + (-6p^3 + 12p^2 - 7p + 1)]$$

$$= np[(3p^3 - 6p^2 + 3p)n + (1 - 7p + 12p^2 - 6p^3)]$$

$$= np[3p(1-p)^2 n + (1-p)(1 - 6p + 6p^2)]$$

$$= 3n^2p^2(1-p)^2 + np(1-p)[1 - 6p(1-p)]$$

$$= 3n^2p^2q^2 + npq(1 - 6pq) \qquad \Big|\because\ 1-p=q$$

Now $\beta_1 = \dfrac{\mu_3^2}{\mu_2^3} = \dfrac{[npq(q-p)]^2}{(npq)^3} = \dfrac{(q-p)^2}{npq} = \dfrac{(1-2p)^2}{npq}$ (*D.U. 1992*)

$$\beta_2 = \frac{\mu_4}{\mu_2^2} = \frac{3n^2p^2q^2 + npq(1-6pq)}{(npq)^2} = 3 + \frac{1-6pq}{npq}$$

Hence $\gamma_1 = \sqrt{\beta_1} = \dfrac{q-p}{\sqrt{npq}} = \dfrac{1-2p}{\sqrt{npq}}$ (*D.U. 1992*)

$$\gamma_2 = \beta_2 - 3 = \frac{1-6pq}{npq}.$$

Note. $\gamma_1 = \dfrac{q-p}{\sqrt{npq}}$ gives **a measure of skewness** of the binomial distribution.

(*D.U. 1982*)

If $p < \frac{1}{2}$, skewness is positive, if $p > \frac{1}{2}$, skewness is negative and if $p = \frac{1}{2}$, it is zero.

$\beta_2 = 3 + \dfrac{1-6pq}{npq}$ gives **a measure of the kurtosis** of the binomial distribution.

(*D.U. 1982*)

Example 1. *Bring out the fallacy, if any, in the following statement : The mean of a binomial distribution is 5 and its standard deviation is 3.*

Sol. Let the binomial distribution be

$$(q+p)^n, \; q+p=1$$

Mean = 5 $\Rightarrow$ $np = 5$ ...(*i*)

$\sigma = 3$ $\Rightarrow$ $\sqrt{npq} = 3$

Squaring $npq = 9$ ...(*ii*)

Dividing (*ii*) by (*i*) $q = \frac{9}{5} > 1.$

But $0 \le q \le 1$. Hence the given statement is false.

Example 2. *Comment on the following :*

For a binomial distribution, mean is 6 and variance is 9.

Sol. Please try yourself. [**Ans.** False statement]

Example 3. *Find the binomial distribution whose mean is 3 and variance 2.* (M.D.U. 1984)

Sol. Let the binomial distribution be $(q+p)^n$.

Mean = 3 $\Rightarrow$ $np = 3$

Variance = 2 $\Rightarrow$ $\sigma^2 = npq = 2$

Dividing $q = \frac{2}{3}$ $\therefore$ $p = 1 - q = \frac{1}{3}$

and therefore $n = 9$.

Hence the required binomial distribution is $(\frac{2}{3} + \frac{1}{3})^9$.

Example 4. *For a binomial distribution, the mean is 6 and the standard deviation is $\sqrt{2}$. Write out all the terms of the distribution.*

Sol. Let the binomial distribution be $(q+p)^n, q+p=1$

Mean = 6 $\Rightarrow$ $np = 6$

$\sigma = \sqrt{2}$ $\Rightarrow$ $\sqrt{npq} = \sqrt{2}$ $\Rightarrow$ $npq = 2$

Dividing $q = \frac{1}{3}$ $\therefore$ $p = 1 - q = \frac{2}{3}$ whence $n = 9$

$\therefore$ The binomial distribution is $(\frac{1}{3} + \frac{2}{3})^9$ and its expansion is

$$(\tfrac{1}{3})^9 + {}^9C_1(\tfrac{1}{3})^8(\tfrac{2}{3}) + {}^9C_2(\tfrac{1}{3})^7(\tfrac{2}{3})^2 + {}^9C_3(\tfrac{1}{3})^6(\tfrac{2}{3})^3 + {}^9C_4(\tfrac{1}{3})^5(\tfrac{2}{3})^4$$
$$+ {}^9C_5(\tfrac{1}{3})^4(\tfrac{2}{3})^5 + {}^9C_6(\tfrac{1}{3})^3(\tfrac{2}{3})^6 + {}^9C_7(\tfrac{1}{3})^2(\tfrac{2}{3})^7 + {}^9C_8(\tfrac{1}{3})(\tfrac{2}{3})^8 + (\tfrac{2}{3})^9.$$

Example 5. *Find the mode of the binomial distribution.*

(M.D.U. 1981)

Or

Find the most probable number of success in a series of n independent trials, the probability of success in each trial being p.

Sol. The mode of the binomial distribution $(q+p)^n$ is the value of r such that

$$T_{r+1} \geq T_r \quad \text{and also} \quad T_{r+1} \geq T_{r+2}$$

where T_r denotes the rth term in the binomial expansion of $(q+p)^n$.

Hence (*i*) $\quad {}^nC_r\, q^{n-r} p^r \geq {}^nC_{r-1}\, q^{n-r+1} p^{r-1}$

$$\Rightarrow \quad \frac{n!}{(n-r)!\,r!}\,p \geq \frac{n!}{(n-r+1)!\,(r-1)!}\,q$$

$$\Rightarrow \quad \frac{p}{r} \geq \frac{q}{n-r+1}$$

$$\Rightarrow \quad (n-r+1)p \geq rq = r(1-p) = r - rp$$

$$\Rightarrow \quad (n+1)p \geq r \qquad \ldots(i)$$

and (*ii*) $\quad {}^nC_r q^{n-r} p^r \geq {}^nC_{r+1}\, q^{n-r-1}\, p^{r+1}$

$$\Rightarrow \quad \frac{n!}{(n-r)!\,r!}\,q \geq \frac{n!}{(n-r-1)!\,(r+1)!}\,p$$

$$\Rightarrow \quad \frac{q}{n-r} \geq \frac{p}{r+1}$$

$$\Rightarrow \quad q(r+1) \geq p(n-r)$$

$$\Rightarrow \quad (1-p)(r+1) \geq p(n-r)$$

$$\Rightarrow \quad r+1-pr-p \geq pn-pr$$

$$\Rightarrow \quad r \geq (n+1)p - 1 \qquad \ldots(ii)$$

From (*i*) and (*ii*)

$$(n+1)p - 1 \leq r \leq (n+1)p.$$

Case I. If $(n+1)p = k$, where k is an integer, then r will increase till $r = k - 1$ and will have the same maximum value when $r = k$ and after that it will begin to decrease.

Case II. If $(n+1)p = k + f$, where k is an integer and f, a proper fraction, then r is maximum when $r = k$.

Note. For the binomial distribution $\left(\frac{2}{5}+\frac{3}{5}\right)^9$,

$$(n+1)p = (9+1) \times \frac{3}{5} = 6$$

$\therefore$ The mode $= (n+1)p - 1$ and $(n+1)p = 5$ and 6. The distribution is bimodal.

For the binomial distribution $\left(\frac{2}{3}+\frac{1}{3}\right)^9$

$$(n+1)p = (9+1)\,\frac{1}{3} = \frac{10}{3} = 3 + \frac{1}{3} \text{ and hence the mode} = 3.$$

↓

integral part

Example 6. *Compute the mode of a binomial distribution with $p = \frac{1}{4}$ and $n = 7$.*

Sol. $p = \frac{1}{4}, \quad n = 7$

$$(n+1)p = (7+1)\frac{1}{4} = 2, \text{ an integer.}$$

The mode $= (n+1)p - 1$ and $(n+1)p = 1$ and 2.

The distribution is bimodal.

Example 7. *Find the mode of a binomial distribution with*

$$p = \frac{1}{2} \quad \text{and} \quad n = 7.$$

Sol. Please try yourself. **[Ans.** 3, 4]

Example 8. *Show that if np be a whole number, the mean of the binomial distribution coincides with the greatest term.*

Sol. Mean of the binomial distribution = np.

If T_{r+1} is the greatest term, then r is the mode of the binomial distribution.

Also $(n+1)p = np + p$ and np is a whole number

$$\therefore \quad r = np$$

Hence the greatest term occurs when $r = np$ = mean.

Example 9. *Find the most probable number of heads in 99 tossings of a biased coin, given that the probability of a head in a single tossing is $\frac{3}{5}$*

Sol. Here $n = 99$, $p = \frac{3}{5}$.

$\therefore \quad (n+1)p = 100 \times \frac{3}{5} = 60$, which is an integer.

Hence $(n+1)p - 1$ and $(n+1)p = 59$ and 60 are the most probable number of heads.

Example 10. *Show that if two symmetrical binomial distributions of degree n (and of the same number of observations) are so superimposed that the rth term of one coincides with the (r + 1)th term of the other, the distribution formed by adding superimposed terms is a symmetrical binomial of degree (n + 1).*

Sol. For symmetrical binomial distribution, $p = q = \frac{1}{2}$

$\therefore$ The symmetrical binomial distribution of degree n is

$$N\left(\frac{1}{2} + \frac{1}{2}\right)^n$$

$$\therefore \quad T_r = N \cdot {}^nC_{r-1}\left(\frac{1}{2}\right)^{n-r+1} \cdot \left(\frac{1}{2}\right)^{r-1} = N \cdot {}^nC_{r-1}\left(\frac{1}{2}\right)^n$$

The $(r+1)$th term of the other symmetrical binomial distribution $N\left(\frac{1}{2} + \frac{1}{2}\right)^n$ is given by

$$T_{r+1} = N \cdot {}^nC_r \left(\frac{1}{2}\right)^{n-r} \cdot \left(\frac{1}{2}\right)^r = N \cdot {}^nC_r \left(\frac{1}{2}\right)^n$$

On the first distribution, the second is so superimposed that T_r of first distribution coincides with T_{r+1} of second distribution.

Now $\quad T_r + T_{r+1} = N\left({}^nC_{r-1} + {}^nC_r\right)\left(\frac{1}{2}\right)^n$

$$= N \,.\, {}^{n+1}C_r \,.\, (\tfrac{1}{2})^n = 2N \,.\, {}^{n+1}C_r (\tfrac{1}{2})^{n+1}$$

$$= 2N \,.\, {}^{n+1}C_r \,.\, (\tfrac{1}{2})^{n+1-r} \,.\, (\tfrac{1}{2})^r$$

$$= (r+1)\text{th term of the symmetrical binomial } 2N\,(\tfrac{1}{2} + \tfrac{1}{2})^{n+1}$$

which is of degree $(n + 1)$. Hence the result.

Example 11. *Ten coins are thrown simultaneously. Find the probability of getting at least seven heads.*

Sol. When one coin is thrown,

p, the probability of getting a head $= \frac{1}{2}$

$$q = 1 - \frac{1}{2} = \frac{1}{2}, \quad n = 10$$

The binomial distribution is $(q + p)^n = (\frac{1}{2} + \frac{1}{2})^{10}$

The probability of getting at least seven heads is given by

$$P(7) + P(8) + P(9) + P(10)$$

$$= {}^{10}C_7(\tfrac{1}{2})^3(\tfrac{1}{2})^7 + {}^{10}C_8(\tfrac{1}{2})^2(\tfrac{1}{2})^8 + {}^{10}C_9(\tfrac{1}{2})(\tfrac{1}{2})^9 + {}^{10}C_{10}(\tfrac{1}{2})^{10}$$

$$= \frac{1}{2^{10}}\,[\,{}^{10}C_3 + {}^{10}C_2 + {}^{10}C_1 + {}^{10}C_0] = \frac{120 + 45 + 10 + 1}{1024} = \frac{176}{1024} = \frac{11}{64}.$$

Example 12. *Eight coins are thrown simultaneously. Find the chance of obtaining at least six heads.*

Sol. Please try yourself. $\left[\textbf{Ans. } \dfrac{37}{256}\right]$

Example 13. *The probability of any ship of a company being destroyed on a certain voyage is 0.02. The company owns 6 ships for the voyage. What is the probability of*

(i) losing one ship.

(ii) losing at most two ships.

(iii) losing none ?

Sol. p, the probability of losing a ship = 0.02

$$q = 1 - 0.02 = 0.98, \qquad n = 6$$

The binomial distribution is $(0.98 + 0.02)^6$

(*i*) The probability of losing one ship $= {}^6C_1(0.98)^5(0.02)$

$$= 6 \times 0.9039 \times 0.02$$

$$= 0.108468 = 0.11$$

(*ii*) The probability of losing at most 2 ships

$$= P(0) + P(1) + P(2)$$

$$= {}^6C_0(0.98)^6 + {}^6C_1(0.98)^5(0.02) + {}^6C_2(0.98)^4(0.02)^2$$

$$= 0.8858 + 0.1084 + 0.0055 = 0.9997$$

(*iii*) The probability of losing none $= {}^nC_0\,(0.98)^6 = 0.8858.$

Example 14. *During war, 1 ship out of 9 was sunk on an average in making a certain voyage. What was the probability that exactly 3 out of a convoy of 6 ships would arrive safely ?*

Sol. p, the probability of a ship arriving safely $= 1 - \frac{1}{9} = \frac{8}{9}$

$$q = \frac{1}{9}, \qquad n = 6$$

Binomial distribution is $(\frac{1}{9} + \frac{8}{9})^6$

The probability that exactly 3 ships arrive safely

$$= {}^6C_3 \left(\frac{1}{9}\right)^3 \left(\frac{8}{9}\right)^3 = \frac{10240}{9^6}.$$

Example 15. *In the long run, 3 vessels out of every 100 are sunk. If 10 vessels are out, what is the probability that*

(*i*) *exactly 6 will arrive safely ?*

(*ii*) *at least 6 will arrive safely ?*

Sol. Please try yourself.

Example 16. *Assume that on the average one telephone number out of fifteen called between 2 P.M. and 3 P.M. on week days is busy. What is the probability that if 6 randomly selected telephone numbers are called (i) not more than three, (ii) at least three of them will be busy ?*

Sol. p, the probability of a telephone number being busy between 2 P.M. and 3 P.M. on week-days $= \frac{1}{15}$

$$q = 1 - \frac{1}{15} = \frac{14}{15}, n = 6$$

Binomial distribution is $\left(\frac{14}{15} + \frac{1}{15}\right)^6$

The probability that not more than three will be busy

$$= P(0) + P(1) + P(2) + P(3)$$

$$= {}^6C_0 \left(\frac{14}{15}\right)^6 + {}^6C_1 \left(\frac{14}{15}\right)^5 \left(\frac{1}{15}\right) + {}^6C_2 \left(\frac{14}{15}\right)^4 \left(\frac{1}{15}\right)^2 + {}^6C_3 \left(\frac{14}{15}\right)^3 \left(\frac{1}{15}\right)^3$$

$$= \frac{(14)^3}{(15)^6} [2744 + 1176 + 210 + 20] = \frac{2744 \times 4150}{(15)^6}$$

$$= 0.9997$$

The probability that at least three of them will be busy

$$= P(3) + P(4) + P(5) + P(6)$$

$$= {}^6C_3 \left(\frac{14}{15}\right)^3 \left(\frac{1}{15}\right)^3 + {}^6C_4 \left(\frac{14}{15}\right)^2 \left(\frac{1}{15}\right)^4 + {}^6C_5 \left(\frac{14}{15}\right) \left(\frac{1}{15}\right)^5 + {}^6C_6 \left(\frac{1}{15}\right)^6$$

$$= 0.005.$$

Example 17. *A machine manufacturing screws is known to produce 5% defectives. In a random sample of 15 screws, what is the probability that there are*

(i) exactly three defectives, *(ii) not more than three defectives ?*

Sol. Please try yourself.

Here $p = 5\% = \dfrac{5}{100} = 0.05, q = 0.95, n = 15.$

Example 18. *The incidence of occupational disease in an industry is such that the workers have a 20% chance of suffering from it. What is the probability that out of six workers chosen at random four or more will suffer from the disease ?*

Sol. Please try yourself. $\left[\textbf{Ans.}\ \dfrac{53}{3125}\right]$

Here $p = 20\% = \dfrac{20}{100} = 0.2, q = .8, n = 6.$

Example 19. *If on the average rain falls on ten days in every thirty, find the probability that rain falls on at least three days of a given week.*

Sol. Please try yourself. $\left[\textbf{Ans.}\ \dfrac{317}{729}\right]$

Example 20. *Six dice are thrown 720 times. How many times do you expect at least three dice to show a five or a six ?*

(K.U. 1981 S ; M.D.U. 1983 S ; D.U. 1988)

Sol. p = the chance of getting 5 or 6 with one die $= \frac{2}{6} = \frac{1}{3}$

$$q = 1 - \tfrac{1}{3} = \tfrac{2}{3}, \qquad n = 6, \qquad N = 729$$

since dice are in sets of 6 and there are 729 sets.

The binomial distribution is $N(q+p)^n$

$$= 729\left(\frac{2}{3} + \frac{1}{3}\right)^6$$

The expected number of times *at least three* dice showing five or six

$$= 729\left[{}^6C_3\left(\frac{2}{3}\right)^3\left(\frac{1}{3}\right)^3 + {}^6C_4\left(\frac{2}{3}\right)^2\left(\frac{1}{3}\right)^4 + {}^6C_5\left(\frac{2}{3}\right)\left(\frac{1}{3}\right)^5 + {}^6C_6\left(\frac{1}{3}\right)^6\right]$$

$$= \frac{729}{3^6}[160 + 60 + 12 + 1] = 233.$$

Example 21. *Six dice are thrown 1458 times. How many times do you expect at least three dice to show a 5 or a 6 ?* (M.D.U. 1982)

Sol. Please try yourself. **[Ans.** 466]

Example 22. *In 256 sets of twelve tosses of a coin, in how many cases may one expect eight heads and four tails ?*

Sol. p = probability of getting a head in a single toss $= \frac{1}{2}$

$$q = 1 - \tfrac{1}{2} = \tfrac{1}{2}. \quad n = 12, \quad N = 256$$

The binomial distribution is $256\left(\frac{1}{2}+\frac{1}{2}\right)^{12}$

The expected number of getting *eight heads* and four tails

$$= 256 \times {}^{12}C_8 \left(\frac{1}{2}\right)^4 \left(\frac{1}{2}\right)^8 = \frac{2^8}{2^{12}} \times {}^{12}C_4$$

$$= \frac{1}{16} \times \frac{12 \times 11 \times 10 \times 9}{4 \times 3 \times 2 \times 1} = \frac{495}{16} = 31 \text{ approx.}$$

Example 23. *In 100 sets of ten tosses of an unbiased (i.e., perfect) coin, in how many cases do you expect to get :*

(i) 7 heads and 3 tails. *(ii) at least 7 heads ?*

Sol. As in example 22

$p = q = \frac{1}{2}$, $n = 10$, $N = 100$

The expected number of getting 7 heads and 3 tails

$$= 100 \times {}^{10}C_7 \left(\frac{1}{2}\right)^3 \left(\frac{1}{2}\right)^7$$

$$= \frac{100}{2^{10}} \times {}^{10}C_3 = \frac{100}{1024} \times \frac{10 \times 9 \times 8}{3 \times 2 \times 1} = \frac{375}{32} = 12 \text{ approx.}$$

The expected number of getting at least 7 heads

$$= 100 \times \left[{}^{10}C_7 \left(\frac{1}{2}\right)^3 \left(\frac{1}{2}\right)^7 + {}^{10}C_8 \left(\frac{1}{2}\right)^2 \left(\frac{1}{2}\right)^8 + {}^{10}C_9 \left(\frac{1}{2}\right) \left(\frac{1}{2}\right)^9 + {}^{10}C_{10} \left(\frac{1}{2}\right)^{10} \right]$$

$$= \frac{100}{2^{10}} [{}^{10}C_3 + {}^{10}C_2 + {}^{10}C_1 + {}^{10}C_0]$$

$$= \frac{100}{1024} (120 + 45 + 10 + 1)$$

$$= \frac{100 \times 176}{1024} = \frac{275}{16} = 17 \text{ approx.}$$

Example 24. *A perfect cubical die is thrown a large number of times in sets of 8. The occurrence of 5 or 6 is called a success. In what proportion of the sets would you expect 3 successes ?*

Sol. p = the chance of getting 5 or 6 with one die $= \frac{2}{6} = \frac{1}{3}$

$q = 1 - \frac{1}{3} = \frac{2}{3}$, $n = 8$

∴ The binomial distribution is $N\left(\frac{2}{3} + \frac{1}{3}\right)^8$, where N is the total number of sets.

The number of sets in which 3 successes are expected

$$= N \cdot {}^8C_3 \left(\frac{2}{3}\right)^5 \left(\frac{1}{3}\right)^3 = N \times \frac{56 \times 32}{243 \times 27} = \frac{1792}{6561} N$$

Hence the proportion of the sets in which three successes are expected

$$= \frac{1792}{6561} N \Big/ N = \frac{1792}{6561} \times 100\% = 27.31\%.$$

Example 25. *Out of 800 families with 4 children each, how many families would be expected to have (i) 2 boys and 2 girls (ii) at least one boy (iii) no girl (iv) at most two girls ? Assume equal probabilities for boys and girls.* (K.U. 1982 S)

Sol. Since probabilities for boys and girls are equal.

p = probability of having a boy = $\frac{1}{2}$

q = " " " " girl = $\frac{1}{2}$

$n = 4$, " N = 800

The binomial distribution is $800\left(\frac{1}{2}+\frac{1}{2}\right)^4$

(*i*) The expected number of families having 2 boys and 2 girls

$$= 800 \cdot {}^4C_2\left(\frac{1}{2}\right)^2\left(\frac{1}{2}\right)^2 = 800 \times 6 \times \frac{1}{16} = 300$$

(*ii*) The expected number of families having *at least one boy*

$$= 800\left[{}^4C_1\left(\frac{1}{2}\right)^3\left(\frac{1}{2}\right) + {}^4C_2\left(\frac{1}{2}\right)^2\left(\frac{1}{2}\right)^2 + {}^4C_3\left(\frac{1}{2}\right)\left(\frac{1}{2}\right)^3 + {}^4C_4\left(\frac{1}{2}\right)^4\right]$$

$$= 800 \times \frac{1}{16}[4+6+4+1] = 750$$

(*iii*) The expected number of families having no girl *i.e.* having 4 boys

$$= 800 \cdot {}^4C_4\left(\frac{1}{2}\right)^4 = 50$$

(*iv*) The expected number of families having *at most two girls i.e.* having *at least 2 boys*

$$= 800\left[{}^4C_2\left(\frac{1}{2}\right)^2\left(\frac{1}{2}\right)^2 + {}^4C_3\left(\frac{1}{2}\right)\left(\frac{1}{2}\right)^3 + {}^4C_4\left(\frac{1}{2}\right)^4\right]$$

$$= 800 \times \frac{1}{16}[6+4+1] = 550.$$

Example 26. *In 800 families with 5 children each, how many families would be expected to have (i) 3 boys and 2 girls (ii) 2 boys and 3 girls (iii) no girl (iv) at the most two girls. (Assume probabilities for boys and girls to be equal)* (K.U. 1983 S)

Sol. Please try yourself. [**Ans.** (*i*) 250, (*ii*) 250, (*iii*) 25, (*iv*) 400]

Example 27. *Assuming that half the population are consumers of rice, so that the chance of an individual being a consumer is $\frac{1}{2}$ and assuming that 100 investigators each take 10 individuals to see whether they are consumers, how many investigators would you expect to report that three people or less were consumers ?*

Sol. Here $p = \frac{1}{2}$ $\therefore$ $q = 1 - \frac{1}{2} = \frac{1}{2}$, $n = 10$, $N = 100$

The binomial distribution is $100\left(\frac{1}{2} + \frac{1}{2}\right)^{10}$.

Number of investigators to report that three people or less (*i.e.* 3 or 2 or 1 or none) are consumers

$$= 100\left[\,{}^{10}C_0\left(\frac{1}{2}\right)^{10} + {}^{10}C_1\left(\frac{1}{2}\right)^{9}\left(\frac{1}{2}\right) + {}^{10}C_2\left(\frac{1}{2}\right)^{8}\left(\frac{1}{2}\right)^{2} + {}^{10}C_3\left(\frac{1}{2}\right)^{7}\left(\frac{1}{2}\right)^{3}\right]$$

$$= \frac{100}{2^{10}}(1 + 10 + 45 + 120) = \frac{17600}{1024} = 17 \text{ nearly.}$$

Example 28. *In a certain town 20% of the population is literate, and assume that 200 investigators take a sample of ten individuals to see whether they are literate. How many investigators would you expect to report that three people or less are literates in the sample ?*

Sol. Please try yourself. **[Ans.** 176]

Example 29. *An irregular six faced die is thrown and the expectation that in 10 throws it will give five even numbers is twice the expectation that it will give four even numbers. How many times in 10,000 sets of 10 throws would you expect to give no even number ?* (D.U. 1991)

Sol. Let p be the probability of getting an even number.

The probability of 5 even numbers in 10 throws

$$= {}^{10}C_5 p^5 q^5, \quad q = 1 - p$$

The probability of 4 even numbers in 10 throws

$${}^{10}C_4 p^4 q^6.$$

By the given condition

$${}^{10}C_5 p^5 q^5 = 2 \times {}^{10}C_4 p^4 q^6$$

$$\Rightarrow \quad \frac{10\,!}{5\,!\,5\,!}\,p = 2 \times \frac{10\,!}{6\,!\,4\,!}\,q$$

$$\Rightarrow \quad \frac{p}{5\,!\,.\,5\,(4)\,!} = 2 \times \frac{q}{5\,!\,.\,4\,!}$$

$$\Rightarrow \quad \frac{3}{5}p = q = 1 - p$$

This gives $p = \frac{3}{8}$ and $\therefore$ $q = \frac{3}{8}$

The binomial distribution is $10000\left(\frac{3}{8} + \frac{5}{8}\right)^{10}$,

Hence the expected number of times where, we get no even number

$$= 10000\left(\frac{3}{8}\right)^{10} = 1 \text{ nearly.}$$

Example 30. *In 103 liters of 4 mice, the number of liters which contained 0, 1, 2, 3, 4 females were noted. The figures are given in the table below :*

Number of female mice (x) :	*0*	*1*	*2*	*3*	*4*
Number of liters (f) :	*8*	*32*	*34*	*24*	*5*

If the chance of obtaining a female in a single trial is assumed constant, estimate this constant of unknown probability. Find also expected frequencies.

Sol. Since the chance of obtaining a female in a single trial is assumed constant, say equal to p, we expect the distribution of the number of female mice to be the binomial distribution

$$103(q+p)^4, \qquad q = 1-p$$

Expected value of the mean of the distribution

$$= np = 4p, \text{ since } n = 4$$

Actual mean of the distribution

$$= \frac{1}{N}\sum fx$$

$$= \frac{1}{103}(0 \times 8 + 1 \times 32 + 2 \times 34 + 3 \times 24 + 4 \times 5)$$

$$= \frac{192}{103}$$

$$\therefore \quad 4p = \frac{192}{103}, \qquad p = \frac{48}{103} = 0.466$$

$\therefore$ Required constant = 0.466

$$q = 1 - p = 0.534$$

Hence expected frequencies are the respective terms of the binomial expansion of $103(0.534 + 0.466)^4$.

Example 31. *Five dice were thrown together 96 times. The number of times 4, 5 or 6 was actually thrown in the experiment is given below. Calculate the expected frequencies :*

No. of dice showing 4, 5 or 6 :	*0*	*1*	*2*	*3*	*4*	*5*
Observed Frequency :	*1*	*10*	*24*	*35*	*18*	*8*

Sol. p = the chance of getting 4, 5 or 6 with one die

$$= \frac{3}{6} = \frac{1}{2} \qquad \therefore \quad q = \frac{1}{2}$$

Hence the expected frequencies of getting 0, 1, 2, 3, 4, 5 successes with 5 dice in throwing 96 times are the successive terms in the binomial expansion of $96\left(\frac{1}{2}+\frac{1}{2}\right)^5$ which are respectively.

$$96\left(\tfrac{1}{2}\right)^5,\ 96.\,{}^5C_1\left(\tfrac{1}{2}\right)^5,\ 96.\,{}^5C_2\left(\tfrac{1}{2}\right)^5,\ 96.\,{}^5C_3\left(\tfrac{1}{2}\right)^5,\ 96.\,{}^5C_4\left(\tfrac{1}{2}\right)^5,\ 96\left(\tfrac{1}{2}\right)^5,$$

$\Rightarrow$ the expected frequencies are

3, 15, 30, 15, 3.

Example 32. *Ten coins are tossed 1024 times and the following frequencies are observed. Compare these frequencies with the expected frequencies :*

No. of heads :	*0*	*1*	*2*	*3*	*4*	*5*	*6*	*7*	*8*	*9*	*10*
Frequencies :	*2*	*10*	*38*	*106*	*188*	*257*	*226*	*128*	*59*	*7*	*3*

Sol. Here $n = 10$ $\quad$ N = 1024

p = the chance of getting a head in one toss = $\frac{1}{2}$

$\therefore \quad q = 1 - p = \frac{1}{2}$

The expected frequencies are the respective terms of the binomial

$$1024\left(\tfrac{1}{2}+\tfrac{1}{2}\right)^{10}$$

The frequency of r heads

$$(0 \le r \le 10) \text{ is } 1024.\ {}^{10}C_r \left(\tfrac{1}{2}\right)^{10-r} . \left(\tfrac{1}{2}\right)^r$$

$$= 1024 \times {}^{10}C_r \left(\tfrac{1}{2}\right)^{10} = {}^{10}C_r.$$

Hence, we have the following comparison

No. of heads :	0	1	2	3	4	5	6	7	8	9	10
Observed frequency :	2	10	38	106	188	257	226	128	59	7	3
Expected frequency :	1	10	45	120	210	252	210	120	45	10	1

(${}^{10}C_0$, ${}^{10}C_1$, ${}^{10}C_2$ and so on).

Example 33. *The following data are the number of seeds germinating out of 10 on damp filter for 80 sets of seeds. Fit a binomial distribution to this data :*

x :	*0*	*1*	*2*	*3*	*4*	*5*	*6*	*7*	*8*	*9*	*10*	*Total*
f :	*6*	*20*	*28*	*12*	*8*	*6*	*0*	*0*	*0*	*0*	*0*	*80*

Sol. Here $\quad n = 10, \qquad N = 80$

Mean $\quad \bar{x} = \dfrac{\Sigma fx}{\Sigma f}$

$$= \frac{1}{80}(0 \times 6 + 1 \times 20 + 2 \times 28 + 3 \times 12 + 4 \times 8 + 5 \times 6 + 0 + 0 + 0 + 0)$$

$$= \frac{174}{80} = 2.175$$

But mean $\quad = np - 10p \qquad \therefore \quad 10p = 2.175$

whence $\quad p = 0.2175$

$\quad q = 1 - p = 0.7825.$

Hence the binomial distribution to be fitted to the data is

$$N(q+p)^n = 80(0.7825 + 0.2175)^{10}.$$

Example 34. *A set of 8 symmetrical coins was tossed 256 times and the frequencies of throws observed were as follows :*

Number of heads :	*0*	*1*	*2*	*3*	*4*	*5*	*6*	*7*	*8*
Frequency of throws :	*2*	*6*	*24*	*63*	*64*	*50*	*36*	*10*	*1*

Fit a binomial distribution to the data.

Sol. Here $n = 8$, $N = 256$, $\Sigma f = 256$

Mean $\bar{x} = \frac{\Sigma fx}{\Sigma f}$

$$= \frac{1}{256}(0 + 6 + 48 + 189 + 256 + 250 + 216 + 70 + 8)$$

$$= \frac{1043}{256}$$

But mean $= np = 8p$

$\therefore \quad 8p = \frac{1043}{256}$ whence $p = \frac{1043}{2048} = 0.5093$

$$q = 1 - p = 0.4907$$

Hence the binomial distribution to be fitted to the data is

$$N(q + p)^n = 256(0.4907 + 0.5093)^8.$$

Example 35. *Prove that for the binomial distribution with parameters n and p, the moments obey the recurrence relation.*

(Renovsky Formula)

$$\mu_{r+1} = pq\left[nr\mu_{r-1} + \frac{d\mu_r}{dp}\right]$$

where the letters have their usual meaning.

Hence deduce the values μ_2, μ_3 and μ_4.

(D.U. 1983, 85, 87, 89, 93 ; Agra 1987 ; M.U. 1985, 89)

Sol. rth moment about the mean is given by

$$\mu_r = \frac{1}{N}\Sigma f(x - \bar{x})^r$$

For the binomial distribution $N(q + p)^n$, $\bar{x} = np$

$$\therefore \quad \mu_r = \frac{1}{N}\Sigma N\, f(x - np)^r$$

$= \Sigma f(x - np)^r$, f being the frequency of x successes

$$= \sum_{x=0}^{n} (x - np)^r \,.\, {}^nC_x\, q^{n-x} p^x$$

$$= \sum_{x=0}^{n} {}^nC_x (x - np)^r (q^{n-x} p^x) \qquad \text{where } q = 1 - p$$

Differentiating w.r.t. p, we get

$$\frac{d\mu_r}{dp} = \sum_{x=0}^{n} {}^nC_x[-rn(x - np)^{r-1} q^{n-x} p^x$$

$$+ (x - np)^r\{-(n - x)q^{n-x-1}p^x + q^{n-x} \,.\, xp^{n-1}\}]$$

$$= -rn \sum_{x=0}^{n} {}^nC_x q^{n-x} p^x (x - np)^{r-1}$$

$$+ \frac{1}{pq} \sum_{x=0}^{n} {}^nC_x (x - np)^r \{-(n-x) q^{n-r} p^r + x q^{n-x+1} p^x\}$$

$$= -rn\mu_{r-1} + \frac{1}{pq} \sum_{x=0}^{n} {}^nC_x (x - np)^r p^x q^{n-x} \{(-n + x)p + xq\}$$

$$= -rn\mu_{r-1} + \frac{1}{pq} \sum_{x=0}^{n} {}^nC_x (x - np)^r p^x q^{n-x} . (x - np)$$

$$(\because \quad p + q = 1)$$

$$= -rn\mu_{r-1} + \frac{1}{pq} \sum_{x=0}^{n} {}^nC_x q^{n-x} p^x (x - np)^{r+1}$$

$$= -rn\mu_{r-1} + \frac{1}{pq} \mu_{r+1}$$

$$\Rightarrow \mu_{r+1} = pq \left[nr\mu_{r-1} + \frac{d\mu_r}{dp} \right] \quad ...(1)$$

Putting $r = 1$ in (1), we get

$$\mu_2 = pq \left[n\mu_0 + \frac{d\mu_1}{dp} \right] = pq[n + 0] \quad [\because \quad \mu_0 = 1, \mu_1 = 1]$$

$$= npq$$

Putting $r = 2$ in (1), we get

$$\mu_3 = pq \left[2n\mu_1 + \frac{d\mu_2}{dp} \right] = pq \left[0 + \frac{d}{dp} \{np(1-p)\} \right]$$

$$= pqn(1 - 2p) = npq(1 - p - p) = npq(q - p)$$

Putting $r = 3$ in (1), we get

$$\mu_4 = pq \left[3n\mu_2 + \frac{d\mu_3}{dp} \right]$$

$$= pq \left[2n . npq + \frac{d}{dp} \{np(1-p)(1-2p)\} \right]$$

$$= pq \left[3n^2pq + n \frac{d}{dp} (p - 3p^2 + 2p^3) \right]$$

$$= npq[2npq + (1 - 6p + 6p^2)]$$

$$= npq[3npq + 1 - 6p(1 - p)]$$

$$= 3n^2p^2q^2 + npq(1 - 6pq).$$

II. POISSON DISTRIBUTION

1. Poisson distribution is a limiting form of Binomial Distribution as $n \to \infty$ and $p \to 0$ in such a manner that $np \to$ a finite constant m.

(D.U. 1983, 88, 89, 90)

In the case of binomial distribution, the probability of r successes is given by

$$P(r) = {}^nC_r q^{n-r} p^r$$

$$= \frac{n(n-1)(n-2) \ldots\ldots (n-r+1)}{r\,!} \cdot q^{n-r} p^r$$

Putting $\quad np = m$, *i.e.*, $p = \dfrac{m}{n}$ and $\quad \therefore \quad q = 1 - p = 1 - \dfrac{m}{n}$

$$P(r) = \frac{n(n-1)(n-2) \ldots\ldots (n-r+1)}{r\,!} \cdot \left(1 - \frac{m}{n}\right)^{n-r} \left(\frac{m}{n}\right)^r$$

$$= \frac{n(n-1)(n-2) \ldots\ldots (n-r+1)}{n^r} \cdot \frac{m^2}{r\,!} \cdot \frac{\left(1 - \frac{m}{n}\right)^n}{\left(1 - \frac{m}{n}\right)^r}$$

$$= \frac{n}{n} \cdot \frac{n-1}{n} \cdot \frac{n-2}{n} \ldots\ldots \frac{n-r+1}{n} \cdot \frac{m^r}{r\,!} \cdot \frac{\left(1 - \frac{m}{n}\right)^n}{\left(1 - \frac{m}{n}\right)^r}$$

$$= \left(1 - \frac{1}{n}\right)\left(1 - \frac{2}{n}\right) \ldots\ldots \left(1 - \frac{r-1}{n}\right) \cdot \frac{m^r}{r\,!} \cdot \frac{\left[\left(1 - \frac{m}{n}\right)^{-n/m}\right]^{-m}}{\left(1 - \frac{m}{n}\right)^r}$$

For fixed r, as $n \to \infty$

$$\left(1 - \frac{1}{n}\right), \left(1 - \frac{2}{n}\right) \ldots\ldots \left(1 - \frac{r-1}{n}\right), \left(1 - \frac{m}{n}\right)^r \text{ all} \to 1$$

and $$\left[\left(1 - \frac{m}{n}\right)^{-n/m}\right]^{-m} \to e^{-m}$$

$$\therefore \quad \boxed{P(r) = \frac{m^r}{r\,!} e^{-m}}$$

2. Definition. The probability distribution of a random variable x is said to be a *Poisson Distribution* if the random variable assumes only non-negative values and if its distribution is given by

$$P(X = r) = \frac{m^r e^{-m}}{r!}$$

where m is called the parameter of the distribution and r takes the values 0, 1, 2 ......

Note 1. Like binomial distribution, the variate of the Poisson Distribution is also a discrete one *i.e.,* it takes only integral values. The probabilities of 0, 1, 2, ... r, ... successes are given by the successive terms of the expansion

$$\sum_{r=0}^{\infty} \frac{m^r e^{-m}}{r!} = e^{-m}\left(1 + \frac{m}{1!} + \frac{m^2}{2!} + \frac{m^3}{3!} + \ldots\ldots + \frac{m^r}{r!} + \ldots\ldots\right)$$

Note 2. In a Poisson Distribution, our interest lies only in the number of occurrences of the event, not in its non-occurrences.

Note 3. A Poisson Distribution may be expected in cases where the chance of any individual event being a success is small. The distribution is used to describe the behaviour of rare events such as serious floods, accidental release of radiation from a nuclear reactor and the like.

Note 4. $\sum_{r=0}^{\infty} \frac{m^r}{r!} = \sum_{r=1}^{\infty} \frac{m^{r-1}}{(r-1)!} = \sum_{r=2}^{\infty} \frac{m^{r-2}}{(r-2)!} = \ldots\ldots = e^m$

$\because$ each $= 1 + \frac{m}{1!} + \frac{m^2}{2!} + \frac{m^3}{3!} + \ldots\ldots$

3. Constants of the Poisson Distribution

Obtain the value of the first four moments for the Poisson distributions. Show that for Poisson distribution, mean and variance are equal.

(K.U. 1982)

Also prove that (i) $m\sigma\gamma_1\gamma_2 = 1$

(ii) $\sqrt{\beta_1}\,(\beta_2 - 3)m\sigma = 1.$

The expected frequencies of 0, 1, 2, ..., r, ... successes of the Poisson Distribution are given by

$$e^{-m}, me^{-m}, \frac{m^2e^{-m}}{2!}, \ldots\ldots, \frac{m^re^{-m}}{r!}, \ldots\ldots$$

Take an arbitrary origin at 0 successes, then the values of the deviations are given by 0, 1, 2, ... , r, ...

Since $\mu_n' = \frac{1}{N}\Sigma f(x - A)^n$ where $N = \Sigma f$

$$\therefore \quad \mu_1' = \sum_{r=0}^{\infty} \frac{m^re^{-m}}{r!}\,.\,r \qquad | \text{ Here } A = 0$$

$$= e^{-m} \sum_{r=0}^{\infty} \frac{m^r}{r!} \,.\, r = me^{-m} \sum_{r=1}^{\infty} \frac{m^{r-1}}{(r-1)!}$$

$$= me^{-m} \,.\, e^m = m$$

Since "$\mu_1' = M - A$"

$\therefore$ **mean** $= \mu_1' = \mathbf{m}$ (*M.D.U. 1983 S ; D.U. 1988*)

$$\mu_2' = \sum_{r=0}^{\infty} \frac{m^r e^{-m}}{r!} \,.\, r^2$$

Writing $\mathbf{r^2 = r(r-1) + r}$

$$\mu_2' = \sum_{r=0}^{\infty} \frac{m^r e^{-m}}{r!} [r(r-1) + r]$$

$$= e^{-m} \sum_{r=0}^{\infty} \frac{m^r}{r!} \,.\, r(r-1) + e^{-m} \sum_{r=0}^{\infty} \frac{m^r}{r!} \,.\, r$$

$$= e^{-m} \,.\, m^2 \sum_{r=0}^{\infty} \frac{m^{r-2}}{(r-2)!} + e^{-m} \,.\, m \sum_{r=1}^{\infty} \frac{m^{r-1}}{(r-1)!}$$

$$= e^{-m} \,.\, m^2 e^m + e^{-m} \,.\, m\, e^m$$

$$= m^2 + m$$

Now $\mu_2 = \mu'_2 - \mu'^2_1 = m^2 + m - m^2 = m$

$\therefore$ **Variance** $\sigma^2 = \mu_2 = \mathbf{m}$ **= mean** (*M.D.U. 1983 S ; D.U. 1988*)

Standard Deviation $\sigma = \sqrt{\mathbf{m}}$

$$\mu_3' = \sum_{r=0}^{\infty} \frac{m^r e^{-m}}{r!} \,.\, r^3$$

Writing $\mathbf{r^3 = r(r-1)(r-2) + 3r(r-1) + r}$

$$\mu_3' = e^{-m} \sum_{r=0}^{n} \frac{m^r}{r!} [r(r-1)(r-2) + 3r(r-1) + r]$$

$$= e^{-m} \,.\, m^3 \sum_{r=3}^{\infty} \frac{m^{r-3}}{(r-3)!} + 3e^{-m} \,.\, m^2 \sum_{r=2}^{\infty} \frac{m^{r-2}}{(r-2)!}$$

$$+ e^{-m} \,.\, m \sum_{r=1}^{\infty} \frac{m^{r-1}}{(r-1)!}$$

$$= e^{-m} . m^3 . e^m + 3e^{-m} . m^2 . e^m + e^{-m} m . e^m$$
$$= m^3 + 3m^2 + m$$
$$\mu_3 = \mu'_3\, 3\mu'_2\mu'_1 + 2\mu'^3_1$$
$$= m^3 + 3m^2 + m - 3m(m^2 + m) + 2m^3 = m$$
$$\mu'_4 = \sum_{r=0}^{\infty} \frac{m^r e^{-m}}{r\,!} . r^4$$

Writing $\mathbf{r^4 = r(r-1)(r-2)(r-3) + 6r(r-1)(r-2) + 7r(r-1) + r}$

$$\mu'_4 = e^{-m} \sum_{r=0}^{\infty} \frac{m^r}{r\,!} \; [r(r-1)(r-2)(r-3) + 6r(r-1)(r-2) + 7r(r-1) + r]$$

$$= e^{-m} . m^4 \sum_{r=4}^{\infty} \frac{m^{r-4}}{(r-4)\,!} + 6e^{-m} . m^3 \sum_{r=3}^{\infty} \frac{m^{r-3}}{(r-3)\,!} + 7e^{-m} . m^2 \sum_{r=2}^{\infty} \frac{m^{r-2}}{(r-2)\,!} + e^{-m} . m \sum_{r=1}^{\infty} \frac{m^{r-1}}{(r-1)\,!}$$

$$= m^4 + 6m^3 + 7m^2 + m \qquad [\because \text{ each sigma } (\Sigma) = e^m]$$

$$\therefore \quad \mu_4 = \mu'_4 - 4\mu'_3\mu'_1 + 6\mu'_2\mu'^2_1 - 3\mu'^4_1$$
$$= (m^4 + 6m^3 + 7m^2 + m) - 4m(m^3 + 3m^2 + m) + 6m^2(m^2 + m) - 3m^4$$
$$= 3m^2 + m$$

Values of Pearson's β and γ co-efficients

$$\beta_1 = \frac{\mu_3^2}{\mu_2^3} = \frac{m^2}{m^3} = \frac{1}{m} \qquad \therefore \quad \gamma_1 = +\sqrt{\beta_1} = \frac{1}{\sqrt{m}}$$

Now $\quad \beta_1 = \frac{1}{m} \Rightarrow \quad m = \frac{1}{\beta_1}$

$\Rightarrow$ mean and variance are reciprocal of β_1

$$\beta_2 = \frac{\mu_4}{\mu_2^2} = \frac{3m^2 + m}{m^2} = 3 + \frac{1}{m} \quad \therefore \quad \gamma_2 = \beta_2 - 3 = \frac{1}{m}.$$

(*i*) $m\sigma\gamma_1\gamma_2 = m\sqrt{m} . \frac{1}{\sqrt{m}} . \frac{1}{m} = 1.$ (*D.U. 1987*)

(*ii*) $\sqrt{\beta_1}\,(\beta_2 - 3)\, m\sigma = \gamma_1\gamma_2\, m\sigma = \frac{1}{\sqrt{m}} . \frac{1}{m} . m\sqrt{m} = 1.$

Example 1. *Deduce the values of the first four moment about the mean of the Poisson Distribution from those of the binomial distribution.* (D.U. 1988)

Sol. For the binomial distribution $(q+p)^n, q+p=1$

$$\mu_1 = 0, \qquad \mu_2 = npq$$

$$\mu_3 = npq\,(q+p), \qquad \mu_4 = 3\,(npq)^2 + npq\,(1-6pq)$$

Since the Poisson Distribution is a limiting case of the binomial distribution when $n \to \infty, p \to 0$ (and $\therefore$ $q = 1 - p \to 1$) in such a manner that $np = m$, the moments of the Poisson Distribution are

$$\mu_1 = 0$$

$$\mu_2 = \text{Lt}\,(npq) = m = \text{Lt}\,(mq) = m$$

$$\mu_3 = \text{Lt}\,npq\,(q-p) = \text{Lt}\,mq(q-p) = m\,1\,.\,(1-0) = m$$

$$\mu_4 = \text{Lt}\,[3(npq)^2 + npq(1-6pq)]$$

$$= \text{Lt}\,[3(mq)^2 + mq(1-6pq)]$$

$$= 3m^2 + m.$$

Example 2. *If X follows the Poisson Law such that* $P(X = 1) = P(X = 2)$, *find* $P(X = 4)$. (M.D.U. 1982 S)

Sol. Since $\quad P(X = r) = \dfrac{m^r e^{-m}}{r\,!}$

$\therefore \quad P(X = 1) = P(X = 2)$

$\Rightarrow \quad \dfrac{me^{-m}}{1\,!} = \dfrac{m^2 e^{-m}}{2\,!}$

$\Rightarrow \quad 1 = \dfrac{m}{2} \quad \therefore \quad m = 2$

$\therefore \quad P(X = 4) = \dfrac{m^4 e^{-m}}{4\,!} = \dfrac{2^4 e^{-2}}{24} = \dfrac{2}{3e^2}.$

Example 3. *Suppose that X has a Poisson Distribution. If*

$$P(X = 2) = \tfrac{2}{3} P(X = 1)$$

find (*i*) $P(X = 0)$ (*ii*) $P(X = 3)$.

Sol. Please try yourself. **[Ans.** (*i*) $e^{-4/3}$ (*ii*) $\frac{32}{81} e^{-4/3}$]

Example 4. *The probabilities of a Poisson variate taking the values 3 and 4 are equal. Calculate the probabilities of the variate taking the values 0 and 1.* (M.D.U. 1984)

Sol. Since $\quad P(X = r) = \dfrac{m^r e^{-m}}{r\,!}$

$\therefore \quad P(X = 3) = P(X = 4)$

$\Rightarrow \quad \dfrac{m^3 e^{-m}}{3\,!} = \dfrac{m^4 e^{-m}}{4\,!}$

$\Rightarrow \quad \dfrac{4\,!}{3\,!} = m \quad \Rightarrow \quad m = 4$

$$\therefore \quad P(X = 0) = \frac{m^0 e^{-m}}{0!} = e^{-4}$$

$$P(X = 1) = \frac{m^1 e^{-m}}{1!} = 4e^{-4}.$$

Example 5. *Criticise the following statement :*
The mean of a Poisson Distribution is 5 while the standard deviation of 4.

Sol. Let m be the parameter of the Poisson distribution.
Then mean = m and $\sigma = \sqrt{m}$

$\therefore \quad \mu = \sqrt{\text{mean}}$

Here $\sigma = 4$ and mean = 5 $\therefore$ $4 = \sqrt{5}$ which is not possible.
Hence the statement is false.

Example 6. *If X is a Poisson variate such that*

$$P(X = 2) = 9\,P(X = 4) + 90\,P(X = 6).$$

Find (i) mean (ii) standard deviation. (M.D.U. 1981)

Sol. Since $\quad P(X = r) = \dfrac{m^r e^{-m}}{r!}$

$$\therefore \quad P(X = 2) = 9\,P(X = 4) + 90\,P(X = 6)$$

$$\Rightarrow \quad \frac{m^2 e^{-m}}{2!} = 9 \cdot \frac{m^4 e^{-m}}{4!} + 90 \cdot \frac{m^6 e^{-m}}{6!}$$

$$\Rightarrow \quad \frac{1}{2!} = 9 \cdot \frac{m^2}{4.3.2!} + 90 \cdot \frac{m^4}{6.5.4.3.2!}$$

$$\Rightarrow \quad 1 = \frac{3m^2}{4} + \frac{m^4}{4} \quad \Rightarrow \quad m^4 + 3m^2 - 4 = 0$$

$$\Rightarrow \quad (m^2 + 4)(m^2 - 1) = 0 \quad \Rightarrow \quad m^2 = -4, 1$$

$$\therefore \quad m = 1 \qquad (\because\ m^2 \text{ can't be } -\text{ve})$$

Hence (*i*) mean = m = 1 (*ii*) S.D. = $\sqrt{m}$ = 1.

Example 7. *Find the mode of the Poisson Distribution.* (D.U. 1991)

Sol. The Poisson distribution with parameter m is the discrete distribution in which the probability of r successes is given by

$$\frac{m^r e^{-m}}{r!}, \; r = 0, 1, 2, \ldots\ldots$$

The mode is that value of r for which $\dfrac{m^r e^{-m}}{r!}$ is greater than or equal to the term that precedes it and the term that follow it.

$$\therefore \quad \frac{m^{r-1} e^{-m}}{(r-1)!} \le \frac{m^r e^{-m}}{r!} \qquad \ldots(i)$$

$$\text{and} \quad \frac{m^{r+1} e^{-m}}{(r+1)!} \le \frac{m^r e^{-m}}{r!} \qquad \ldots(ii)$$

(*i*) gives $r \leq m$ ...(*iii*)

(*ii*) gives $m \leq r + 1$ or $m - 1 \leq r$ ...(*iv*)

Combining (*iii*) and (*iv*) $m - 1 \leq r \leq m$

If m is an integer, then there are two modes m – 1 and m. Thus the distribution is bimodal.

If m is not an integer, the mode is the integral value between m – 1 and m.

Example 8. *A Poisson distribution has a double mode at x = 1 and x = 2. What is the probability that x will have one or the other of these two values ? ($e^{-2} = .1353$).*

Sol. If a Poisson distribution, with parameter *m*, is bimodal then the two modes are at the points $x = m - 1$ and $x = m$. Since we are given that the two modes are at the points $x = 1$ and $x = 2$, we find that $m = 2$.

Since $P(x = r) = \dfrac{m^r \cdot e^{-m}}{r\,!}$

$\therefore$ $P(x = 1) = 2e^{-2}$ and $P(x = 2) = 2e^{-2}$

Required probability $= P(x = 1) + P(x = 2)$

| By Additive law of probability

$= 4e^{-2} = 4 \times .1353 = .5412.$

Example 9. *If a Poisson distribution has a double mode at X = 1 and X = 2, find P(X = 1).*

Sol. Please try yourself. **[Ans.** $2e^{-2}$]

Example 10. *Give some examples of the occurrence of Poisson distribution in different fields.*

Sol. The following are some examples of Poisson variates :

(*i*) The number of cars passing through a certain street in time *t*.

(*ii*) The number of defective screws per box of 100 screws.

(*iii*) The number of deaths in a district by rare disease.

(*iv*) The number of air accidents in some unit of time.

(*v*) The number of printing mistakes at each page of the book.

(*vi*) The emission of radioactive (alpha) particles.

(*vii*) The number of pieces of a certain merchandise sold by a store in time *t*.

(*viii*) The number of telephone calls received at particular telephone exchange in some unit of time.

Example 11. *Fit a Poisson distribution to the following and calculate theoretical frequencies :*

Deaths :	*0*	*1*	*2*	*3*	*4*
Frequency :	*122*	*60*	*15*	*2*	*1*

Sol. Mean of given distribution

$$= \frac{\Sigma fx}{\Sigma f}$$

$$= \frac{122 \times 0 + 60 \times 1 + 15 \times 2 + 2 \times 3 + 1 \times 4}{122 + 60 + 15 + 2 + 1}$$

$$= \frac{60 + 30 + 6 + 4}{200} = \frac{1}{2} = 0.5$$

This is the parameter (m) of the Poisson distribution.

Now $\quad e^{-m} = e^{-0.5} = 1 - (0.5) + \frac{(0.5)^2}{2!} - \frac{(0.5)^3}{3!} + \frac{(0.5)^4}{4!} \ldots\ldots$

$$= 1 - 0.5 + 0.125 - 0.0208 + 0.0026 \ldots\ldots$$

$$= 1.1276 - 0.5208 = 0.6068$$

$\therefore$ Required Poisson distribution is $N \cdot \frac{m^r e^{-m}}{r!}$

$$= 200\, e^{-0.5} \cdot \frac{(0.5)^r}{r!}. \qquad | \because \; N = \Sigma f = 200$$

$$= 200 \times 0.6068 \frac{(0.5)^r}{r!} = 121.36 \times \frac{(0.5)^r}{r!}$$

To find theoretical frequencies, since frequencies are always integers, they are to be converted to the nearest integer.

r	$P(r)$	*Theoretical Frequency*
0	121.36	121
1	$121.36 \times 0.5 = 60.68$	61
2	$121.36 \times \frac{(0.5)^2}{2!} = 15.17$	15
3	$121.36 \times \frac{(0.5)^3}{3!} = 2.53$	3
4	$121.36 \times \frac{(0.5)^4}{4!} = 0.32$	0
		Total = 200

Example 12. *Data was collected over a period of 10 years, showing number of deaths from horse kicks in each of the 20 army corps. The distribution of deaths was as follows :*

No. of deaths	:	*0*	*1*	*2*	*3*	*4*	*Total*
Frequency	:	*109*	*65*	*22*	*3*	*1*	*200*

Fit a Poisson distribution to the data and calculate the theoretical frequencies.

Sol. Mean of given distribution

$$= \frac{\Sigma fx}{\Sigma f} = \frac{65 + 44 + 9 + 4}{200} = \frac{122}{200} = 0.61$$

This is the parameter (m) of the Poisson distribution.

Now $\quad e^{-m} = e^{-0.61} = 1 - 0.61 + \frac{(0.61)^2}{2!} - \frac{(0.61)^3}{3!} + \frac{(0.61)^4}{4!} \ldots\ldots$

$$= 1 - 0.61 + 0.1860 - 0.0379 + 0.0054 = 0.5435$$

$\therefore$ Required Poisson distribution is $N . \dfrac{m^r e^{-m}}{r!}$

$$= 200\, e^{-0.61} \frac{(0.61)^r}{r!} \qquad | \because \quad N = \Sigma f = 200$$

$$= 200 \times 0.5435 \frac{(0.61)^r}{r!} = 108.7 \times \frac{(0.61)^r}{r!}$$

r	$P(r)$	Theoretical Frequency
0	108.7	109
1	$108.7 \times 0.61 = 66.3$	66
2	$108.7 \times \dfrac{(0.61)^2}{2!} = 20.2$	20
3	$108.7 \times \dfrac{(0.61)^3}{3!} = 4.1$	4
4	$108.7 \times \dfrac{(0.61)^4}{4!} = 0.7$	1
		Total 200

Example 13. *For the following data, fit a Poisson distribution and calculate the theoretical frequencies ?*

No. of births :	*0*	*1*	*2*	*3*	*4*
Frequency :	*105*	*66*	*24*	*4*	*1*

(K.U. 1983)

Sol. Please try yourself.

[**Ans.** Poisson distribution is $104.58 \times \dfrac{(0.65)^r}{r!}$]

Theoretical frequencies are 105, 68, 22, 4, 1.

Example 14. *Fit a Poisson distribution to the following data which gives the number of yeast cells per square for 400 squares.*

($e^{-1.32} = 0.2674$)

No. of cells per square (x) :	*0*	*1*	*2*	*3*	*4*	*5*	*6*	*7*	*8*	*9*	*10*	*Total*
No. of squares (f) :	*103*	*143*	*98*	*42*	*8*	*4*	*2*	*0*	*0*	*0*	*0*	*400*

Find the expected frequencies.

Sol. Mean $= \dfrac{\Sigma fx}{\Sigma f} = \dfrac{529}{400} = 1.32$

This is parameter (m) of the Poisson distribution.

Poisson distribution is $N . \dfrac{m^r e^{-m}}{r!} = 400\, e^{-1.32} . \dfrac{(1.32)^r}{r!}$

$$= 400 \times 0.2674 \times \frac{(1.32)^r}{r\,!} = 106.96 \times \frac{(1.32)^r}{r\,!}$$

Theoretical frequencies are the values of $106.96 \times \frac{(1.32)^r}{r\,!}$ when $r = 0$, 1, 2,......, 10 (converted to the nearest integer).

$\therefore$ Theoretical frequencies are :

107, 141, 93, 41, 14, 4, 1, 0, 0, 0, 0.

Example 15. *Find the mean and standard deviation for the table of deaths of women over 85 years old recorded in a three year period.*

No. of deaths recorded in a day	:	*0*	*1*	*2*	*3*	*4*	*5*	*6*	*7*
No. of days	:	*364*	*376*	*218*	*89*	*33*	*13*	*2*	*1*

Find the expected number of days with one death recorded from the Poisson series fitted to the data.

Sol.

x	f	fx	fx^2
0	364	0	0
1	376	376	376
2	218	436	872
3	89	267	801
4	33	132	528
5	13	65	325
6	2	12	72
7	1	7	49
Total	1096	1295	3023

Mean $m = \frac{\Sigma fx}{\Sigma f} = \frac{1295}{1096} = 1.18$

$$\sigma^2 = \frac{\Sigma fx^2}{\Sigma f} - \left(\frac{\Sigma fx}{\Sigma f}\right)^2 = \frac{3023}{1096} - (1.18)^2$$

$$= 2.758 - 1.392 = 1.366$$

$\therefore$ $\sigma = 1.17$ (nearly).

Now $e^{-m} = e^{-1.18} = 1 - 1.18 + \frac{(1.18)^2}{2\,!} - \frac{(1.18)^3}{3\,!} + \frac{(1.18)^4}{4\,!} - \frac{(1.18)^5}{5\,!} + ...$

$$= 1 - 1.18 + 0.696 - 0.273 + 0.08 - 0.009 = 0.316$$

Poisson distribution is $N . \frac{m^r e^{-m}}{r\,!} = 1096 \times 0.316 \frac{(1.18)^r}{r\,!}$

Expected number of days with one death per day

$$= 1096 \times 0.316 \times 1.18 \qquad \textit{(by putting } r = 1\textit{)}$$
$$= 397.$$

Example 16. *Letters were received in an office on each of 100 days. Assuming the following data to form a random sample from a Poisson distribution, find the expected frequencies, correct to the nearest unit, taking* $e^{-4} = .0183$.

No. of letters	:	*0*	*1*	*2*	*3*	*4*	*5*	*6*	*7*	*8*	*9*	*10*
Frequency	:	*1*	*4*	*15*	*22*	*21*	*20*	*8*	*6*	*2*	*0*	*1*

(D.U. 1987)

Sol. $N = \Sigma f = 100$

$$\text{Mean } m = \frac{\Sigma fx}{\Sigma f} = \frac{0 + 4 + 30 + 66 + 84 + 100 + 48 + 42 + 16 + 0 + 10}{100}$$

$$= \frac{400}{100} = 4$$

$$\text{Poisson distribution is N}\frac{m^r e^{-m}}{r!} = 100\, e^{-4} \cdot \frac{4^r}{r!}$$

$$= 100 \times .0183 \frac{4^r}{r!} = 1.83 \times \frac{4^r}{r!}$$

r	$P(r)$	*Expected Frequency*
0	1.83	2
1	$1.83 \times 4 = 7.32$	7
2	$1.83 \times \frac{4^2}{2!} = 14.64$	15
3	$1.83 \times \frac{4^3}{3!} = 19.52$	20
4	$1.83 \times \frac{4^4}{4!} = 19.52$	20
5	$1.83 \times \frac{4^5}{5!} = 15.62$	16
6	$1.83 \times \frac{4^6}{6!} = 10.41$	10
7	$1.83 \times \frac{4^7}{7!} = 5.95$	6
8	$1.83 \times \frac{4^8}{8!} = 2.97$	3
9	$1.83 \times \frac{4^9}{9!} = 1.32$	1
10	$1.83 \times \frac{4^{10}}{10!} = 0.53$	1

Example 17. *The following mistakes per page were observed in a book :*

No. of mistakes per page :	*0*	*1*	*2*	*3*	*4*	*Total*
Frequency :	*211*	*90*	*19*	*5*	*0*	*325*

Fit a Poisson distribution to the data and find the expected frequencies.

Sol. Please try yourself. **[Ans.** 209, 92, 20, 3, 0]

Example 18. *Below are given the number of vacancies of judges occurring in a High Court over a period of 96 years :*

No. of vacancies :	*0*	*1*	*2*	*3*	*Total*
Frequency :	*59*	*27*	*9*	*1*	*96*

Fit a Poisson distribution to represent the frequencies of vacancies per year and find the expected frequencies.

Sol. Please try yourself. **[Ans.** 58, 29, 7, 1]

Example 19. *Find the probability that at most 5 defective fuses will be found in a box of 200 fuses if experience shows that 2% of such fuses are defective.* (D.U. 1985)

Sol. p = the probability of finding defective fuses = $\frac{2}{100}$

$\therefore$ Average number of defective fuses in a box of 200 fuses

$$= np = 200 \times \frac{2}{100} = 4$$

$\therefore$ The mean of the Poisson distribution is given by

$m = np = 4$

$$e^{-4} = 1 - 4 + \frac{4^2}{2!} - \frac{4^3}{3!} + \frac{4^4}{4!} \ldots\ldots = 0.0183$$

Required probability $P(r \le 5)$

$$= \sum_{r=0}^{5} \frac{4^r e^{-4}}{r!}$$

$$= e^{-4}\left(1 + 4 + \frac{4^2}{2!} + \frac{4^3}{3!} + \frac{4^4}{4!} + \frac{4^5}{5!}\right)$$

$$= 0.0183\,(1 + 4 + 8 + 10.6667 + 10.6667 + 8.5334)$$

$$= 0.0183 \times 42.8668 = 0.7845.$$

Example 20. *Find the probability that no defective fuse will be found in a box of 200 fuses if experience shows that 2% such fuses are defective.*

Sol. Please try yourself. **[Ans.** 0.0183]

Example 21. *In a certain factory turning razor blades, there is a small chance $\frac{1}{500}$ for any blade to be defective. The blades are in packets of 10. Use Poisson's distribution to calculate the approximate number of packets containing no defective, one defective and two defective blades respectively in a consignment of 10000 packets.* (D.U. 1986, 92)

Sol. p = the probability of getting a defective blade $= \frac{1}{500}$

$n = 10, \qquad N = 10000$

$\therefore$ Mean $m = np = 10 \times \frac{1}{500} = 0.02$

$$e^{-m} = e^{-0.02} = 1 - 0.02 + \frac{(0.02)^2}{2!} \ldots\ldots = .9802 \text{ approx.}$$

Poisson distribution is $N \cdot \frac{m^r e^{-m}}{r!} = 10000 \times .9802 \times \frac{(0.02)^r}{r!}$

$$= 9802 \times \frac{(0.02)^r}{r!}$$

Putting $r = 0$, the number of packets containing no defective blade $= 9802$

Putting $r = 1$, the number of packets containing one defective blade $= 9802 \times 0.02 = 196$ approx.

Putting $r = 2$, the number of packets containing two defective blades $= 9802 \times \frac{(0.02)^2}{2!} = 2$ approx.

Example 22. *A car-hire firm has two cars, which it hires out day by day. The number of demands for a car on each day is distributed as a Poisson distribution with mean 1.5. Calculate the proportion of days on which neither car is used and the proportion of days on which some demand is refused.*

($e^{-1.5} = 0.2231$) (Agra 1983)

Sol. Since the number of demands for a car is distributed as a Poisson distribution with mean $m = 1.5$.

$\therefore$ Proportion of days on which neither car is used

= Probability of there being no demand for the car

$$= \frac{m^0 e^{-m}}{0!} = e^{-0.5}. = 0.2231$$

Proportion of days on which some demand is refused

= probability for the number of demands to be more than two

$= 1 - P(x \le 2)$

$$= 1 - \left(e^{-m} + \frac{me^{-m}}{1!} + \frac{m^2 e^{-m}}{2!} \right)$$

$$= 1 - e^{-1.5} \left(1 + 1.5 + \frac{(1.5)^2}{2} \right)$$

$= 1 - 0.2231\,(1 + 1.5 + 1.125)$

$= 1 - 0.2231 \times 3.625 = 1 - 0.8087375$

$= 0.1912625.$

Example 23. *Six coins are tossed 6400 times. Using the Poisson distribution, what is the approximate probability of getting six heads x times ?*

Sol. Probability of getting one head with one coin = $\frac{1}{2}$.

$\therefore$ The probability of getting six heads with six coins

$$= \left(\frac{1}{2}\right)^6 = \frac{1}{64}$$

$\therefore$ Average number of six heads with six coins in 6400 throws

$$= np = 6400 \times \frac{1}{64} = 100$$

$\therefore$ The mean of the Poisson distribution = 100

Approximate probability of getting six heads x times when the distribution is Poisson

$$= \frac{m^x e^{-m}}{x!} = \frac{(100)^x \cdot e^{-100}}{100!}.$$

Example 24. *The probability that a man aged 50 years will die within a year is 0.01125. What is the probability that of 12 such men at least 11 will reach their fifty first birthday ?*

Sol. Since the probability of death is very small, we use Poisson distribution.

Here $p = 0.01125, n = 12$

$\therefore$ $m = np = 12 \times 0.01125 = 0.135$

$$e^{-m} = e^{-0.135} = 1 - 0.35 + \frac{(0.135)^2}{2!} - \frac{(0.35)^3}{3!} + ...$$

$$= 1 - 0.135 + 0.00906 - 0.00041$$

$$= 0.87366.$$

The probability that at least 11 persons will survive

= The probability that at most one person dies

$$= P_{(0)} + P_{(1)}$$

$$= \frac{m^0 e^{-m}}{0!} + \frac{m^1 e^{-m}}{1!}$$

$$= e^{-m}(1 + m) = 0.87366 \times 1.135 = 0.9916.$$

Example 25. *A manufacture r of cotter pins knows that 5% of his product of defective. If he sells pins in boxes of 100 and guarantees that not more than 4 pins will be defective, what is the approximate probability that a box will fail to meet the guaranteed quality ? ($e^{-5} = 0.0067$)*

Sol. Here $p = \frac{5}{100}$, $n = 100$

$\therefore$ $np = 5.$

Hence the mean m of the Poisson distribution = 5 and the Poisson distribution is

$$P_{(r)} = \frac{m^r e^{-m}}{r!} \quad r = 0, 1, 2, \ldots\ldots$$

Probability that a box will fail to meet the guaranteed quality

= probability of there being more than 4 defective pins in a box of 100 pins

$$= 1 - P(r \le 4)$$

$$= 1 - [P_{(0)} + P_{(1)} + P_{(2)} + P_{(3)} + P_{(4)}]$$

$$= 1 - \left[e^{-m} + \frac{m\,e^{-m}}{1!} + \frac{m^2 e^{-m}}{2!} + \frac{m^3 e^{-m}}{3!} + \frac{m^4 e^{-m}}{4!} \right]$$

$$= 1 - e^{-m} \left(1 + m + \frac{m^2}{2} + \frac{m^3}{6} + \frac{m^4}{24} \right)$$

$$= 1 - e^{-5} \left(1 + 5 + \frac{25}{2} + \frac{125}{6} + \frac{625}{24} \right)$$

$$= 1 - .0067\,(1 + 5 + 12.5 + 20.83 + 26.04)$$

$$= 1 - .0067 \times 65.37 = 1 - .437979 = .5620 \text{ approx.}$$

Example 26. *It is known that the probability that an item produced by a certain machine will be defective is 0.01. By applying Poisson's approximations, show that the probability that a random sample of 100 items selected at random from the total output will contain not more than one defective item is 2/e.* (D.U. 1983)

Sol. Please try yourself.

[**Hint.** $p = 0.01, \quad n = 100, \quad m = np = 1$

Required probability $= P(r \le 1) = P_{(0)} + P_{(1)}$.]

Example 27. *Show that in a Poisson distribution with unit mean, mean deviation about mean is* $\left(\frac{2}{e}\right)$ *times the standard deviation*

(M.D.U. 1982 ; D.U. 1982, 93 ; Kanpur 1990)

Sol. Here $m = 1, \quad f = \frac{m^r e^{-m}}{x!} = \frac{e^{-1}}{x!}$

$$N = \Sigma f = \sum_{x=0}^{\infty} \frac{e^{-1}}{x!} = e^{-1} \sum_{x=0}^{\infty} \frac{1}{x!}$$

$$= e^{-1} \left(1 + \frac{1}{1!} + \frac{1}{2!} + \frac{1}{3!} + \ldots\ldots \text{ to } \infty \right)$$

$$= e^{-1} . e = 1$$

Standard deviation $(\sigma) = \sqrt{m} = \sqrt{1} = 1$

Hence mean deviation about the mean

$$= \frac{1}{N} \Sigma f \,|\, x - m \,| = \Sigma f \,|\, x - 1 \,| \qquad \Big|\because \quad m = 1, N = 1$$

$$= \sum_{x=0}^{\infty} \frac{e^{-1}}{x\,!} \cdot |\, x - 1 \,|$$

$$= e^{-1}\left(1 + 0 + \frac{1}{2\,!} + \frac{2}{3\,!} + \frac{3}{4\,!} \ldots\ldots\right)$$

$$= e^{-1}\left[1 + \frac{2-1}{2\,!} + \frac{3-1}{3\,!} + \frac{4-1}{4\,!} + \ldots\ldots\right]$$

$$= e^{-1}\left[1 + \left(1 - \frac{1}{2\,!}\right) + \left(\frac{1}{2\,!} - \frac{1}{3\,!}\right) + \left(\frac{1}{3\,!} - \frac{1}{4\,!}\right) + \ldots\ldots\right]$$

$$= e^{-1}\,[1 + 1] = 2e^{-1}$$

$$= \frac{2}{e} \times 1 = \frac{2}{e} \times \text{standard deviation.}$$

Example 28. *Prove the following recurrence formula for a Poisson distribution with parameter m*

$$\mu_{r+1} = m\left(r\,\mu_{r-1} + \frac{du_r}{dm}\right)$$

(D.U. 1985, 90 ; Meerut 1988, 90)

Sol. We have $\mu_r = \Sigma f(x - m)^r$

$$= \sum_{x=0}^{\infty} \frac{m^x\, e^{-m}}{x\,!}\,(x - m)^r \text{ for a Poisson distribution.}$$

Differentiating both sides w.r.t. parameter m

$$\frac{d\mu_r}{dm} = \sum_{x=0}^{\infty} \frac{1}{x\,!}\,[x m^{x-1} e^{-m}\,(x - m)^r + m^x(-e^{-m})(x - m)^r + m^x e^{-m}\, r(x - m)^{r-1}(-1)]$$

$\Big|$ using $\frac{d}{dm}$ (uvw) formula

$$= \sum_{x=0}^{\infty} \frac{1}{x\,!}\,[m^{x-1}\, e^{-m}\,(x - m)^r\,(x - m) - r m^r\, e^{-m}\,(x - m)^{r-1}]$$

$$= \sum_{x=0}^{\infty} \frac{m^{x-1}\, e^{-m}}{x\,!}\,(x + m)^{r+1} - r\sum_{x=1}^{\infty} \frac{m^x\, e^{-m}}{x\,!}\,(x - m)^{r-1}$$

$$m\frac{d\mu_r}{dm}=\sum_{x=0}^{\infty}\frac{m^x e^{-m}}{x!}(x-m)^{r+1}-rm\sum_{x=0}^{\infty}\frac{m^x e^{-m}}{x!}(x-m)^{r-1}$$

$$=\mu_{r+1}-rm\mu_{r-1}$$

Hence $\mu_{r+1}=m\left(r\mu_{r-1}+\frac{d\mu_r}{dm}\right)$.

III. NORMAL (or Gaussian) DISTRIBUTION

1. Normal distribution is another limiting form of the binomial distribution under the following conditions :

(*i*) n, the number of trials is indefinitely large *i.e.*, $n \to \infty$ and

(*ii*) neither p nor q is very small.

2. To derive the normal distribution as a limiting case of binomial distribution when p = q.

Let the binomial distribution be $N(q+p)^n$. Since $p+q=1$, $\therefore\ p=q=\frac{1}{2}$. The binomial distribution is symmetrical. Let us assume that n is an even integer *i.e.*, let $n=2k$, $k\in Z$. (There is no loss of generally in this assumption because n is to tend to infinity ultimately).

The frequencies of r and $(r+1)$ successes can be written as

$$f_r = N\,.\,{}^{2k}C_r(\tfrac{1}{2})^{2k},\qquad f_{r+1}=N\,.\,{}^{2k}C_{r+1}(\tfrac{1}{2})^{2k}$$

$\therefore$ $$f_r > f_{r+1}$$

if $$N\,.\,{}^{2k}C_r\,.\,(\tfrac{1}{2})^{2k} > N\,.\,{}^{2k}C_{r+1}\,.\,(\tfrac{1}{2})^{2k}.$$

or if $${}^{2k}C_r > {}^{2k}C_{r+1}$$

or if $$\frac{2k\,!}{(2k-r)\,!\,.\,r\,!} > \frac{2k\,!}{(2k-r-1)\,!\,.\,(r+1)\,!}$$

or if $$\frac{1}{(2k-r)\,(2k-r-1)\,!\,r\,!} > \frac{1}{(2k-r-1)\,!\,.\,(r+1)\,.\,r\,!}$$

or if $$\frac{1}{2k-r} > \frac{1}{r+1}$$

or if $$2k-r < r+1$$

or if $$2k-1 < 2r$$

or if $$r > k-\tfrac{1}{2}$$

Similarly $f_r > f_{r-1}$ if ${}^{2k}C_r > {}^{2k}C_{r-1}$

i.e., if $r < k+\frac{1}{2}$.

Hence if $k-\frac{1}{2} < r < k+\frac{1}{2}$, f_r is greatest.

Evidently when $r=k$, f_r is maximum, say y_0. Then

$$y_0 = N\,.\,{}^{2k}C_k\,(\tfrac{1}{2})^{2k} = N\,(\tfrac{1}{2})^{2k}\frac{2k\,!}{k\,!\,k\,!}$$

The frequency curve tails off symmetrically on either side of this greatest ordinate. The frequency of $k + x$ successes is given by (x successes more than k)

$$y_x = \text{N} \cdot {}^{2k}C_{k+x} \left(\tfrac{1}{2}\right)^{2k} = \text{N} \cdot \left(\tfrac{1}{2}\right)^{2k} \cdot \frac{2k\,!}{(k-x)\,!\,(k+x)\,!}$$

$$\therefore \quad \frac{y_x}{y_0} = \frac{k\,!\;k\,!}{(k+x)\,!\,(k-x)\,!}$$

$$= \frac{k(k-1)(k-2)\,......\,(k-x+1)\;k-x\,!\;k\,!}{(k+x)(k+x-1)\,......\,(k+2)(k+1)\,.\,k\,!\;k-x\,!}$$

$$= \frac{k(k-1)(k-2)\,......\,(k-x+1)}{(k+1)(k+2)\,......\,(k+x)}$$

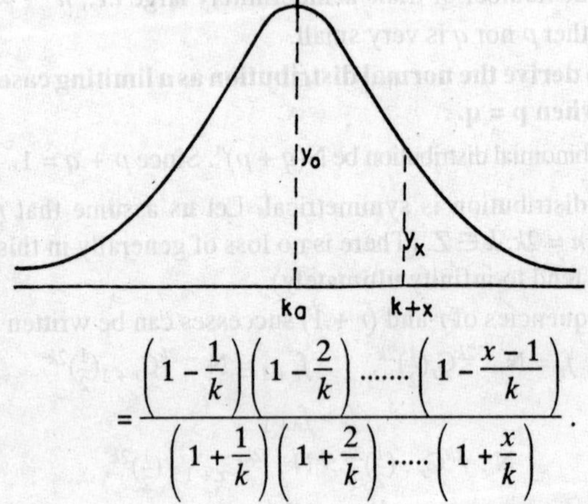

$$= \frac{\left(1-\frac{1}{k}\right)\left(1-\frac{2}{k}\right)\,......\,\left(1-\frac{x-1}{k}\right)}{\left(1+\frac{1}{k}\right)\left(1+\frac{2}{k}\right)\,......\,\left(1+\frac{x}{k}\right)}.$$

Since n is large, k is large $\quad \therefore \quad \frac{1}{k}, \frac{2}{k}$, ... are small (*i.e.*, < 1)

Taking logarithms

$$\log \frac{y_x}{y_0} = \log\left(1-\frac{1}{k}\right) + \log\left(1-\frac{2}{k}\right) + \,.....\, \log\left(1-\frac{x-1}{k}\right)$$
$$-\log\left(1+\frac{1}{k}\right) - \log\left(1+\frac{2}{k}\right) \,......\, \log\left(1+\frac{x}{k}\right)$$

$$= \left(-\frac{1}{k}\,......\right) + \left(-\frac{2}{k}\,......\right) + \,......\, + \left(-\frac{x-1}{k} - \,......\right)$$
$$-\left(\frac{1}{k} - \,......\right) - \left(\frac{2}{k} - \,......\right) - \,......\, - \left(\frac{x}{k} - \,......\right)$$

$$\because \quad \log(1-x) = -x - \frac{x^2}{2} - \,.......$$
$$\log(1+x) = x - \frac{x^2}{2} + \,.......$$

$\left(\text{Since } k \text{ is very small, } \frac{x}{k} \text{ is very small and } \therefore \text{ higher powers of } \frac{x}{k} \text{ may be neglected}\right)$

$$= -\frac{1}{k}(1 + 2 + \ldots\ldots + \overline{x - 1}) - \frac{1}{k}(1 + 2 + \ldots\ldots + x)$$

$$= -\frac{2}{k}(1 + 2 + \ldots. + \overline{x - 1}) - \frac{x}{k}$$

$$= -\frac{2}{k} \cdot \frac{(x-1)(x-1+1)}{2} - \frac{x}{2}$$

$$\left| \because \quad 1 + 2 + \ldots\ldots + n = \frac{n(n+1)}{2} \right.$$

$$= -\frac{x}{k}(x-1) - \frac{x}{k} = -\frac{x^2}{k}$$

$$\therefore \quad \frac{y_x}{y_0} = e^{-\frac{x^2}{k}} \quad \text{or} \quad y_x = y_0\, e^{-\frac{x^2}{k}}$$

But in the case of binomial distribution

$$\sigma^2 = npq = n \cdot \frac{1}{2} \cdot \frac{1}{2} = \frac{1}{4} \cdot 2k = \frac{k}{2}$$

$$\therefore \quad k = 2\sigma^2$$

$$\therefore \quad y_x = y_0\, e^{-\frac{x^2}{2\sigma^2}}$$

This is thē **normal distribution.**

Note 1. Even if $p \neq q$, the normal distribution is given by

$$y_x = y_0\, e^{-\frac{x^2}{2\sigma^2}}$$

Note 2. The curve $y = y_0\, e^{-\frac{x^2}{2\sigma^2}}$ is called the **normal curve.**

3. Standard Form of the Normal Curve.

To make the curve $y = y_0\, e^{-\frac{x^2}{2\sigma^2}}$ normal probability curve, the value of y_0 is determined in such a way that the total frequency may be 1.

$$\Rightarrow \quad \int_{-\infty}^{\infty} y_0\, e^{-\frac{x^2}{2\sigma^2}}\, dx = 1$$

$$\text{or} \quad 2y_0 \int_0^{\infty} e^{-\frac{x^2}{2\sigma^2}}\, dx = 1$$

($\because$ the integrand is an even function of x)

Put $\frac{x}{\sqrt{2}\sigma} = t$ so that $dx = \sqrt{2}\,\sigma\,dt$

When $x = 0$, $t = 0$; when $x \to \infty, t \to \infty$

$$\therefore \quad 2y_0 \int_0^{\infty} e^{-t^2} . \sqrt{2}\,\sigma dt = 1$$

or
$$2\sqrt{2}\, y_0\, \sigma \int_0^{\infty} e^{-t^2}\, dt = 1$$

or
$$2\sqrt{2}\, \sigma y_0 . \tfrac{1}{2}\sqrt{\pi} = 1 \quad \therefore \quad y_0 = \frac{1}{\sigma\sqrt{2\pi}}$$

Hence $y_0 = \frac{1}{\sigma\sqrt{2\pi}} e^{-\frac{x^2}{2\sigma^2}}$ is the **Standard Form** of the normal curve.

Note. If the total frequency is N, the **general form** of the normal curve is

$$y = \frac{N}{\sigma\sqrt{2\pi}} e^{-\frac{(x-m)^2}{2\sigma^2}}$$

where $x - m$ is the excess of the mean over the value chosen as origin.

4. To obtain the first four moments of the normal distribution and hence deduce the value of β_1 and β_2.

(M.D.U. 1982 S ; D.U. 1985, 86)

Equation to the normal curve with the origin at the mean is

$$y = -\frac{1}{\sigma\sqrt{2\pi}} e^{-\frac{x^2}{2\sigma^2}}$$

The nth moment about the origin is

$$\mu'_n = \int_{-\infty}^{\infty} \frac{1}{\sigma\sqrt{2\pi}} e^{-\frac{x^2}{2\sigma^2}} . x^n\, dx$$

Putting $\frac{x}{\sigma} = t$, $dx = \sigma dt$

$$\therefore \quad \mu'_n = \frac{1}{\sigma\sqrt{2\pi}} \int_{-\infty}^{\infty} e^{-\frac{1}{2}t^2} . \sigma^n t^n . \sigma dt$$

$$= \frac{\sigma^n}{\sqrt{2\pi}} \int_{-\infty}^{\infty} e^{-\frac{1}{2}t^2} . t^n\, dt$$

If n is odd $e^{-\frac{1}{2}t^2} . t^n$ is an odd function of t and

$$\therefore \quad \mu'_n = 0$$

Hence all odd order moments about the mean vanish ($\because$ the mean is the origin) *(D.U. 1992, 93)*

$\therefore \quad \mu'_1 = 0 = \mu'_3 = \mu'_5 = ...$

If n is even, say $2p$, then

$$\mu'_{2p} = \frac{1}{\sigma\sqrt{2\pi}}\int_{-\infty}^{\infty} e^{-\frac{x^2}{2\sigma^2}} . x^{2p}\, dx$$

$$= \frac{2}{\sigma\sqrt{2\pi}}\int_{0}^{\infty} e^{-\frac{x^2}{2\sigma^2}} . x^{2p}\, dx$$

$|\because$ the integrand is an even function of x

$$= \frac{2}{\sigma\sqrt{2\pi}}\int_{0}^{\infty} e^{-\frac{x^2}{2\sigma^2}} . x^{2p-1} . x\, dx$$

Putting $\frac{x^2}{2\sigma^2} = t, \quad xdx = \sigma^2 dt$

$$\therefore \quad \mu'_{2p} = \frac{2}{\sigma\sqrt{2\pi}}\int_{0}^{\infty} e^{-t} . (2\sigma^2 t)^{\frac{2p-1}{2}} . \sigma^2\, dt$$

$$= \frac{2^{1/2} . 2^{\frac{2p-1}{2}}}{\sigma\sqrt{\pi}} . \sigma^{2p-1+2}\int_{0}^{\infty} e^{-t}\, t^{p-1/2}\, dt$$

$$= \frac{2^p \sigma^{2p}}{\sqrt{\pi}}\int_{0}^{\infty} e^{-t}\, t^{p+1/2-1}\, dt$$

$$= \frac{2^p \sigma^{2p}}{\sqrt{\pi}}\, \Gamma\left(p + \tfrac{1}{2}\right) \qquad \Big|\ \because \int_{0}^{\infty} x^{n-1} e^{-x}\, dx = \Gamma(n)$$

$$= \frac{(\sqrt{2})^{2p}\sigma^{2p}}{\sqrt{\pi}}\, \Gamma\left(\frac{2p+1}{2}\right)$$

or $$\mu'_n = \frac{(\sqrt{2})^n\sigma^n}{\sqrt{\pi}}\, \Gamma\left(\frac{n+1}{2}\right)$$

$$\therefore \quad \mu'_2 = \frac{2\sigma^2}{\sqrt{\pi}}\, \Gamma\left(\frac{3}{2}\right) = \frac{2\sigma^2}{\sqrt{\pi}} . \frac{1}{2}\sqrt{\pi} = \sigma^2$$

$$\mu'_4 = \frac{4\sigma_4}{\sqrt{\pi}}\, \Gamma\left(\frac{5}{2}\right) = \frac{4\sigma^4}{\sqrt{\pi}} . \frac{3}{2} . \frac{1}{2} . \sqrt{\pi} = 3\sigma^4$$

Since the mean is at the origin itself.

$$\mu'_n = \mu_n$$

$$\therefore \quad \mu_1 = 0 = \mu_3$$

And $\mu_2 = \mu'_2 = \sigma^2$

$\mu_4 = \mu'_4 = 3\sigma^4$

Now $\beta_1 = \dfrac{\mu_3^2}{\mu_2^3} = 0$ $(\because \quad \mu_3 = 0)$

$$\beta_2 = \frac{\mu_4}{\mu_2^2} = \frac{3\sigma^4}{\sigma^4} = 3$$

Note. $\gamma_1 = \sqrt{\beta_1} = 0, \quad \gamma_2 = \beta_2 - 3 = 0.$

Example 1. *Prove that the total area under normal probability curve is unity.*

Sol. Normal probability curve is given by

$$y = \frac{1}{\sigma\sqrt{2\pi}} e^{-\frac{(x-m)^2}{2\sigma^2}}$$

Area under this curve

$$= \int_{-\infty}^{\infty} y dx = \frac{1}{\sigma\sqrt{2\pi}} \int_{-\infty}^{\infty} e^{-\frac{(x-m)^2}{2\sigma^2}} dx$$

Put $\dfrac{x-m}{\mu\sqrt{2}} = t, \quad dx = \sigma\sqrt{2}\,dt$

$$\text{Required area} = \frac{1}{\sigma\sqrt{2\pi}} \int_{-\infty}^{\infty} e^{-t^2} \sigma\sqrt{2}\, dt$$

$$= \frac{1}{\sqrt{\pi}} \int_{-\infty}^{\infty} e^{-t^2} dt = \frac{1}{\sqrt{\pi}} \times \sqrt{\pi} = 1.$$

Example 2. *Find the mean of the normal distribution.*

(K.U. 1981 S, 82 S)

Sol. The general form of the normal curve is

$$y = \frac{N}{\sigma\sqrt{2\pi}} e^{-\frac{(x-m)^2}{2\sigma^2}}$$

$$\text{Mean} = \frac{1}{N} \Sigma fx = \frac{1}{N} \int_{-\infty}^{\infty} yx\, dx$$

$$= \frac{1}{\sigma\sqrt{2\pi}} \int_{-\infty}^{\infty} x e^{-\frac{(x-m)^2}{2\sigma^2}} dx$$

Putting $\dfrac{x-m}{\sigma\sqrt{2}} = t, \quad dx = \sigma\sqrt{2}\, dt$

$$\therefore \quad \text{Mean} = \frac{1}{\sigma\sqrt{2\pi}}\int_{-\infty}^{\infty}(m+\sigma\sqrt{2}t)\,e^{-t^2}\,\sigma\sqrt{2}\,dt$$

$$= \frac{1}{\sqrt{\pi}}\int_{-\infty}^{\infty}(m+\sigma\sqrt{2}\,t)\,e^{-t^2}\,dt$$

$$= \frac{m}{\sqrt{\pi}}\int_{-\infty}^{\infty}e^{-t^2}\,dt + \frac{\sigma\sqrt{2}}{\sqrt{\pi}}\int_{-\infty}^{\infty}te^{-t^2}\,dt$$

$$= \frac{m}{\sqrt{\pi}}\sqrt{\pi} + \frac{\sigma\sqrt{2}}{\sqrt{\pi}}(0)$$

$\Big|\because \; te^{-t^2}$ is an odd function of t

$= m.$

Example 3. *Prove that for a normal curve, the ordinate at the mean is the maximum ordinate.*

Sol. The equation of the normal curve is

$$y = \frac{N}{\sigma\sqrt{2\pi}}\,e^{-\frac{(x-m)^2}{2\sigma^2}}, \text{ Mean} = m$$

$$\frac{dy}{dx} = \frac{N}{\sigma\sqrt{2\pi}}\,e^{-\frac{(x-m)^2}{2\sigma^2}} \cdot -\frac{2(x-m)^2}{2\sigma^2}$$

$$= -\frac{N(x-m)}{\sigma^3\sqrt{2\pi}}\,e^{-\frac{(x-m)^2}{2\sigma^2}}$$

$$\frac{d^2y}{dx^2} = -\frac{N}{\sigma^3\sqrt{2\pi}}\left[1\,.\,e^{-\frac{(x-m)^2}{2\sigma^2}} + (x-m)\,.\,e^{-\frac{(x-m)^2}{2\sigma^2}}\,.\,\frac{-2(x-m)}{2\sigma^2}\right]$$

$$= -\frac{N}{\sigma^3\sqrt{2\pi}}\,.\,e^{-\frac{(x-m)^2}{2\sigma^2}}\left[1-\frac{(x-m)^2}{\sigma^2}\right]$$

Now $\dfrac{dy}{dx} = 0$ when $x = m$ and

$$\left[\frac{d^2y}{dx^2}\right]_{x=m} = -\frac{N}{\sigma^3\sqrt{2\pi}} < 0$$

Hence y, the ordinate is maximum when $x = m$ *i.e.,* the ordinate at the mean is the maximum ordinate.

Example 4. *Find the mode of the normal distribution.*

Sol. The equation of the normal curve is

$$y = \frac{N}{\sigma\sqrt{2\pi}}\,e^{-\frac{(x-m)^2}{2\sigma^2}}$$

Mode is the value of x corresponding to $y = y_0$, where y_0 is the maximum frequency.

Proceeding as in Example 3, y is maximum when $x = m$.

Hence **the mode = the mean = m.**

Example 5. *Find the median of the normal distribution.*

Sol. If M is the median of the normal distribution, we have

$$\int_{-\infty}^{M} \frac{N}{\sigma\sqrt{2\pi}} e^{-\frac{(x-m)^2}{2\sigma^2}} dx = \frac{N}{2}$$

$$\Rightarrow \quad \frac{1}{\sigma\sqrt{2\pi}} \int_{\infty}^{M} e^{-\frac{(x-m)^2}{2\sigma^2}} dx = \frac{1}{2}$$

$$\Rightarrow \quad \frac{1}{\sigma\sqrt{2\pi}} \int_{-\infty}^{M} e^{-\frac{(x-m)^2}{2\sigma^2}} dx + \frac{1}{\sigma\sqrt{2\pi}} \int_{-\infty}^{M} e^{-\frac{(x-m)^2}{2\sigma^2}} dx = \frac{1}{2} \qquad ...(1)$$

Now $\displaystyle \frac{1}{\sigma\sqrt{2\pi}} \int_{-\infty}^{m} e^{-\frac{(x-m)^2}{2\sigma^2}} dx$

$$= \frac{1}{\sigma\sqrt{2\pi}} \int_{\infty}^{0} e^{-t^2} (-\sigma\sqrt{2})dt \qquad \text{where} \quad -\frac{x-m}{\sigma\sqrt{2}} = t$$

$$= \frac{1}{\sqrt{\pi}} \int_{0}^{\infty} e^{-t^2} dt = \frac{1}{\sqrt{\pi}} \cdot \tfrac{1}{2}\sqrt{\pi} = \tfrac{1}{2}$$

$\therefore$ From (1)

$$\frac{1}{2} + \frac{1}{\sigma\sqrt{2\pi}} \int_{m}^{M} e^{-\frac{(x-m)^2}{2\sigma^2}} dx = \tfrac{1}{2}$$

$$\Rightarrow \quad \int_{m}^{M} e^{-\frac{(x-m)^2}{2\sigma^2}} dx = 0 \quad \Rightarrow \quad M = m$$

☞ **(Hence for the normal distribution mean, median, and mode coincide).** *(D.U. 1988, 89)*

Example 6. *Find the variance and standard deviation of a normal distribution.*

Sol. The equation of the normal curve is

$$y = \frac{N}{\sigma\sqrt{2\pi}} e^{-\frac{(x-m)^2}{2\sigma^2}}, \qquad \text{Mean} = m$$

$$\text{Variance} = \frac{1}{N} \int_{-\infty}^{\infty} \frac{N}{\sigma\sqrt{2\pi}} e^{-\frac{(x-m)^2}{2\sigma^2}} \cdot (x-m)^2 \, dx$$

$$= \frac{1}{\sigma\sqrt{2\pi}} \int_{-\infty}^{\infty} (x-m)^2 \,.\, e^{-\frac{(x-m)^2}{2\sigma^2}} dx$$

Putting $\frac{x-m}{\sigma\sqrt{2}} = t, \qquad dx = \sigma\sqrt{2}\, dt$

$$\text{Variance} = \frac{1}{\sigma\sqrt{2\pi}} \int_{-\infty}^{\infty} 2\sigma^2 t^2 \,.\, e^{-t^2} \,.\, \sigma\sqrt{2}\, dt$$

$$= \frac{2\sigma^2}{\sqrt{\pi}} \int_{-\infty}^{\infty} t^2 e^{-t^2} dt = \frac{4a^2}{\sqrt{\pi}} \int_0^{\infty} t^2 e^{-t^2} dt$$

Putting $t^2 = z, \qquad 2t\, dt = dz$

or $\qquad dt = \frac{dz}{2\sqrt{z}}$

$$\therefore \quad \text{Variance} = \frac{4\sigma^2}{\sqrt{\pi}} \int_0^{\infty} ze^{-z} \,.\, \frac{dz}{2\sqrt{z}} = \frac{2\sigma^2}{\sqrt{\pi}} \int_0^{\infty} z^{1/2} e^{-z} dz$$

$$= \frac{2\sigma^2}{\sqrt{\pi}} \int_0^{\infty} z^{3/2-1} e^{-2} dz = \frac{2\sigma^2}{\sqrt{\pi}} \Gamma\left(\tfrac{3}{2}\right)$$

$$= \frac{2\sigma^2}{\sqrt{\pi}} \,.\, \tfrac{1}{2}\sqrt{\pi} = \sigma^2.$$

Standard deviation $= \sqrt{\text{Variance}} = \sigma$.

Example 7. *Find the points of inflexion of the normal curve and show that they occur at a distance equal to standard deviation from the mean ordinate.* (D.U. 1989)

Sol. Let the equation of the normal curve be

$$y = \frac{1}{\sigma\sqrt{2\pi}} e^{-\frac{(x-m)^2}{2\sigma^2}}$$

Taking logarithms

$$\log y = \log \frac{1}{\sigma\sqrt{2\pi}} - \frac{-(x-m)^2}{\sigma^2}$$

Differentiating w.r.t. x

$$\frac{1}{y} \cdot \frac{dy}{dx} = -\frac{x-m}{\sigma^2}$$

Differentiating again

$$\frac{1}{y} \cdot \frac{d^2y}{dx^2} - \frac{1}{y^2} \cdot \left(\frac{dy}{dx}\right)^2 = -\frac{1}{\sigma^2}$$

$$\frac{d^2y}{dx^2} = \frac{1}{y} \cdot \left(\frac{dy}{dx}\right)^2 - \frac{y}{\sigma^2}$$

$$= \frac{1}{y} \cdot \frac{y^2 (x-m)^2}{\sigma^4} - \frac{y}{\sigma^2}$$

$$= \frac{y}{\sigma^2} [(x-m)^2 - \sigma^2]$$

At a point of inflexion.

$$\frac{d^2y}{dx^2} = 0 \qquad \left(\text{and } \frac{d^3y}{dx^3} \neq 0 \text{ which can be shown}\right)$$

$$\therefore \qquad (x-m)^2 = \sigma^2$$

or $\qquad x - m = \pm\sigma \quad$ or $\quad x = m \pm \sigma.$

Thus, the curve has two points of inflexion, one at $m - \sigma$ and the other at $m + \sigma$, *i.e.,* at a distance from the mean, equal to the standard deviation.

Example 8. *Show that the mean deviation from the mean of the normal distribution is about $\frac{4}{5}$ of its standard deviation.* (D.U. 1983, 86, 91)

OR

Prove that the mean deviation from the mean of the general normal distribution is $\sigma\sqrt{\frac{2}{\pi}}$. (D.U. 1985)

Sol. Let the equation of the normal curve be

$$y = \frac{1}{\sigma\sqrt{2\pi}} e^{-\frac{(x-m)^2}{2\sigma^2}}$$

Standard deviation = σ.

Mean deviation from the mean

$$= \int_{-\infty}^{\infty} y\,|x-m|\,dx$$

$$= \frac{1}{\sigma\sqrt{2\pi}} \int_{-\infty}^{\infty} |x-m| \,.\, e^{-\frac{(x-m)^2}{2\sigma^2}}\, dx$$

$$= \frac{\sigma^2}{\sigma\sqrt{2\pi}} \int_{-\infty}^{\infty} |z|\, e^{-\frac{1}{2}z^2}\, dz \quad \text{where } z = \frac{x-m}{\sigma}$$

$$= \frac{\sigma}{\sqrt{2\pi}} \left[\int_{-\infty}^{\infty} -z \,.\, e^{-\frac{1}{2}z^2}\, dz + \int_{0}^{\infty} z\, e^{-\frac{1}{2}z^2}\, dz\right]$$

$$\left|\because \quad |z| = \int_{z}^{-z} \right. \qquad \begin{matrix}\text{if } z < 0 \\ \text{if } z \geq 0\end{matrix}$$

$$= \frac{\sigma}{\sqrt{2\pi}} \left[-\int_{\infty}^{0} t\, e^{-\frac{1}{2}t^2} dt + \int_{0}^{\infty} z\, e^{-\frac{1}{2}z^2} dz \right]$$

where $t = -z$

$$= \frac{\sigma}{\sqrt{2\pi}} \left[\int_{0}^{\infty} z\, e^{-\frac{1}{2}z^2} dz + \int_{0}^{\infty} z\, e^{-\frac{1}{2}z^2} dz \right]$$

$$\left(\because \int_a^b f(x)\, dx = -\int_b^a f(x)\, dx \text{ and } \int_a^b f(x)\, dx = \int_a^b f(z)\, dz \right)$$

$$= \sigma \sqrt{\frac{2}{\pi}} \int_0^{\infty} z\, e^{-\frac{1}{2}z^2} dz$$

$$= \sigma \sqrt{\frac{2}{\pi}} \int_0^{\infty} \tfrac{1}{2}\, e^{-\frac{1}{2}t} dt$$

where $t = z^2$

$$= \sigma \sqrt{\frac{2}{\pi}} \left[-e^{-\frac{1}{2}t} \right]_0^{\infty} = -\sigma \sqrt{\frac{2}{\pi}}\, [0 - 1]$$

$$= \sqrt{\frac{2}{\pi}} \,.\, \sigma = 0.7979\, \sigma = \tfrac{4}{5}\, \sigma \text{ (approx.)}$$

$$= \tfrac{4}{5} \times \text{standard deviation (approx.)}.$$

Example 9. *For a certain normal distribution, the first moment about 10 is 40 and the fourth moment about 50 is 48. What is the arithmetic mean and standard deviation of the distribution ?* (M.D.U. 1982)

Sol. First moment about A is given by $\mu'_1 = M - A$ where M is the mean.

Here $\mu'_1 = 40$, $A = 10$

$\therefore$ $M = \mu'_1 + a = 40 + 10 = 50$

Fourth moment about 50, the mean, is 48

$\Rightarrow$ $\mu_4 = 48$.

But for a normal distribution with standard deviation σ,

$$\mu_4 = 3\sigma^4 \quad \Rightarrow \quad 48 = 3\sigma^4 \quad \Rightarrow \quad \sigma^4 = 16 \quad \Rightarrow \quad \sigma = 2$$

Hence mean = 50 S.D. = 2.

Example 10. *Prove that for a normal distribution,*

$$\mu_{2n} = (2n - 1)\, \sigma^2\, \mu_{2n-2}. \qquad \text{(D.U. 1986, 92, 93)}$$

Sol. μ_{2n} = 2nth moment about the mean

$$= \int_{-\infty}^{\infty} \frac{1}{\sigma\sqrt{2\pi}} e^{-\frac{(x-m)^2}{2\sigma^2}} . (x-m)^{2n} dx$$

Putting $\frac{x-m}{\sigma} = z, dx = \sigma dz$

$$\mu_{2n} = \int_{-\infty}^{\infty} \frac{1}{\sigma\sqrt{2\pi}} e^{-\frac{1}{2}z^2} . (\sigma z)^{2n} . \sigma dz$$

$$= \frac{\sigma^{2n}}{\sqrt{2\pi}} \int_{-\infty}^{\infty} z^{2n} e^{-\frac{1}{2}z^2} dz$$

$$= \frac{2\sigma^{2n}}{\sqrt{2\pi}} \int_{0}^{\infty} z^{2n} e^{-\frac{1}{2}z^2} dz$$

(since the integrand is an even function of z)

Putting $\frac{z^2}{2} = t, \quad z\, dz = dt \quad$ or $\quad dz = \frac{dt}{\sqrt{2t}}$

$$\mu_{2n} = \frac{2\sigma^{2n}}{\sqrt{2\pi}} \int_{0}^{\infty} (2t)^n . e^{-t} . \frac{dt}{\sqrt{2t}}$$

$$= \frac{2^n\sigma^{2n}}{\sqrt{\pi}} \int_{0}^{\infty} e^{-t} . t^{n-\frac{1}{2}} dt = \frac{2^n \sigma^{2n}}{\sqrt{\pi}} \int_{0}^{\infty} e^{-t} . t^{(n+\frac{1}{2})-1} dt$$

$$\Rightarrow \quad \mu_{2n} = \frac{2^n \sigma^{2n}}{\sqrt{\pi}} \Gamma\left(n - \tfrac{1}{2}\right)$$

Changing n to $(n-1)$, we get

$$\mu_{2n-2} = \frac{2^{n-1} \sigma^{2n-2}}{\sqrt{\pi}} \Gamma\left(n - \tfrac{1}{2}\right)$$

$$\therefore \quad \frac{\mu_{2n}}{\mu_{2n-2}} = 2\sigma^2 . \frac{\Gamma\left(n+\frac{1}{2}\right)}{\Gamma\left(n-\frac{1}{2}\right)}$$

$$= 2\sigma^2 . \frac{\left(n-\frac{1}{2}\right)\Gamma\left(n-\frac{1}{2}\right)}{\Gamma\left(n-\frac{1}{2}\right)} \quad \Big| \because \Gamma(r) = (r-1)\,\Gamma(r-1)$$

$$= \sigma^2 (2n-1)$$

Hence $\mu_{2n} = \sigma^2 (2n-1) \mu_{2n-2}$.

☞Note. $\mu_{2n} = \sigma^2 (2n-1) \mu_{2n-2}$ ...(1)

Changing n to $(n-1)$ in (1)

$$\mu_{2n-2} = \sigma^2 (2n-3) \mu_{2n-4}$$

Putting this value of μ_{2n-2} in (1), we have

$$\mu_{2n} = (\sigma^2)^2 (2n-1)(2n-3)\, \mu_{2n-4} \qquad ...(2)$$

Generalising from (1) and (2)

$$\mu_{2n} = (\sigma^2)^n (2n-1)(2n-3) \,.....\, 3.1\,.\,\mu_0$$

$$= (2n-1)(2n-3) \,......\, 3.1\,\sigma^{2n} \qquad \Big|\because \quad \mu_0 = 1$$

(M.D.U. 1983 S)

Example 11. *Obtain the equation of the normal curve that may be fitted to the following data :*

Class :	*60—65*	*65—70*	*70—75*	*75—80*	*80—85*
Frequency :	*3*	*21*	*150*	*335*	*326*
Class :	*85—90*	*90—95*	*95—100*		
Frequency :	*135*	*26*	*4*		

(K.U. 1982)

Sol. Let the equation of the normal curve be

$$y = \frac{N}{\sigma\sqrt{2\pi}}\, e^{-\frac{(x-m)^2}{2\sigma^2}}$$

Now from the given data, we shall find $N = \Sigma f$; m, the mean and σ, the S.D.

Class	*Mid-values (x)*	*f*	$u = \frac{x - 77.5}{5}$	*fu*	fu^2
60—65	62.5	3	−3	− 9	27
65—70	67.5	21	−2	− 42	84
70—75	72.5	150	−1	−150	150
75—80	77.5	335	0	0	0
80—85	82.5	326	1	326	326
85—90	87.5	135	2	270	540
90—95	92.5	26	3	78	234
95—100	97.5	4	4	16	64
		N = 1000		489	1425

Mean $\quad m = A + \frac{i}{N}\Sigma fu$

$$= 77.5 + \frac{5}{1000} \times 489 = 77.5 + 2.445 = 79.945$$

$$\sigma = i\sqrt{\frac{1}{N}\Sigma fu^2 - \left(\frac{\Sigma fu}{N}\right)^2}$$

$$= 5\sqrt{\frac{1425}{1000} - \frac{489 \times 489}{1000 \times 1000}}$$

$$= \frac{5}{1000}\sqrt{1425000 - 239121} = \frac{5 \times 1088.9}{1000}$$

$$= 5.4445.$$

Hence the equation of the normal curve fitted to the given data is

$$y = \frac{1000}{5.4445\sqrt{2\pi}} e^{-\frac{1}{2}\left(\frac{x - 79.945}{5.4445}\right)^2}$$

Example 12. *Obtain the equation of the normal curve that may be fitted to the following data :*

Variable (x) :	*0*	*1*	*2*	*3*	*4*	*5*
Frequency (f) :	*10*	*14*	*19*	*8*	*5*	*4*

Sol. Please try yourself. $\left[\textbf{Ans.} \; y = \frac{60}{1.45\sqrt{2\pi}} e^{-\frac{1}{2}\left(\frac{x - 1.93}{2.09}\right)^2}\right]$

Example 13. *A factory turns out an article by mass production methods. From past experience it appears that 10 articles on the average are rejected out of every batch of 100. Find the standard deviation of the number of rejections in a batch and write down the equation to the normal curve which may be taken to represent the distribution of the number of rejects in a large series of batches of 100.*

Sol. Here $N = 100$, Mean $m = 10$

$$p = \frac{10}{100} = \frac{1}{10} \qquad q = 1 - p = \frac{9}{10}$$

$$\sigma = \sqrt{Npq} = \sqrt{100 \times \frac{1}{10} \times \frac{9}{10}} = 3$$

Equation of the normal curve is

$$y = \frac{N}{\sigma\sqrt{2\pi}} e^{-\frac{1}{2}\left(\frac{x - m}{\sigma}\right)^2} = \frac{100}{3\sqrt{2\pi}} e^{-\frac{(x - 10)^2}{18}}$$

Example 14. *If two normal universes A and B have the same total frequency but the standard deviation of universe A is k times that of the universe B, show that the maximum frequency of universe A is $\frac{1}{k}$ times that of universe B.*

Sol. Let N be the same total frequency for each of the two universes A and B.

If σ is the standard deviation of universe B, then the standard deviation of universe A is $k\sigma$. Let m and m' be the means of the universes A and B respectively.

The equations of the normal universes A and B are

$$y = \frac{N}{k\sigma\sqrt{2\pi}} e^{-\frac{1}{2}\left(\frac{x - m}{k\sigma}\right)^2} = y_0 \, e^{-\frac{1}{2}\left(\frac{x - m}{k\sigma}\right)^2}$$

and $$y = \frac{N}{\sigma\sqrt{2\pi}} e^{-\frac{1}{2}\left(\frac{x-m'}{\sigma}\right)^2} = y_0' e^{-\frac{1}{2}\left(\frac{x-m'}{\sigma}\right)^2}$$

where $$y_0 = \frac{N}{k\sigma\sqrt{2\pi}} \qquad y_0' = \frac{N}{\sigma\sqrt{2\pi}}$$

so that $$y_0 = \frac{1}{k} y_0'$$

Since, for a normal distribution, the maximum frequency occurs at the point x = mean.

$\therefore$ Maximum frequency of universe

$$A = \left[y_0 e^{-\frac{1}{2}\left(\frac{x-m}{k\sigma}\right)^2} \right]_{x = m} = y_0$$

Maximum frequency of universe

$$B = \left[y_0' e^{-\frac{1}{2}\left(\frac{x-m'}{\sigma}\right)^2} \right]_{x = m'} = y_0'$$

Since $y_0 = \frac{1}{k} y_0'$

$\therefore$ Maximum frequency of universe A

$$= \frac{1}{k} \times \text{Maximum frequency of universe B.}$$

Example 15. *Assuming the mean height of soldiers to be 68.22 inches with a variance of 10.8 (inches)^2, find how many soldiers in a regiment of 1000 would you expect to be over 6 feet tall. (Given : area under the standard normal curve between z = 0 and z = 0.35 is 0.1368 and between z = 0 and z = 1.15 is 0.3746).* (D.U. 1990)

Sol. Assume that the distribution of height is normal.

Standard normal variate

$$z = \frac{x - m}{\sigma}$$

Here $m = 68.22$

$$\sigma = \sqrt{10.8} = 3.286$$

$$x = 6 \text{ ft.} = 72 \text{ inches}$$

$$\therefore \quad z = \frac{72 - 68.22}{3.286}$$

$$= 1.15.$$

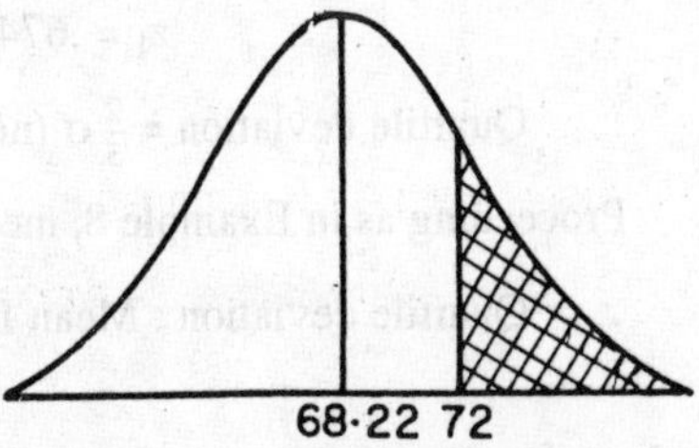

$\therefore$ From $z = 0$ to $z = 1.15$, area under the curve is 0.3746.

Since total area under normal curve is 1.

$\therefore$ area lying to the right of $z = 0$ is .5 and hence, the area lying to the right of $z = 1.15$ is $.5 - 0.3746 = .1254$

Hence the number of soldiers out of 1000 who are over 6 ft. tall

$= 1000 \times .1254 = 125.4$ or 125 nearly.

Example 16. *Prove that for the normal distribution, the quartile deviation, the mean deviation and standard deviation are approximately in the ratio 10 : 12 : 15.*

$$\left[\frac{1}{\sqrt{2\pi}}\int_0^{.6745} e^{-z^2/2}\,dz = 0.25\right]$$

(D.U. 1981, 88)

Sol. Standard normal variate

$$z = \frac{x-m}{\sigma}$$

when $x = Q_3,\ z = \dfrac{Q_3 - m}{\sigma} = z_1$ (say)

when $x = Q_1,\ z = \dfrac{Q_1 - m}{\sigma} = -z_1$

(which is obvious from the fig.)

Subtracting $\dfrac{Q_3 - Q_1}{\sigma} = 2z_1$.

$\therefore$ Quartile deviation $= \dfrac{Q_3 - Q_1}{2} = \sigma z_1$

Area under the curve between $z = 0$ and $z = z_1$ is 0.25.

$$\Rightarrow \quad \frac{1}{\sqrt{2\pi}}\int_0^{z_1} e^{-z^2/2}\,dz = 0.25$$

But $\dfrac{1}{\sqrt{2\pi}}\displaystyle\int_0^{.6745} e^{-z^2/2}\,dz = 0.25$ (given)

$\therefore \quad z_1 = .6745 = \frac{2}{3}$ (nearly)

$\therefore$ Quartile deviation $= \frac{2}{3}\sigma$ (nearly)

Proceeding as in Example 8, mean deviation $= \frac{4}{5}\sigma$

$\therefore$ Quartile deviation : Mean Deviation : S.D. $= \frac{2}{3}\sigma : \frac{4}{5}\sigma : \sigma$

$= 10 : 12 : 15.$

8

Random and Simple Sampling

(*ELEMENTARY IDEA*)

8.1. POPULATION OR UNIVERSE

An aggregate of objects (animate or inanimate) under study is called **population or universe.** It is thus a collection of individuals or of their attributes (qualities) or of results of operations which can be numerically specified.

A universe containing a finite number of individuals or members is called a **finite universe.** For example, the universe of the weights of students in a particular class or the universe of smokers in Rohtak district.

A universe with infinite number of members is known as an **infinite universe.** For example, the universe of pressures at various points in the atmosphere.

In some cases, we may be even ignorant whether or not a particular universe is infinite, *e.g.*, the universe of stars.

The universe of concrete objects is an **existent universe.** The collection of all possible ways in which a specified event can happen is called a **hypothetical universe.** The universe of heads and tails obtained by tossing a coin an infinite number of times (provided that it does not wear out) is a hypothetical one.

8.2. SAMPLING

The statistician is often confronted with the problem of discussing universe of which he cannot examine every member *i.e.*, of which complete enumeration is impracticable. For example, if we want to have an idea of the average per capita income of the people of India, enumeration of every earning individual in the country is a very difficult task. Naturally, the question arises : What can be said about a universe of which we can examine only a limited number of members ? This question is the origin of the Theory of Sampling.

A finite sub-set of a universe is called a **sample.** A sample is thus a small portion of the universe. The number of individual in a sample is called the **sample size**. The process of selecting a sample from a universe is called **sampling.**

The theory of sampling is a study of relationship existing between a population and samples drawn from the population. The fundamental object of sampling is to get as much information as possible of the whole universe by examining only a part of it. An attempt is thus made through sampling to give the maximum information about the parent universe with the minimum effort.

Sampling is quite often used in our day-to-day practical life. For example, in a shop we asses the quality of sugar, rice or any other commodity by taking only a handful of it from the bag and then decide whether to purchase it or not. A housewife normally tests the cooked products to find if they are properly cooked and contain the proper quantity of salt or sugar, by taking a spoonful of it.

8.3. RANDOM SAMPLING

The selection of an individual from the universe in such a way that each individual of the universe has the same chance of being selected is called **random sampling.**

Suppose we have a finite universe of size N and we take a sample of size n. Then there are $^{N}C_n$ possible samples. Random sampling is that technique in which each of the $^{N}C_n$ samples has an equal chance of being selected. A sample obtained by random sampling is called a **random sample.** A random sample is free from favouritism and human bias.

The simplest method which is normally used for random sampling is the **lottery system** in which a miniature universe is created. It is illustrated by an example.

Suppose we want to select r individuals out of n. Assign the number 1 to n, one number to each individual. Write the numbers 1 to n on n slips and make the slips as homogeneous as possible in shape, size etc. Put the slips in a bag. Shuffle thoroughly and draw r slips one by one. The 'r' individuals corresponding to the numbers on the slips drawn, will constitute the random sample.

8.4. SAMPLING OF ATTRIBUTES

The sampling of attributes may be regarded as the drawing of samples from a universe whose numbers possess the attribute A or α (not A). The universe is thus divided into two mutually exclusive and collectively exhaustive classes—one class possessing the attribute A and the other class not possessing the attribute A. The **presence** of a particular attribute in sampled unit may be termed as **success** and its **absence as failure.**

8.5. SIMPLE SAMPLING

By simple sampling we mean random sampling in which each event has the same probability of success and the probability of event is independent of the success or failure of events in the preceding trails. Thus **simple sampling is a special case of random sampling in which trials are independent and probability of success is constant.**

For example, counting the number of successes in the throwing of a dice or tossing of a coin is a case of simple sampling since the probability of getting heads with a coin is unaffected by the previous trials and remains constant irrespective of the number of trials made provided the coin remains unbiased.

Sampling, though random, need not be simple. For example, suppose in an urn there are 2 white balls and 3 red balls. Now the probability of drawing a white ball is $\frac{2}{5}$. Suppose at the first trial, we draw a white ball. The probability of getting a white ball at the second trial, **when the ball is not replaced,** is $\frac{1}{4}$ which is clearly not the same as the probability in the first trial. Thus the sampling though random, is not simple. However, if the ball drawn at the first trial was put back in the urn before the next trial, success or failure having been noted, the random sampling would become simple sampling.

Random sampling from an infinite universe is always simple but random sampling from a finite universe may or may not be simple according as the members drawn are replaced or not.

8.6. MEAN AND STANDARD DEVIATION IN SIMPLE SAMPLING OF ATTRIBUTES

The conditions of simple sampling *viz.* constant probability p and independent events satisfy the basic assumptions of the binomial distribution.

Suppose we take N samples, each having n members. Let p be the probability of success of each member and q of failure so that $p + q = 1$. The frequencies of samples with 0, 1, 2, ..., n successes are the terms of the binomial expansion $N(q + p)^n$.

The binomial probability distribution thus determined is called the **sampling distribution** of the number of successes in the sample.

Expected value of (*i*) mean is np

(*ii*) variance is npq

(*iii*) standard deviation is $\sqrt{npq}$

Instead of recording the number of successes in each sample, we might record the proportion of successes *i.e.*, $\frac{1}{n}$th of the number in each sample.

$$\therefore \quad \text{Mean of proportion of success} = \frac{np}{n} = p$$

$$\text{S.D. of proportion of success} = \frac{\sqrt{npq}}{n} = \sqrt{\frac{pq}{n}}$$

(Standard deviation of a sampling distribution is also called the **standard error** and abbreviated as S.E.)

8.7. TESTS OF SIGNIFICANCE FOR LARGE SAMPLES

Suppose a large number of samples is classified according to the frequencies of an attribute. This gives rise to a binomial distribution which tends to a normal distribution for large values of n, the sample size. Therefore a great majority of its members lie within a range $\pm 3\sigma$ on each ride of the mean *i.e.*, within a range $\pm 3\sqrt{npq}$ on each side of np.

☞ **If the number of successes in a large sample of size n differs from the expected value np by more than $3\sqrt{npq}$, we call the difference highly significant and the truth of the hypothesis is very improbable.**

To test a statistical hypothesis, we accept the hypothesis as correct and then calculate np and $\sqrt{npq}$ and apply the above test to decide whether to accept or reject the hypothesis.

Example 1. *A coin is tossed 400 times and it turns up head 216 times. Discuss whether the coin may be unbiased one.* (M.D.U. 1983 S)

Sol. Suppose the coin is unbiassed one. Then p, the probability of getting head in one toss $= \frac{1}{2}$, $q = 1 - p = \frac{1}{2}$

Expected number of heads in 400 tosses

$$= np = 400 \times \frac{1}{2} = 200$$

Actual number of heads in 400 tosses = 216

The deviation of the actual number of heads from expected number of heads

$$= 216 - 200 = 16$$

Standard deviation $= \sqrt{npq} = \sqrt{400 \times \frac{1}{2} \times \frac{1}{2}} = 10$

$\therefore$ The deviation 16 is only $\frac{16}{10} = 1.6$ times the standard deviation

Hence deviation is likely to appear as a result of fluctuations of simple sampling. The hypothesis is therefore accepted and the coin may be taken unbiassed one.

Example 2. *A coin is tossed 1000 times and the head comes out 550 times. Can the deviation from expected value be due to fluctuations of simple sampling ?*

Sol. On the assumption of an unbiassed coin, the chance of turning up a head in one toss is $\frac{1}{2}$

$$\therefore \quad p = \frac{1}{2}, \qquad q = \frac{1}{2}.$$

Expected number of heads is 1000 tosses

$$= np = 1000 \times \frac{1}{2} = 500$$

Actual number of heads in 1000 tosses = 550.

The deviation of the actual number of heads from expected number of heads

$$= 550 - 500 = 50$$

Standard deviation

$$= \sqrt{npq} = \sqrt{1000 \times \frac{1}{2} \times \frac{1}{2}} = 15.81.$$

The deviation 50 is $\frac{50}{15.81} = 3.1 \; (> 3)$ times the standard deviation and is most unlikely to appear as a result of fluctuations of simple sampling.

Example 3. *A coin is tossed 10000 times and it turns up head 5195 times. Is it reasonable to think that the coin is unbiased?* (K.U. 1982)

Sol. Suppose the coin is unbiased.

p, the chance of turning up a head in one toss

$$= \tfrac{1}{2} \qquad \therefore \quad q = \tfrac{1}{2}$$

Expected number of heads in 10000 tosses

$$= np = 10000 \times \tfrac{1}{2} = 5000$$

Actual number of heads in 10000 tosses = 5195.

The deviation of the actual number of heads from expected number of heads $= 5195 - 5000 = 195$

Standard deviation

$$= \sqrt{npq} = \sqrt{10000 \times \frac{1}{2} \times \frac{1}{2}}$$

$$= \sqrt{2500} = 50$$

The deviation 195 is $\frac{195}{50} = 3.9 \; (> 3)$ times the standard deviation and is most unlikely to appear as a result of fluctuations of simple sampling.

$\therefore$ It is not reasonable to think that the coin is unbiased.

Example 4. *A certain cubical die was thrown 9000 times and a 5 or 6 was obtained 3240 times. On the assumption of random throwing, do the data indicate an unbiased die?* (M.D.U. 1982)

Sol. On the assumption of an unbiased die, the chance p of getting a 5 or 6 in one throw

$$= \frac{2}{6} = \frac{1}{3} \qquad \therefore \quad q = 1 - \frac{1}{3} = \frac{2}{3}.$$

Expected number of successes in 9000 throws

$$= np = 9000 \times \tfrac{1}{3} = 3000$$

Actual number of successes in 9000 throws

$$= 3240$$

Difference $= 3240 - 3000 = 240$

Standard deviation

$$= \sqrt{npq} = \sqrt{9000 \times \frac{1}{3} \times \frac{2}{3}} = 44.72.$$

The deviation of the actual number of successes from expected number of successes = 240 is $\frac{240}{44.72} = 5.4 \; (> 3)$ times the standard deviation. This is most unlikely to appear as a result of fluctuations of simple sampling. Therefore our assumption is incorrect and the die is certainly biased one.

Example 5. *In some dice throwing experiments, Weldon threw dice 49152 times and of these 25145 yielded a 4, 5 or 6. Is this consistent with the hypothesis that the dice were unbiased ?*

Sol. On the assumption that the dice were unbiased, the chance p of getting a 4, 5, or 6 in a single throw of one die

$$= \frac{3}{6} = \frac{1}{2} \quad \therefore \quad q = 1 - \frac{1}{2} = \frac{1}{2}$$

Expected number of successes $= np = 49152 \times \frac{1}{2} = 24576$

Actual " " " $= 25145$

Difference $= 25145 - 24576 = 569$

Standard deviation $= \sqrt{npq}$

$$= \sqrt{49152 \times \frac{1}{2} \times \frac{1}{2}} = \sqrt{12288} = 110.8$$

The deviation of the actual number of successes from expected number of successes = 569 is $\frac{569}{110.8} > 3$ times the standard deviation. This is most unlikely to appear as a result of fluctuation of simple sampling.

Therefore our assumption is incorrect and the given results are inconsistent with the hypothesis that the dice were unbiased.

Example 6. *A die is thrown 9000 times and a throw of 3 or 4 observed 3240 times. Show that the die cannot be regarded as an unbiased one.*

Sol. Please try yourself.

Example 7. *In 324 throws of a six-faced die, odd points appeared 181 times. Is it reasonable to think that the die is unbiased ?*

Sol. Please try yourself. **[Ans.** Yes]

Example 8. *In a sample of 500 people from Andhra Pradesh, 280 are found to be rice eaters and the rest wheat eaters. Can we assume that both the food articles are equally popular ?* (D.U. 1986)

Sol. Let us take the hypothesis that both the food articles are equally popular.

Then $p = q = \frac{1}{2}$, $n = 500$

Expected number of rice eaters $= np = 500 \times \frac{1}{2} = 250$

Acutal " " " " $= 280$

Difference $= 280 - 250 = 30$

$$\text{S.D.} = \sqrt{npq} = \sqrt{500 \times \frac{1}{2} \times \frac{1}{2}} = 11.2$$

$$\frac{\text{Difference}}{\text{S.D.}} = \frac{30}{11.2} < 3.$$

$\therefore$ The hypothesis is correct.

Example 9. *In a hospital 480 female and 520 male bodies were born in a week. Do these figures confirm the hypothesis that males and females are born in equal number ?*

Sol. Let us take the hypothesis that the male and female babies are born in **equal number.** Then $p = q = \frac{1}{2}$

Total number of births,

$$m = 480 + 520 = 1000$$

Expected number of female babies

$$= np = 1000 \times \frac{1}{2} = 500$$

Actual number of female babies

$$= 480$$

Difference = 500 – 480 = 20

$$\text{S.D.} = \sqrt{npq} = \sqrt{1000 \times \frac{1}{2} \times \frac{1}{2}} = \sqrt{250} = 15.8$$

$$\frac{\text{Difference}}{\text{S.D.}} = \frac{20}{15.8} < 3$$

$\therefore$ The hypothesis is correct. Hence the male and female babies are born in equal number.

Example 10. *Of 10000 babies born in U.P. 5200 are male children. Taking this to be a random sample of the births in U.P. show that it throws considerable doubt on the hypothesis that the sexes are born in equal proportion.*

Sol. Total number of births, $n = 10000$

Observed **proportion** of males $= \frac{5200}{10000} = .52$

Assuming that the hypothesis is correct, the chance of birth of a male

$$= \frac{1}{2}$$

$$\therefore \qquad p = \frac{1}{2} = .5, \qquad q = 1 - \frac{1}{2} = \frac{1}{2}$$

Deviation between the proportions

$$= .52 - .5 = .02$$

S.D. of proportion of male

$$= \sqrt{\frac{pq}{n}} = \sqrt{\frac{1}{2 \times 2 \times 10000}}$$

$$= \frac{1}{200} = .005$$

$$\frac{\text{Difference}}{\text{S.D.}} = \frac{.02}{.005} = 4 > 3.$$

$\therefore$ The difference is not due to fluctuations of sampling and the hypothesis is considerably doubtful.

Example 11. *The records of a certain hospital showed the birth of 723 males and 617 females in a certain week. Do these confirm to the hypothesis that the sexes are born in equal proportions ?*

Sol. Total number of births, $n = 723 + 617 = 1340$

Observed **proportion** of males $= \dfrac{732}{1340} = .54$

Assuming that the hypothesis is correct, the chance of the birth of a male $= \frac{1}{2}$

$$\therefore \quad p = \tfrac{1}{2} = .5, \quad q = 1 - \tfrac{1}{2} = \tfrac{1}{2}.$$

Difference between the proportions

$$= .54 - .5 = .04$$

S.D. of proportion of males

$$= \sqrt{\frac{pq}{n}} = \sqrt{\frac{1}{2 \times 2 \times 1340}} = \frac{1}{\sqrt{5360}} = .01$$

$$\frac{\text{Difference}}{\text{S.D.}} = \frac{.04}{.01} = 4 > 3$$

Hence the figures do not confirm to the hypothesis.

Example 12. *In a hospital, 230 females and 270 males were born in a week. Do these figures confirm to the hypothesis that the sexes are born in equal proportions ?*

Sol. Please try yourself. **[Ans.** Yes]

Example 13. *A sample of 900 days is taken from meteorological records of a certain district and 100 of them are found to be foggy. What are the probable limits to the percentage of foggy days in the district ?*

(D.U. 1987)

Sol. p, the proportion of foggy days in the sample of 900 days

$$= \frac{100}{900} = \frac{1}{9} \qquad \therefore \qquad q = 1 - \frac{1}{9} = \frac{8}{9}$$

S.E. of the proportion of foggy days

$$= \sqrt{\frac{pq}{n}} = \sqrt{\frac{1}{9} \times \frac{8}{9} \times \frac{1}{900}} = 0.0105$$

Probable limits are $p \mp 3\sqrt{\dfrac{pq}{n}} = 0.1111 \mp 3 \times 0.0105$

$$= 0.1111 \mp 0.0315 = 0.9796 \text{ and } 0.1426$$

$$= 0.0796 \times 100\% \text{ and } 0.1426 \times 100\%$$

$$= 7.96\% \text{ and } 14.26\%$$

Hence the percentage of foggy days lies between 7.96 and 14.26.

Example 14. *A random sample of 500 pine-apples was taken from a large consignment and 65 were found to be bad. Show that the S.E. of the proportion of bad ones in a sample of this size is 0.015 and deduce that the percentage of bad pine-apples in the consignment almost certainly lies between 8.5 and 17.5.*

Sol. p, the proportion of bad pine-apples in the given sample

$$= \frac{65}{500} = 0.13$$

$$\therefore \quad q = 1 - p = 0.87$$

S.E. of the proportion of bad pine-apples

$$= \sqrt{\frac{pq}{n}} = \sqrt{0.13 \times 0.87 \times \frac{1}{500}}$$

$$= 0.015$$

Probable limits of the bad pine-apples in the consignment are

$$p \mp 3\sqrt{\frac{pq}{n}} = 0.13 \mp 3 \times 0.015$$

$$= 0.13 \mp 0.045 = 0.085 \text{ and } 0.175$$

$$= 0.085 \times 100\% \text{ and } 0.175 \times 100\%$$

$$= 8.5\% \text{ and } 17.5\%.$$

Hence the percentage of bad pine-apples lies between 8.5 and 17.5.

Example 15. *400 oranges are taken from a large consignment and 50 are found to be bad. Estimate the percentage of bad oranges in the consignment and assign limits within which the percentage lies.*

Sol. p, the proportion of bad oranges in the given sample

$$= \frac{50}{400} = \frac{1}{8} = .125$$

$$\therefore \quad q = 1 - \frac{1}{8} = \frac{7}{8}$$

Percentage of bad oranges in the consignment

$$= 0.125 \times 100\% = 12.5\%.$$

S.E. of the proportion of bad oranges

$$= \sqrt{\frac{pq}{n}} = \sqrt{\frac{1}{8} \times \frac{7}{8} \times \frac{1}{400}}$$

$$= .0165$$

Probable limits of bad oranges in the consignment are

$$p \mp 3\sqrt{\frac{pq}{n}} = 0.125 \mp 0.0495$$

$$= 0.0755 \text{ and } 0.1745$$

$$= 0.6755 \times 100\% \text{ and } 0.1745 \times 100\%$$

$$= 7.55\% \text{ and } 17.45\%.$$

Example 16. *In a certain maternity home during a year, there were 1600 births of which 840 were males. Test the hypothesis that male and female births are equally likely. Supposing the null hypothesis is not given, determine the* $\pm 3\sigma$ *confidence limits for the proportion of male births.*

Sol. Please try yourself.

Example 17. *A sample of 100 days is taken from mateorological records of a certain district and of them 10 are found to be foggy. What are the probable limits of the percentage of foggy days in the district ?*

Sol. Please try yourself. **[Ans.** 5.98% and 14.02%]

Example 18. *500 apples are taken at random from a large basket and 50 are found to be bad. Estimate the proportion of bad apples in the basket and assign limits within which the percentage most probable lies.*

Sol. Please try yourself. **[Ans.** 0.1%, 6.1% and 13.9%]

Example 19. *Out of a simple sample of 1000 individuals from the inhabitants of a country we find that 36% of them have blue eyes and the remainder have eyes of some other colour. What can we infer about the proportion of blue eyed individuals in the whole population ?*

Sol. p, the proportion of individuals having blue eyes

$$= \frac{36}{100} = .36$$

$$\therefore \quad q = 1 - p = .64.$$

S.E. of the proportion of individuals having blue eyes

$$= \sqrt{\frac{pq}{n}} = \sqrt{\frac{.36 \times .64}{1000}}$$

$$= .015$$

Probable limits of blue eyed individuals are

$$p \mp 3\sqrt{\frac{pq}{n}} = .36 \mp .045$$

$$= .315 \text{ and } .405$$

$$= .315 \times 100\% \text{ and } .405 \times 100\%$$

$$= 31.5\% \text{ and } 40.5\%.$$

Example 20. *1000 ladies are chosen at random from the inhabitants of Bombay State. and it is found that 55% of them have dark eyes and the remainder have eyes of some other colour. What can be inferred about the proportion of dark eyed ladies in the whole population of Bombay State ?*

Sol. Please try yourself. **[Ans.** 50.29% and 59.71%]

Example 21. *In a locality of 18000 families, a sample of 840 families was selected. Of these 840 families, 206 families were found to have a monthly income of Rs. 50 or less. It is desired to estimate how many out of the 18000 families have a monthly income of Rs. 50 or less. Within what limits would you place your estimate ?* (D.U. 1983)

Sol. p, the proportion of families having monthly income of Rs. 50 or less

$$= \frac{206}{840} = .245$$

so that $q = 1 - p = .755$

S.E. of the proportion of families having monthly income of Rs. 50 or less

$$= \sqrt{\frac{pq}{n}} = \sqrt{\frac{.245 \times .755}{840}}$$

$$= 0.015$$

Probable limits of families having monthly income of Rs. 50 or less are

$$p \mp 3\sqrt{\frac{pq}{n}} = .245 \mp .045$$

$$= .200 \text{ and } .290$$

$$= .200 \times 100\% \text{ and } .290 \times 100\%$$

$$= 20\% \text{ and } 29\%$$

$$= 18000 \times \frac{20}{100} \text{ and } 18000 \times \frac{29}{100}$$

$$= 3600 \text{ and } 5220.$$

Example 22. *Given that on the average 4% of insured men of age 65 die within a year, and that 60 of a particular group of 1000 such men died within a year. Can this group be regarded as a representative sample ?*

Sol. p, the proportion of deaths in the sample

$$= \frac{4}{100} = .04$$

so that $q = .96$

Observed proportion of deaths

$$= \frac{60}{1000} = .06$$

Deviation of proportion of deaths

$$= .06 - .04 = .02$$

S.E. of proportion of deaths

$$= \sqrt{\frac{pq}{n}} = \sqrt{\frac{.04 \times .96}{1000}}$$

$$= .0062$$

$$\frac{\text{Deviation}}{\text{S.E.}} = \frac{.02}{.0062} > 3$$

Hence the group chosen is not a representative sample.

9

Tests of Significance

I. CHI-SQUARE (χ^2) TEST

1. When a coin is tossed 200 times, the theoretical considerations lead us to expect 100 heads and 100 tails. But in practice, these results are rarely achieved. The quantity χ^2 (a Greek letter, pronounced as chi-square) describes the magnitude of discrepancy between theory and observation. If χ 0, the observed and expected frequencies completely coincide. The greater the discrepancy between the observed and expected frequencies, the greater is the value of χ^2. Thus χ^2 **affords a measure of the correspondence between theory and observation.**

If O_i ($i = 1, 2, ..., n$) is a set of observed (experimental) frequencies and E_i ($i = 1, 2, ..., n$) is the corresponding set of expected (theoretical or hypothetical) frequencies, then χ^2 **is defined as**

$$\chi^2 = \sum_{i=1}^{n} \left[\frac{(O_i - E_i)^2}{E_i} \right]$$

and degrees of freedom ($d.f.$) = $(n - 1)$.

2. Degree of Freedom.

While comparing the calculated value of χ^2 with the table value, we have to determine the degrees of freedom.

If we have to choose any four numbers whose sum is 50, we can exercise our independent choice for any three numbers only, the fourth being 50 minus the total of the three numbers selected. Thus, though we were to choose any four numbers, our choice was reduced to three because of one condition imposed. There was only one restraint on our freedom and our degrees of freedom were $4 - 1 = 3$. If two restrictions are imposed, our freedom to choose will be further curtailed and degrees of freedom will be $4 - 2 = 2$.

In general, the number of degrees of freedom is the total number of observations less the number of independent constraints imposed on the observations. Degrees of freedom ($d.f.$) are usually denoted by ν (the letter 'nu' of the Greek alphabet).

Thus, $\nu = n - k$ where k is the number of independent constraints in a set of data of n observations.

Note. (*i*) For a $p \times q$ contingency table (p columns and q rows).

$$v = (p - 1)(q - 1)$$

(*ii*) In the case of a contingency table, the expected frequency of any class

$$= \frac{\text{Total of row in which it occurs} \times \text{Total of col. in which it occurs}}{\text{Total number of observations}}$$

3. χ^2 test is one of the simplest and the most general test known. It is applicable to a very large number of problems in practice which can be summed up under the following heads :

(*i*) as a test of goodness of fit.

(*ii*) as a test of independence.

(*i*) **χ^2 Test as a test of goodness of fit.** χ^2 test enables us to ascertain how well the theoretical distributions such as Binomial, Poisson or Normal etc. fit empirical distributions, *i.e.,* distributions obtained from sample data. If the **calculated value of χ^2 is less than the table value** at a specified level (generally 5%) of significance, the **fit is considered to be good,** *i.e.,* the divergence between actual and expected frequencies is attributed to fluctuations of simple sampling. If the calculated value of χ^2 is greater than the table value, the fit is considered to be poor.

(*ii*) **χ^2 Test as a test of independence.** With the help of χ^2 test, we can find whether or not two attributes are associated. We take the null hypothesis that there is no association between the attributes under study *i.e.,* **we assume that the two attributes are independent. If the calculated value of χ^2 is less than the table value** at a specified level (generally 5%) of significance, the hypothesis holds good, *i.e.,* **the attributes are independent** and do not bear any association. On the other hand, if the calculated value of χ_2 is greater than the table value at a specified level of significance, we say that the results of the experiment do not support the hypothesis. In other words, the attributes are associated.

4. Conditions for applying χ^2 Test

Following are the conditions which should be satisfied before χ^2 test can be applied.

(*a*) N, the total number of frequencies should be large. It is difficult to say what constitutes largeness, but as an arbitrary figure, we may say that **N should be at least 50,** however, few the cells.

(*b*) No theoretical cell-frequency should be small. Here again, it is difficult to say what constitutes smallness, but 5 should be regarded as the very minimum and **10 is better**. If small theoretical frequencies occur (*i.e.,* < 10), the difficulty is overcome by grouping two or more classes together before calculating (O – E). **It is important to remember that the number of degrees of freedom is determined with the number of classes after regrouping.**

(*c*) The constraints on the cell frequencies, if any, should be linear.

Example 1. *Calculate χ^2 for the following data :*

Class	:	*A*	*B*	*C*
Observed Frequency	:	*37*	*44*	*19*
Expected Frequency	:	*31*	*38*	*31*

Sol. $\chi^2 = \sum \frac{(O-E)^2}{E}$

$$= \frac{(37-31)^2}{31} + \frac{(44-38)^2}{38} + \frac{(19-31)^2}{31}$$

$$= \frac{36}{31} + \frac{36}{38} + \frac{144}{31}$$

$$= 1.16 + 0.96 + 4.64 = 6.76.$$

Example 2. *Find the value of χ^2 for the following data :*

Class	:	*A*	*B*	*C*	*D*	*E*
Observed Frequency	:	*8*	*29*	*44*	*15*	*4*
Theoretical Freq.	:	*7*	*24*	*38*	*24*	*7*

(M.U.D. 1982 S)

Sol. The condition for the application of χ^2 test is that the frequencies should not be less than 10 in any class. Since some of the frequencies in the data are less than 10, we regroup the data in order to fulfil the condition. The data on regrouping becomes

Class	:	A and B	C	D and E
Observed frequency	:	37	44	19
Theoretical freq.	:	31	38	31

Now proceeding as in Ex. 1,

$$\chi^2 = 6.76.$$

Example 3. *Find the value of χ^2 for the following data :*

Observed Freq. :	*10*	*4*	*15*	*18*	*20*	*15*	*5*	*2*	*3*
Expected Freq. :	*10*	*7*	*10*	*15*	*25*	*10*	*5*	*5*	*5*

So. Since some of the frequencies are less than 10, we regroup the data. The data on regrouping becomes

Observed Freq. :	14	15	18	20	15	10
Expected Freq. :	17	10	15	25	10	15

$$\chi^2 = \Sigma \frac{(O-E)^2}{E} = \frac{(14-17)^2}{17} + \frac{(15-10)^2}{10} + \frac{(18-15)^2}{15} + \frac{(20-25)^2}{25}$$

$$+ \frac{(15-10)^2}{10} + \frac{(10-15)^2}{15}$$

$$= \frac{9}{17} + \frac{25}{10} + \frac{9}{15} + \frac{25}{25} + \frac{25}{10} + \frac{25}{15}$$

$$= 0.53 + 2.5 + 0.60 + 1 + 2.5 + 1.67$$

$$= 8.80.$$

Example 4. *Find dice were thrown 192 times and the number of times 4, 5 or 6 were obtained are as follows :*

No. of dice throwing 4, 5 or 6	:	*5*	*4*	*3*	*2*	*1*	*0*
f	:	*6*	*46*	*70*	*48*	*20*	*2*

Calculate χ^2.

Sol. Here observed frequencies are given. Let us calculate theoretical frequencies.

Probability p of throwing 4, 5 or 6 with one die

$$= \frac{3}{6} = \frac{1}{2} \quad \therefore \quad q = 1 - \frac{1}{2} = \frac{1}{2}$$

$\therefore$ The theoretical frequencies of getting 5, 4, 3, 2, 1, 0 successes with 5 dice are respectively the successive terms of $N(q+p)^n$

$$= 192(\tfrac{1}{2} + \tfrac{1}{2})^5$$

which are $192 \times (\frac{1}{2})^2$, $192 \times {}^5C_1 (\frac{1}{2})^5$, $192 \times {}^5C_2 (\frac{1}{2})^5$, $192 \times {}^5C_3(\frac{1}{2})^5$,

$$192 \times {}^5C_4(\tfrac{1}{2})^5, 192 \times (\tfrac{1}{2})^5$$

i.e., 6, 30, 60, 60, 30, 6 respectively.

Therefore the data becomes

No. of die throwing 4, 5 or 6	:	5	4	3	2	1	0
Observed Frequency	:	6	46	70	48	20	2
Expected Frequency	:	6	30	60	60	30	6

Since some of the frequencies are less than 10, we regroup the data. The data on regrouping becomes

No. of dice throwing 4, 5 or 6	:	5 or 4	3	2	1 or 0
Observed Frequency	:	52	70	48	22
Expected Frequency	:	36	60	60	36

$$\therefore \quad \chi^2 = \Sigma \frac{(O-E)^2}{E}$$

$$= \frac{(52-36)^2}{36} + \frac{(70-60)^2}{60} + \frac{(48-60)^2}{60} + \frac{(22-36)^2}{36}$$

$$= 7.11 + 1.66 + 2.4 + 5.44 = 16.61.$$

Example 5. *Five dice were thrown 96 times and the numbers 4, 5 or 6 were thrown as given below :*

No. of dice throwing 4, 5 or 6	:	*5*	*4*	*3*	*2*	*1*	*0*
f	:	*7*	*19*	*35*	*24*	*8*	*3*

Calculate χ^2.

Sol. Please try yourself. **[Ans.** 8.3]

Example 6. *A die is thrown 264 times with the following results :*

No. appeared on the die	:	*1*	*2*	*3*	*4*	*5*	*6*
Frequency	:	*40*	*32*	*28*	*58*	*54*	*60*

Show that the die is biased (χ^2 for 5 d.f. at 5% level = 11.07).

(K.U. 1982 S)

Sol. Taking the hypothesis that the die is unbiased and the distribution is uniform *i.e.,* the expected frequency of each of the numbers 1, 2, ..., 6 is $\frac{264}{6} = 44$, the data becomes

No. appeared on the die	:	1	2	3	4	5	6
Observed Frequency	:	40	32	28	58	54	60
Expected Frequency	:	44	44	44	44	44	44

$$\chi^2 = \Sigma \frac{(O - E)^2}{E}$$

$$= \frac{(40-44)^2}{44} + \frac{(32-44)^2}{44} + \frac{(28-44)^2}{44} + \frac{(58-44)^2}{44} + \frac{(54-44)^2}{44} + \frac{(60-44)^2}{44}$$

$$= \frac{1}{44}(16 + 144 + 256 + 196 + 100 + 256) = \frac{968}{44} = 22$$

The number of degrees of freedom

= No. of observation – No. of restrictions

= 6 – 1 = 5

The table value of χ^2 for 5 d.f. at 5% level = 11.07

The calculated value of χ^2 (= 22) is greater than the table value. Therefore the results of the experiment do not support the hypothesis. Hence the die is biased.

Example 7. *In 120 throws of a single die, the following distribution of faces was obtained.*

Faces	:	*1*	*2*	*3*	*4*	*5*	*6*	*Total*
Frequency	:	*30*	*25*	*18*	*10*	*22*	*15*	*120*

Find the value of χ^2 *on the basis that the die was unbiased.*

(M.D.U. 1983 S)

Sol. Please try yourself. **[Ans.** 85.81]

Example 8. *Five coins are tossed 3200 times and the following results are obtained :*

No. of heads	:	*0*	*1*	*2*	*3*	*4*	*5*
Frequency	:	*80*	*570*	*1100*	*900*	*500*	*50*

Test the hypothesis that the coins are biased.

You can make use of the following data in drawing your conclusion :

Degree of freedom	:	*1*	*2*	*3*	*4*	*5*	*6*
χ^2 *value at 5% level of significance*	:	*3.841*	*5.991*	*7.851*	*9.483*	*11.070*	*12.59*

(D.U. 1983, 92, 93)

Sol. Please try yourself.

Example 9. *A die is thrown 132 times with the following results :*

No. turned up :	*1*	*2*	*3*	*4*	*5*	*6*
Frequency :	*16*	*20*	*25*	*14*	*29*	*28*

Test the hypothesis that the die is unbiased. (χ^2 *for 5 d.f. at 5% level* $= 11.07$). (K.U. 1982)

Sol. Please try yourself. **[Ans.** The die is unbiased]

Example 10. *200 digits were chosen at random from a set of tables. The frequencies of the digits were :*

Digit :	*0*	*1*	*2*	*3*	*4*	*5*	*6*	*7*	*8*	*9*
Frequency :	*18*	*19*	*23*	*21*	*16*	*25*	*22*	*20*	*21*	*15*

Use the chi-square test to assess the correctness of the hypothesis that the digits were distributed in the equal number in the tables from which these were chosen. (χ^2 *for 9 d.f. at 5% level* $= 16.919$).

(M.D.U. 1981 ; K.U. 1981 S ; D.U. 1992)

Sol. Taking the hypothesis that the digits were distributed in equal number in the tables from which they were chosen, the expected frequency of each digit $= \frac{200}{20} = 20$. Now

Digit :	0	1	2	3	4	5	6	7	8	9
Observed Freq. :	18	19	23	21	16	25	22	20	21	15
Expected Freq. :	20	20	20	20	20	20	20	20	20	20

$$\chi^2 = \Sigma \frac{(O-E)^2}{E}$$

$$= \frac{(18-20)^2}{20} + \frac{(19-20)^2}{20} + \frac{(23-20)^2}{20} + \frac{(21-20)^2}{20} + \frac{(16-20)^2}{20}$$

$$+ \frac{(25-20)^2}{20} + \frac{(22-20)^2}{20} + \frac{(20-20)^2}{20} + \frac{(21-20)^2}{20} + \frac{(15-20)^2}{20}$$

$$= \frac{1}{20}(4+1+9+1+16+25+4+0+1+25) = \frac{86}{20} = 4.3.$$

Degrees of freedom $= 10 - 1 = 9$

Table value of χ^2 for 9 degrees of freedom $= 16.919$

Since the calculated value of χ^2 is less than the table value, the hypothesis that the digits are distributed in equal number, holds good.

Example 11. *Three hundred digits were chosen at random from a set of tables. The frequencies of the digits were as follows :*

Digit :	*0*	*1*	*2*	*3*	*4*	*5*	*6*	*7*	*8*	*9*
Frequency :	*28*	*29*	*33*	*31*	*26*	*35*	*32*	*30*	*31*	*25*

Using the χ^2 *test assess the hypothesis that the digits were distributed in equal number in the table.* (χ^2 *for 9 d.f. at 5% level* $= 16.92$)

Sol. Please try yourself. **[Ans.** The hypothesis holds good]

Example 12. *The following table gives the number of aircraft accidents that occurred during the various days of the week. Find whether the accidents are uniformly distributed over the week.*

Day :	*Sun.*	*Mon.*	*Tue.*	*Wed.*	*Thu.*	*Fri.*	*Sat.*	*Total*
No. of accidents :	*14*	*16*	*8*	*12*	*11*	*9*	*14*	*84*

(χ^2 *for 4 d.f. at 5% level = 9.41)* (D.U.1990)

Sol. Taking the hypothesis that the accidents are uniformly distributed over the week, the expected frequencies of accidents on any day $= \frac{84}{7} = 12$.

Day :	*Sun.*	*Mon.*	*Tue.*	*Wed.*	*Thu.*	*Fri.*	*Sat.*
Observed No. of accidents :	14	16	8	12	11	9	14
Expected No. of accidents :	12	12	12	12	12	12	12

Since the frequencies in some classes are less than 10, we regroup the data.

Day :	*Sun.*	*Mon.*	*Tue. and Wed.*	*Thu.*	*Fri. and Sat.*
O :	14	16	20	11	23
E :	12	12	24	12	24

$$\chi^2 = \Sigma \frac{(O-E)^2}{E} = \frac{(14-12)^2}{12} + \frac{(16-12)^2}{12} + \frac{(20-24)^2}{24} + \frac{(11-12)^2}{12} + \frac{(23-24)^2}{24}$$

$$= \frac{1}{24}[8 + 32 + 16 + 2 + 1] = \frac{59}{24} = 2.46.$$

Degrees of freedom = 5 – 1 = 4

Table value of χ^2 for 4 d.f. at 5% level = 9.41

Since the calculated value of χ^2 is less than the table value, the hypothesis holds good.

Example 13. *Examine the goodness of fit for the following frequency distribution :*

Observed Frequency :	*15*	*20*	*10*	*25*	*18*	*10*	*7*
Theoretical Frequency :	*15*	*15*	*15*	*15*	*15*	*15*	*15*

(χ^3 *for 6 d.f. at 5% level = 12.59)*

Sol. $\chi^2 = \Sigma \frac{(O-E)^2}{E}$ = 16.53 (Please calculate it yourself.)

Degree of freedom = 7 – 1 = 6

Table value of χ^2 for 6 d.f. at 5% level = 12.59

Since the calculated value of χ^2 is greater than the table value, hence the fit is poor.

Example 14. *Apply the χ^2 test of goodness of fit to the following data :*

Observed frequency :	1	5	20	28	42	22	15	5	2
Theoretical frequency :	1	6	18	25	40	25	18	6	1

(χ^2 *for 6 d.f. at 5% level = 12.59*)

Sol. Since the frequencies are less than 10 in the beginning and end of the series, we regroup as follows :

O	E	$(O-E)^2$	$\frac{(O-E)^2}{E}$
$\left.\begin{matrix}1\\5\end{matrix}\right\} = 6$	$\left.\begin{matrix}1\\6\end{matrix}\right\} = 7$	1	0.143
20	18	4	0.222
28	25	9	0.360
42	40	4	0.100
22	25	9	0.360
15	18	9	0.500
$\left.\begin{matrix}5\\2\end{matrix}\right\} = 7$	$\left.\begin{matrix}6\\1\end{matrix}\right\} = 7$	0	0.000

$$\chi^2 = \Sigma \frac{(O-E)^2}{E} = 1.685$$

Degrees of freedom = 7 – 1 = 6

Table value of χ^2 for 6 d.f. at 5% level = 12.59

Since the calculated value of χ^2 is much less than the table value, the fit is good.

Example 15. *The following mistakes per page were observed in a book.*

No. of mistakes per page	*No. of times the mistake occurred*
0	*211*
1	*90*
2	*19*
3	*5*
4	*0*
Total	*325*

Fit a Poisson distribution to the data and test the goodness of fit. ($e^{-0.44}$ – *0.6444 and* χ^2 *for 1 d.f. at 5% level = 3.84*).

Sol. Here

x :	0	1	2	3	4
f :	211	90	19	5	0

$$N = \Sigma f = 325 \qquad \text{Mean } m = \frac{\Sigma fx}{N} = \frac{90 + 38 + 15}{325}$$

$$= \frac{143}{325} = 44$$

$\therefore$ The Poisson distribution fitted to the data is

$$325 \times \frac{e^{-0.44}(0.44)^r}{r\,!} = 325 \times 0.6444 \times \frac{(0.44)^r}{r\,!}$$

$$= 209.43 \times \frac{(0.44)^r}{r\,!}$$

The expected frequencies of 0, 1, 2, 3, 4 mistakes are (putting $r = 0, 1, 2, 3, 4$) 209.43, 92.15, 20.27, 2.97, 0.33.

$\therefore$ We have

O	E	$(O-E)^2$	$\frac{(O-E)^2}{E}$
211	209.43	2.46	0.012
90	92.15	4.62	0.050
19, 5, 0 = 24	20.27, 2.97, 0.33 = 23.57	0.18	0.008
			0.070

$$\chi^2 = \sum \frac{(O-E)^2}{E} = 0.070$$

Since m is calculated and the totals should agree, degrees of freedom $= 3 - 2 = 1$

Table value of χ^2 for 1 d.f. at 5% level = 3.84

Since the calculated value of χ^2 is much less than the table value, the fit is good.

Example 16. *The following table gives the observed frequencies of a distribution and the frequencies of the normal distribution. Apply χ^2 test of goodness of fit.*

O :	*8*	*27*	*111*	*217*	*270*	*223*	*110*	*29*	*5*
E :	*6*	*31*	*107*	*216*	*280*	*216*	*107*	*31*	*6*

(χ^2 for 4 d.f. at 5% level = 9.49)

Sol. Since frequencies in the beginning and end are less than 10, therefore, regrouping is necessary.

O	E	$(O-E)^2$	$\frac{(O-E)^2}{E}$
8, 27 } = 35	6, 31 } = 37	4	0.108
111	107	16	0.149
217	216	1	0.005
270	280	100	0.357
223	216	49	0.226
110	107	9	0.084
29, 5 } = 34	31, 6 } = 37	9	0.243
			1.172

$$\therefore \qquad \chi^2 = \sum \frac{(O-E)^2}{E} = 1.172$$

Since in a normal distribution, the mean and standard deviation are calculated and the totals should agree, degrees of freedom

$$= 7 - 3 = 4$$

Table value of χ^2 for 4 d.f. at 5% level = 9.49.

The calculated value of χ^2 is much less than the table value, therefore, the fit is good.

Example 17. *The theory predicts the proportion of beans in the four groups A, B, C and D should be 9 : 3 : 3 : 1. In an experiment among 1600 beans, the numbers in the four groups were 882, 313, 287 and 118. Does the experimental result support the theory ? (χ^2 for 3 d.f. at 5% level = 7.815).*

Sol. Taking the hypothesis that the theory fits well into the experiment *i.e.,* the experimental result supports the theory, the expected frequencies are respectively,

$$\frac{9}{16} \times 1600, \frac{3}{16} \times 1600, \frac{3}{16} \times 1600, \frac{1}{16} \times 1600$$

i.e., 900, 300, 300, 100. Thus

O :	882	313	287	118
E :	900	300	300	100

$$\chi^2 = \sum \frac{(O-E)^2}{E} = \frac{(882-900)^2}{900} + \frac{(313-300)^2}{300} + \frac{(287-300)^2}{300} + \frac{(118-100)^2}{100}$$

$$= 0.3600 + 0.5633 + 0.5633 + 3.2400 = 4.7266$$

Degrees of freedom = 4 – 1 = 3

Table value of χ^2 for 3 d.f. as 5% level of significance = 7.815

Since the calculated value of χ^2 is less than the table value, the hypothesis may the accepted. Hence the experimental results support the theory.

Example 18. *In experiments on pea-breeding, Mendel got the following frequencies of seeds :*

315 round and yellow, 101 wrinkled and yellow, 108 round and green, 32 wrinkled and green ; total 556. Theory predicts that the frequencies should be in the proportion 9 : 3 : 3 : 1. Examine the correspondence between theory and experiment. (χ^2 for 3 d.f. at 5% level = 7.815).

Sol. Taking the hypothesis that the theory fits well into the experiment, the expected frequencies are respectively

$$\frac{9}{16} \times 556, \frac{3}{16} \times 556, \frac{3}{16} \times 556, \frac{1}{16} \times 556$$

i.e., 313, 104, 104, 35. Thus

O :	315	101	108	32
E :	313	104	104	35

$$\therefore \quad \chi^2 = \sum \frac{(O-E)^2}{E} = \frac{(315-313)^2}{313} + \frac{(101-104)^2}{104} + \frac{(108-104)^2}{104} + \frac{(32-35)^2}{35}$$

$$= 0.013 + 0.09 + 0.15 + 0.26 = 0.513$$

Degrees of freedom = 4 – 1 = 3

Table value of χ^2 for 3 d.f. at 5% level of significance = 7.815.

Since the calculated value of χ^2 is much less than the table value, the hypothesis may be accepted. Hence there is much correspondence between theory and experiment.

Example 19. *Among 65 off-springs of a certain cross between guinea pigs, 34 were red. 10 were black and 20 were white. According to the genetic model, hence numbers should be in the ratio 9 : 3 : 4. Are the data consistent with the model ? (χ^2 for 2 d.f. at 5% level = 5.99).*

Sol. Please try yourself. **[Ans.** Yes]

Example 20. *Children having one parent of blood-type M and the other type N will always be one of the three types M, MN, N and average proportions of these will be 1 : 2 : 1.*

Out of 300 children having one M parent and one N parent. 30% were found to be of type M, 45% of the type MN and the remaining of type N. Use χ^2 test to test the hypothesis. (χ^2 for 2 d.f. at 5% level = 5.99).

Sol. Please try yourself. **[Ans.** Theory appears to be correct.]

Example 21. *A sample analysis of examination results of 500 students was made. It was found that 220 students failed. 170 had secured a third class, 90 were placed in second class and 20 got a first class. Are these figures commensurate with the general examination result which is the ratio of 4 : 3 : 2 : 1 for the various categories respectively. (χ^2 for 3 d.f. at 5% level to significance = 7.81).*

Sol. Please try yourself. **[Ans.** No]

Example 22. *The following figures show the distribution of digits in numbers chosen at random from a telephone directory :*

Digits :	0	1	2	3	4	5	6	7	8	9	*Total*
Freq. :	*1026*	*1107*	*997*	*966*	*1075*	*933*	*1107*	*972*	*964*	*853*	*10000*

Test whether the digits may be taken to occur equally frequently in the directory. (χ^2 *for 9 d.f. at 5% level = 16.919*). (M.D.U. 1982)

Sol. Please try yorself. **[Ans.** No]

Example 23. *A survey of 320 families with 5 children each revealed the following distribution :*

No. of boys :	5	4	3	2	1	0
No. of girls :	0	1	2	3	4	5
No. of families :	14	56	110	88	40	12

Is this result consistent with the hypothesis that male and female births are equally probable. (χ^2 *for 5 d.f. at 5% level of significance = 11.07*).

(K.U. 1983)

Sol. Taking the hypothesis that male and female births are equally probable, p, the probability of male births $= \frac{1}{2} = q$

The probability of r male births in a family of 5 out of 320 families.

$$= \text{N} \,.\, {}^5C_r\, q^{5-r} p^r = 320\ {}^5C_r \left(\tfrac{1}{2}\right)^5 \quad | \text{ Binomial Dist. N}(q+p)^5$$

$$= 10\ {}^5C_r.$$

$\therefore$ The probability of 5, 4, 3, 2, 1, 0 male births in a family of 5 out of 320 families are respectively.

$$10 \times {}^5C_5,\ 10 \times {}^5C_4,\ 10 \times {}^5C_3,\ 10 \times {}^5C_2,\ 10 \times {}^5C_1,\ 10 \times {}^5C_0$$

i.e., 10, 50, 100, 100, 50, 10. Thus

O :	14	56	100	88	40	12
E :	10	50	100	100	50	10

$$\chi^2 = \sum \frac{(\text{O}-\text{E})^2}{\text{E}} = \frac{(14-10)^2}{10} + \frac{(56-50)^2}{50} + \frac{(110-100)^2}{100}$$

$$+ \frac{(88-100)^2}{100} + \frac{(40-50)^2}{50} + \frac{(12-10)^2}{10}$$

$$= 1.60 + 0.72 + 1 + 1.44 + 2 + 0.40 = 7.16$$

Degrees of freedom = 6 – 1 = 5

Table value of χ^2 for 5 d.f. at 5% level of significance = 11.07

Since the calculated value of χ^2 is less than the table value, the hypothesis may be accepted.

Example 24. *4 coins were tossed 160 times and the following results were obtained :*

No. of heads :	0	1	2	3	4
Observed freq. :	17	52	54	31	6

Under the assumption that the coins are balanced (i.e., unbiased) find the expected frequencies of getting 0, 1, 2, 3 or 4 heads and test the goodness of fit. (χ^2 *for 4 f.d. at 5% level = 9.49*)

Sol. Please try yourself. [**Ans.** 10, 40, 60, 40, 10 ; the fit is poor]

Example 25. *The number of books borrowed from a public library during a particular week is given below. Test the hypothesis that the number of books borrowed does not depend on the day of the week.*

Days	*No. of books borrowed*
Monday	*140*
Tuesday	*132*
Wednesday	*160*
Thursday	*148*
Friday	*134*
Saturday	*150*

(χ^2 for 5 d.f. at 5% level of significance = 11.07)

Sol. Please try yourself. [**Ans.** Hypothesis of correct]

II. STUDENT'S TEST

1. Gosset, who wrote under the pen-name of Student, derived a theoretical distribution which has come to be known as "student's *t*-distribution". The quantity *t* is defined as

$$t = \frac{\bar{x} - \mu}{S} \sqrt{n}$$

where n = the number of observations in the sample

$\bar{x} = \frac{1}{n} \sum_{i=1}^{n} x_i$ is the mean of the sample

μ = the mean of the parent population from which the sample has been drawn

$S = \sqrt{\frac{1}{n-1} \sum_{i-1}^{n} (x_i - \bar{x})^2}$ is the S.D. of sample

and not $\sqrt{\frac{1}{n} \sum_{i-1}^{n} (x_i - \bar{x})^2}$

Degrees of freedom = $n - 1$.

2. Application of the *t*-distribution. *The t*-distribution has a wide number of applications in Statistics, some of which are enumerated below :

(*i*) **To test the significance of a sample mean.** To test if the sample mean $\bar{x}$ differs significantly from the hypothetical value μ of the population mean, compare the calculated value of $|t|$ with the table value at a certain level of significance (usually 5%).

If the calculated value of $|t|$ exceeds the table value of *t* at 5%, we say that the difference between $\bar{x}$ and μ is significant and the sample does not seem to have been drawn from the population with mean μ. However if the calculated value of $|t|$ is less than the table value of *t* at 5%, we say that the difference between $\bar{x}$ and μ is not significant. The deviation may be due to sampling fluctuations and the sample might have been drawn from the population with mean μ.

(*ii*) To test the significance of the difference between two sample means.

(*iii*) To test the significance of the co-efficient of correlation.

Note. On computation of S^2 for numerical problems

$$S^2 = \frac{1}{n-1} \Sigma (x_i - \bar{x})^2$$

$$= \frac{1}{n-1} \Sigma (x_i^2 - 2x_i\bar{x} + \bar{x}^2)$$

$$= \frac{1}{n-1} [\Sigma x_i^2 - 2\bar{x}\, \Sigma x_i + n\bar{x}^2]$$

$$= \frac{1}{n-1} \left[\Sigma x_i^2 - 2 \cdot \frac{1}{n} \Sigma x_i \Sigma x_i + n \left(\frac{1}{n} \Sigma x_i \right)^2 \right]$$

$$= \frac{1}{n-1} \left[\Sigma x_i^2 - \frac{2}{n} (\Sigma x_i)^2 + \frac{1}{n} (\Sigma x_i)^2 \right]$$

$$= \frac{1}{n-1} \left[\Sigma x_i^2 - \frac{(\Sigma x_i)^2}{n} \right]$$

If we take $d_i = x_i - A$ where A is any arbitrary number, knowing that variance is independent of change of origin

$$S^2 = \frac{1}{n-1} \left[\Sigma d_i^2 - \frac{(\Sigma d_i)^2}{n} \right] \text{ Also } \bar{x} = A + \frac{\Sigma d_i}{n}.$$

Example 1. *A machine which produces mica insulating washers of use in electric devices is set to turn out washers having a thickness of 10 mm. A sample of 10 washers has an average thickness of 9.52 mm with a standard deviation of 0.60 mm. Find out t.*

Sol. Here $\bar{x}$ = mean of sample = 9.52 mm.

n = number of observation in sample = 10

S = standard deviation of sample = 0.60

μ = mean of parent population from which sample is taken = 10 mm.

$$\therefore \quad t = \frac{\bar{x} - \mu}{S} \sqrt{n} = \frac{-9.52 - 10}{0.60} \sqrt{10} = \frac{-0.48}{0.60} \sqrt{10}$$

$$= -\frac{4}{5} \times 3.16 = -2.528.$$

Example 2. *Find the student's t for the following variable values in a sample of eight :*

– 4, – 2, – 2, 0, 2, 2, 3, 3

taking the mean of the universe to be zero.

Sol.

x	$d = x - 2$	d^2
− 4	− 6	36
− 2	− 4	16
− 2	− 4	16
0	− 2	4
2	0	0
2	0	0
3	1	1
3	1	1
	$\Sigma d = -14$	$\Sigma d^2 = 74$

Here $n = 8$ assumed mean A = 2

$$\therefore \quad \bar{x} = A + \frac{\Sigma d}{n} = 2 - \frac{14}{8} = \frac{1}{4} = 0.25$$

$$S^2 = \frac{1}{n-1}\left[\Sigma d^2 - \frac{(\Sigma d)^2}{n}\right] = \frac{1}{7}\left[74 - \frac{196}{8}\right]$$

$$= \frac{1}{7}[74 - 24.5] = \frac{1}{7} \times 49.5 = 7.07$$

$$\therefore \quad S = \sqrt{7.07} = 2.66. \text{ Also } \mu = 0$$

$$\therefore \quad t = \frac{\bar{x} - \mu}{S}\sqrt{n} = \frac{0.25 - 0}{2.66}\sqrt{8} = 0.27.$$

Example 3. *Find the student's t for the following variable values in a sample of 10.*

− 6, − 4, − 1, − 1, 0, 1, 1, 3, 4, 5

taking the mean of the universe to be zero. (K.U. 1983 S)

Sol. Please try yourself. **[Ans.** 0.18]

Example 4. *A machinist is making engine parts with axle diameters of 0.700 inch. A random sample of 10 parts shows a mean diameter of 0.742 inch with a standard deviation of 0.040 inch. Compute statistic you would use to test whether the work is meeting the specification. Also state how you would proceed further ?*

Sol. Here $\mu = 0.700$ inches, $\bar{x} = 0.742$ inches, S = 0.040 inches and $n = 10$

Taking the hypothesis that the product is confirming to specifications *i.e.*, there is no significant difference between $\bar{x}$ and μ, the test statistic is

$$t = \frac{\bar{x} - \mu}{S}\sqrt{n} = \frac{0.742 - 0.700}{0.040}\sqrt{10}$$

$$= \frac{0.042}{0.040}\sqrt{10} = \frac{21}{20} \times 3.16 = 3.318$$

Degrees of freedom $= n - 1 = 9$.

How to proceed further ? We will now compare this calculated value of t with the table value of t for 9 d.f. at a certain level of significance (say 5%). Let this table value of t be denoted by t_0.

(*i*) If $t = 3.318 > t_0$, the value of t is significant which implies that $\bar{x}$ differs singinficantly from μ and the hypothesis is rejected. The product is not meeting the specifications.

(*ii*) If $t = 3.318 > t_0$, the value of t is not significant which implies that there is no significance different between $\bar{x}$ and μ. The deviation $(\bar{x} - \mu)$ is due to fluctuations of samples and hypothesis is accepted at 5% level of significance. The product confirms to specifications.

Example 5. *A random sample of nine form men of a large city gave a mean height 68 inch and the unbiased estimate of the population variance from the sample was 4.5 inches. Proceed as far as you can to test for a mean height of 68.5 inches for the mean of the city. Also state how would you proceed further.*

Sol. Please try yourself.

Example 6. *Ten individuals are chosen at random from a population and their heights are found to be in inches 63, 63, 64, 65, 66, 69, 69, 70, 70, 71. Discuss the suggestion that the mean height in the universe is 65 inches given that for 9 d.f. the value of student's that 5% level of significance is 2.262.*

Sol.

x	$d = x - 69$	d^2
63	– 6	36
63	– 6	36
64	– 5	25
65	– 4	16
66	– 3	9
69	0	0
69	0	0
70	1	1
70	1	1
71	2	4
	$\Sigma d = -20$	$\Sigma d^2 = 128$

$$\bar{x} = A + \frac{\Sigma d}{n} = 69 - \frac{20}{10} = 67$$

$$S^2 = \frac{1}{n-1}\left[\Sigma d^2 - \frac{(\Sigma d)^2}{n}\right] = \frac{1}{9}\left[128 - \frac{400}{10}\right]$$

$$= \frac{1}{9} \times 88 \quad \therefore \quad S = \sqrt{\frac{88}{9}} = 3.13$$

Here $\mu = 65$ inches

On the assumption that the mean for the universe is 65 inches,

$$t = \frac{\overline{x} - \mu}{S}\sqrt{n} = \frac{67 - 65}{3.13}\sqrt{10} = 2.02$$

Degrees of freedom

$= 10 - 1 = 9$

Table value of t **for 9 d.f. at 5% level of significance**

$= 2.262.$

Since calculated value of $t = 2.02 < 2.262 =$ **table value, the difference is due to fluctuations of sampling. The hypothesis is, therefore, accepted and the mean height in the universe is 65 inches.**

Example 7. *Ten individuals are chosen at random from a population and their heights are found to be in inches, 63, 63, 66, 67, 68, 69, 70, 70, 71, 71. In the light of this data, discuss the suggestion that the mean height in the universe is 66 inches. (Value of t for 9 d.f. at 5% level of significance = 2.62).*

Sol. Please try yourself. **[Ans.** The suggestion in acceptable]

Example 8. *A certain stimulus administered to each of 12 patients resulted in the following increases of blood pressures :*

5, 2, 8, – 1, 3, 0, 6, – 2, 1, 5, 0, 4.

Can it be concluded that the stimulus will be, in general, accompanied by an increase in blood pressure ? (Value of t for 11 d.f. at 5% level of significance = 2.201). (D.U. 1985, 86)

Sol.

x	$d = x - 3$	d^2
5	2	4
2	– 1	1
8	5	25
– 1	– 4	16
3	0	0
0	– 3	9
6	3	9
– 2	– 5	25
1	– 2	4
5	2	4
0	– 3	9
4	1	1
	$\Sigma d = -5$	$\Sigma d^2 = 107$

$$\overline{x} = A + \frac{\Sigma d}{n} = 3 - \frac{5}{12} = 2.58$$

$$S^2 = \frac{1}{n-1}\left[\Sigma d^2 - \frac{(\Sigma d)^2}{n}\right] = \frac{1}{11}\left[107 - \frac{25}{12}\right]$$

$$= \frac{1}{11} \times \frac{1259}{12} \qquad \therefore \quad S = \sqrt{\frac{1259}{132}} = 2.96$$

Assuming that the stimulus will, in general, not be accompanied by an increase in blood pressure, *i.e.*, the mean of the universe $\mu = 0$, we get

$$t = \frac{\bar{x} - \mu}{S}\sqrt{n} = \frac{2.58 - 0}{2.96} \times \sqrt{12} = 3.01.$$

Degrees of freedom = 12 – 1 = 11.

Table value of t for 11 d.f. at 5% level of significance = 2.201

Since calculated value of $t = 3.01 > 2.201$ = table value, the hypothesis is rejected. Hence the stimulus will be, in general, accompanied by blood pressure.

Example 9. *10 patients to whom a drug was administered, registered the following additional hours of sleep :*

0.7, – 1.1, – 0.2, 1.2, 0.1, 3.4, 3.7, 0.8, 1.8, 2.0.

Compute the statistic you would use to determine whether these data justify the claim that the drug does produce additional sleep and state how you would proceed further.

Sol. Please try yourself.

Example 10. *An experiment was conducted on nine rats. The experiment showed that due to smoking, the pulse rate increased in the following order : 5, 3, 4, – 1, 2, – 3, 4, 3, 1.*

Can you maintain that smoking leads to an increase in the pulse rate ? (t for 8 d.f. at 5% level of significance = 2.31).

Sol. Please try yourself. **[Ans.** Yes]

Example 11. *Nine patients to whom a certain drink was administered registered the following increments in blood pressure :*

7, 3, – 1, 4, – 3, 5, 6, – 4, 1.

Show that the data do not indicate that the drink was responsible for these increments.

Sol. Please try yourself.

Example 12. *The nine itmes of a sample had the following values :*

45, 47, 50, 52, 48, 47, 49, 53, 51.

Does the mean of the nine items differ significantly from the assumed population mean of 47.5 ? (t for 8 d.f. at 5% level of significance = 2.31)

(D.U. 1983)

Sol.

x	$d = x - 49$	d^2
45	– 4	16
47	– 2	4
50	1	1
52	3	9
48	– 1	1
47	– 2	4
49	0	0
53	4	16
51	2	4
	$\Sigma d = 1$	$\Sigma d^2 = 55$

$$\bar{x} = A + \frac{\Sigma d}{n} = 49 + \frac{1}{9} = 49.11$$

$$S^2 = \frac{1}{n-1}\left[\Sigma d^2 - \frac{(\Sigma d)^2}{n}\right]$$

$$= \frac{1}{8}\left[55 - \frac{1}{9}\right] = \frac{54.89}{8} = 6.86$$

$$S = \sqrt{6.86} = 2.62$$

Taking the hypothesis that the mean of the population from which the sample is drawn is 47.5.

$$t = \frac{\bar{x} - \mu}{S}\sqrt{n} = \frac{49.11 - 47.5}{2.62} \times \sqrt{9}$$

$$= \frac{1.61 \times 3}{2.62} = 1.85$$

$$\text{d.f.} = 9 - 1 = 8$$

The calculated value of $t = 1.85 < 2.31$ = table value of t for 8 d.f. at 5% level.

Therefore, the difference between $\bar{x}$ and μ is not significant and the hypothesis holds.

Example 13. *A random sample of 10 boy had the following I. Q.'s :*

70, 120, 110, 101, 88, 83, 95,98, 107, 100

Do these data support the assumption of a population mean I.Q. of 100 ? (t for 9 d.f. at 5% level of significance = 2.262).

Sol.

x	$d = x - 90$	d^2
70	− 20	400
120	30	900
110	20	400
101	11	121
88	− 2	4
83	− 7	49
95	5	25
98	8	64
107	17	289
100	10	100
	$\Sigma d = 72$	$\Sigma d^2 = 2352$

$$\bar{x} = A + \frac{\Sigma d}{n} = 90 + \frac{72}{10} = 97.2$$

$$S^2 = \frac{1}{n-1}\left[\Sigma d^2 - \frac{(\Sigma d)^2}{n}\right]$$

$$= \frac{1}{9}\left[2352 - \frac{(72)^2}{10}\right] = 203.73$$

$$\therefore \quad S = \sqrt{203.73}$$

Taking the hypothesis that the data are consistent with the assumption of a mean I.Q. of 100 in the population, *i.e.*, $\mu = 100$, we have

$$t = \frac{\bar{x} - \mu}{S}\sqrt{n}$$

$$= \frac{97.2 - 100}{\sqrt{203.73}} \times \sqrt{10} = \frac{-2.8}{\sqrt{\frac{203.73}{10}}}$$

$$= \frac{-2.8}{\sqrt{20.373}} = \frac{-2.8}{4.514} = -0.6$$

d.f. = 10 – 1 = 9.

Calculated $|t| = 0.6 < 2.262$ = table value of t for 9 d.f. at 5% level of significance.

Hence the data are consistent with the assumption of mean I.Q. of 100 in the population.

Example 14. *A random sample of size 16 has 53 as mean and the sum of the squares of the deviations taken from mean is 150. Can this sample be regarded as taken from the population having 56 as mean ? (t for 15 d.f. at 5% level of significance = 2.13).*

Sol. Taking the hypothesis that the sample is taken from the population having mean 56, we have

$$\mu = 56$$

$$n = 16, \quad \bar{x} = 53 \quad \Sigma(x - \bar{x})^2 = 150$$

$$\therefore \quad S = \sqrt{\frac{1}{n-1}\Sigma(x - \bar{x})^2}$$

$$= \sqrt{\frac{150}{15}} = \sqrt{10} = 3.162$$

$$t = \frac{\bar{x} - \mu}{S}\sqrt{n}$$

$$= \frac{53 - 56}{3.162} \times \sqrt{16} = \frac{-12}{3.162} = -3.8$$

d.f. = 16 – 1 = 15

Calculated value of $|t| = 3.8 > 2.13$ = table value of t for 15 d.f. at 5% level of significance.

∴ The result of experiment does not support the hypothesis and hence, the sample is not drawn from the population having 56 as mean.

Example 15. *A manufacturer of gunpowder has developed a new powder which is designed to produce a muzzle velocity equal to 3000 ft./sec. Eight shells are loaded with the charge and the muzzle velocities measured. The resulting velocities are as follows :*

3005, 2935, 2965, 2995, 3905, 2935, 2905.

Do these data present sufficient evidence to indicate that the average velocity differs from 3000 ft./sec.

Sol. Please try yourself.

III. SNEDECOR'S F–TEST
(For Equality of Population Variances)

Let x_i (i = 1, 2, ..., n_1) and y_i (j = 1, 2, ... n_2) be the values of two independent random samples drawn from the normal populations with the same variance. Let $\overline{x}$ and $\overline{y}$ be the sample means.

Let

$$S_x^2 = \frac{1}{n_1 - 1}\sum_{i=1}^{n_1} (x_i - \overline{x})^2$$

and

$$S_y^2 = \frac{1}{n_2 - 1}\sum_{j=1}^{n_2} (y_j - \overline{y})^2$$

Then, we define the statistic F by the relation

$$F = \frac{S_x^2}{S_y^2}.$$

where greater of the two variances S_x^2 and S_y^2 is to b aken in the numerator and n_1 corresponds to the greater variance.

If $r_1 = n_1 - 1$ and $r_2 = n_2 - 1$, then F has (r_1, r_2) d.f.

Taking the hypothesis that two samples have been drawn from normal populations with the same variance, we compare the calculated value of F with its table value.

If the calculated value of F > the table value of F with $(n_1 - 1, n_2 - 1)$ d.f. at 5% level of significance, the ratio is significant and the hypothesis may be rejected.

If the calculated value of F < the table value of F with $(n_1 - 1, n_2 - 1)$ d.f. at 5% level of significance , the hypothesis may be accepted and the two samples might have come from two normal populations with the same variance.

Example 1. *Two samples of size 9 and 8 give the sum of squares of deviations from their respective means equal to 160 inches square and 91 inches square respectively. Can they be regarded as drawn from the two normal populations with same variance ? (F for 8 and 7 d.f. = 3.73).*

(D.U. 1987)

Sol. Taking the hypothesis that the two samples have been drawn from two normal populations with same variance.

$$S_x^2 = \frac{\Sigma(x-\bar{x})^2}{n_1-1} = \frac{160}{9-1} = 20$$

$$S_y^2 = \frac{\Sigma(y-\bar{y})^2}{n_2-1} = \frac{91}{8-1} = 13$$

$$\therefore \quad F = \frac{S_x^2}{S_y^2} = \frac{20}{13} = 1.54$$

Table value of F for $(n_1 - 1, n_2 - 1) = (9-1, 8-1) = (8, 7)$ d.f. is 3.73.

Calculated value of F = 1.54 < 3.73 = table value for (8, 7) d.f.

∴ The hypothesis may be accepted. Hence the two samples may be regarded as drawn from two normal populations with same variance.

Example 2. *In one sample of 8 observations, the sum of the squares of the deviations of the sample values from the sample mean was 84.4 and in another sample of 10 observations, it was 102.6. Test whether the two samples have been drawn from two normal populations with the same variance. (F for 7 and 9 d.f. at 5% level of significance = 3.29).* (D.U. 1985)

Sol. Please try yourself. **[Ans.** Yes**]**

Example 3. *Two independent samples of 8 and 7 items respectively had the following values of the variable (weight in ounces) :*

Sample I :	*9*	*11*	*13*	*11*	*15*	*9*	*12*	*14*
Sample II :	*10*	*12*	*10*	*14*	*9*	*8*	*10*	

Do the two estimates of population variance differ significantly ? Given that for 7 and 6 d.f. the value of F at 5% level of significance is 4.20 nearly. (D.U. 1992)

Sol. Taking the hypothesis that the two populations have the same variance, we calculate F.

Sample I (x)	$x-\bar{x}$ $= x - 11.75$	$(x-\bar{x})^2$	*Sample II* (y)	$y-\bar{y}$ $= y - 10.43$	$(y-\bar{y})^2$
9	− 2.75	7.5625	10	− 0.43	0.1849
11	− 0.75	0.5625	12	1.57	2.4649
13	1.25	1.5625	10	− 0.43	0.1849
11	− 0.75	0.5625	14	3.57	12.7449
15	3.25	10.5625	9	− 1.43	2.0449
9	− 2.75	7.5625	8	− 2.43	5.8049
12	0.25	0.0625	10	− 0.43	0.1849
14	2.25	5.0625			
Total : 94		33.5000	73		23.6143

$$\bar{x} = \frac{94}{8} = 11.75 \text{ oz}, \qquad \bar{y} = \frac{73}{7} = 10.43 \text{ oz.}$$

$$S_x^2 = \frac{\Sigma(x-\bar{x})^2}{n_1 - 1} = \frac{33.5}{8-1} = 4.8$$

$$S_y^2 = \frac{\Sigma(y-\bar{y})^2}{n_2 - 1} = \frac{23.6143}{7-1} = 3.94$$

$$\therefore \quad F = \frac{S_x^2}{S_y^2} = \frac{4.8}{3.94} < 2$$

Table value of F for $(n_1 - 1, n_2 - 1) = (7, 6)$ d.f. at 5% level of significance = 4.20 nearly.

Calculated value of F < table value.

The hypothesis may be accepted. Hence the differences are not significant and the samples may well be drawn from the populations having the same variance.

Example 4. *Two random samples drawn from two normal populations are :*

Sample I : *20, 16, 26, 27, 23, 22, 18, 24, 25, 19*
Sample II : *27, 33, 42, 35, 32, 34, 38, 28, 41, 43, 30, 37.*

Obtain the estimates of the variances of the population and test whether the two populations have the same variance.

[F for 11 and 9 d.f. at 5% level of significance = 3.11] (D.U. 1990)

Sol. Taking the hypothesis that the two populations have the same variance, we apply F-test.

Sample I (x)	$x - \bar{x}$ $= x - 22$	$(x-\bar{x})^2$	*Sample II* (y)	$y - \bar{y}$ $= y - 35$	$(y-\bar{y})^2$
20	−2	4	27	−8	64
16	−6	36	33	−2	4
26	4	16	42	7	49
27	5	25	35	0	0
23	1	1	32	−3	9
22	0	0	34	−1	1
18	−4	16	38	3	9
24	2	4	28	−7	49
25	3	9	41	6	36
19	−3	9	43	8	64
			30	−5	25
			37	2	4
Total : 220		120	420		314

$$\bar{x} = \frac{\Sigma x}{n_1} = \frac{220}{10} = 22, \qquad \bar{y} = \frac{420}{12} = 35$$

$$S_x^2 = \frac{\Sigma(x-\bar{x})^2}{n_1 - 1} = \frac{120}{9} = 13.33$$

$$S_y^2 = \frac{\Sigma(y-\bar{y})^2}{n_2-1} = \frac{314}{11} = 28.55$$

$$F = \frac{S_y^2}{S_x^2} = \frac{28.55}{13.33}$$

(Since numerator is always the greater of S_x^2 and S_y^2)

$= 2.14$

d.f. of greater variance = 11

d.f. of lesser variance = 9

Table value of F for 11 and 9 d.f. at 5% level of significance

$= 3.11$

Calculated value of F < table value.

∴ The hypothesis may be accepted. Hence the samples may well we drawn from the populations having the same variance.

Example 5. *Two independent samples of 8 and 7 items respectively have the following values of the variable (weight in ounces) :*

Sample I :	*39*	*41*	*43*	*41*	*45*	*39*	*42*	*44*
Sample II :	*40*	*42*	*40*	*44*	*39*	*38*	*40*	

Do the estimates of the population variance differ significantly ? Given that for 7 and 6 d.f. the value of F at 5% level of significance is 3.87.

(D.U. 1993)

Sol. Please try yourself.

Example 6. *Two independent random samples, each of 8 individuals provide the following data. Estimate the variance ratio and test the significance :*

63	*64*	*65*	*65*	*66*	*66*	*67*	*68*
69	*66*	*67*	*67*	*66*	*68*	*69*	*69*

The value of F at 5% level for 7 d.f. is 3.8 approximately.

Sol. Please try yourself.

IV. Z-TEST

This test is used to test the significance of the correlation co-efficient in small samples. If r is the correlation co-efficient of the sample and ρ, that of the population, calculate the value of

$$\frac{Z-\xi}{\dfrac{1}{\sqrt{n-3}}}$$

where $Z = \frac{1}{2}\tanh^{-1} r = \frac{1}{2}\log_e\left(\frac{1+r}{1-r}\right)$ or $1.1513 \log_{10}\left(\frac{1+r}{1-r}\right)$

$$\xi = \tfrac{1}{2}\tanh^{-1}\rho = \frac{1}{2}\log_e\left(\frac{1+\rho}{1-\rho}\right) \text{ or } 1.1513 \log_{10}\left(\frac{1+\rho}{1-\rho}\right)$$

$$\frac{1}{\sqrt{n-3}} = \text{S.E.}$$

If the absolute value of this $\frac{\text{difference}}{\text{S.E.}}$ exceeds 1.96, the difference is significant at 5% level.

Example 1. *Test the significance of the correlation r = 0.5 from a sample of size 18 against hypothetical correlation* $\rho = 0.7$.

Sol. We have to test the hypothesis that correlation in the population is 0.7.

$$Z = \tfrac{1}{2}\log_e\left(\frac{1+r}{1-r}\right) = 1.1513 \log_{10}\left(\frac{1+0.5}{1-0.5}\right)$$

$$= 1.1513 \log 3 = 1.1513 \times 0.4771$$

$$= 0.549$$

$$\xi = \tfrac{1}{2}\log_e\left(\frac{1+\rho}{1-\rho}\right) = 1.1513 \log_{10}\left(\frac{1+0.7}{1-0.7}\right)$$

$$= 1.1513 \log 5.67 = 1.1513 \times 0.7536$$

$$= 0.868$$

$$Z - \xi = 0.549 - 0.0868 = -0.319$$

$$\text{S.E.} = \frac{1}{\sqrt{n-3}} = \frac{1}{\sqrt{18-3}} = \frac{1}{\sqrt{15}} = 0.26$$

Absolute value of $\frac{Z-\xi}{\text{S.E.}} = \frac{0.319}{0.26} = 1.23$ which is less than 1.96 (5% level of significance) and is, therefore, not significant. Hence the sample may be regarded as coming from population with $\rho = 0.7$.

Example 2. *From a sample of 19 pairs of observations, the correlation is 0.5 and the corresponding population value is 0.3. Is the difference significant ?*

Sol. Here $n = 19$, $r = 0.5$, $\rho = 0.3$

$$Z = \tfrac{1}{2}\log_e\left(\frac{1+r}{1-r}\right) = 1.1513 \log_{10}\left(\frac{1+0.5}{1-0.5}\right)$$

$$= 1.1513 \log 3 = 1.1513 \times 0.4771 = 0.55$$

$$\xi = \tfrac{1}{2}\log_e\left(\frac{1+\rho}{1-\rho}\right) = 1.1513 \log_{10}\left(\frac{1+0.3}{1-0.3}\right)$$

$$= 1.1513 \log 1.857 = 1.1513 \times 0.2695 = 0.31$$